Intelligent Soil Management for Sustainable Agriculture

Kodoth Prabhakaran Nair

Intelligent Soil Management for Sustainable Agriculture

The Nutrient Buffer Power Concept

Springer

Kodoth Prabhakaran Nair
International Agricultural Scientist
c/o Mavila Pankajakshy
Calicut, Kerala, India

ISBN 978-3-030-15529-2 ISBN 978-3-030-15530-8 (eBook)
https://doi.org/10.1007/978-3-030-15530-8

This Springer imprint is published by the registered company Springer Nature Switzerland AG.
The registered company address is: Gewerbestrasse 11, 6330 Cham, Switzerland

I dedicate this book, which was compiled under very trying circumstances, to Pankajam, my wife, who is all to me, and also, to the memory of my late parents, my father, Kuniyeri Pookkalam Kannan Nair, an illustrious Police Officer who served the British Police and who was decorated with the King George V medal for bravery and honesty, and my mother, Kodoth Padinhareveetil Narayani Amma, daughter of the aristocratic Kodoth family of North Malabar, Kerala State, India, both of whom left me an orphan at a very young age, but whose boundless love and blessings made me what I am today.

Photos of my father and mother

Foreword

Global agriculture, in particular, Indian agriculture, is at a cross roads. The highly extractive agriculture has run out of steam. Of the 328.73 mha of geographical area in India, as much as 120.40 mha has degraded soils. One of the principal reasons for the steep decline in soil fertility is the rapid, but, sustained depletion of soil carbon, the bedrock of soil fertility, due to unscientific and nonjudicious use of chemical fertilizers. The soil nutrient management in much of South Asia, in particular, in India, still seems to be rooted in "text book" knowledge. As our experience has shown, more often than not, blanket recommendations of chemical fertilizers have been quite off the mark, and this leads to a lack of faith in these recommendations by millions of Indian farmers. "Soil test" and issue of "Soil Health Cards" still base their recommendations on routine and conventional soil analysis.

It is against this worrisome background that Professor K. P. Prabhakaran Nair's book, *Intelligent Soil Management for Sustainable Agriculture: The Nutrient Buffer Power Concept*, comes as a breath of fresh air. Prof. Nair is an eminent and acclaimed soil scientist with a very rich professional background, spanning more than three decades in Europe, Africa, and Asia. He has been using highly sophisticated analytical techniques like electro ultrafiltration to quantify the bioavailability of a plant nutrient, which suits the resource-rich European and American farmers, and have also developed simple analytical techniques to suit the pocket of a poor Asian, African, or Latin American farmer. The book catalogues several field success stories from Europe, Africa, and Asia.

Prof. Nair's academic credentials are impeccable. A Rockefeller Fellow, a Senior Fellow of the world-renowned Alexander von Humboldt Research Foundation of the Federal Republic of Germany, a former Professor of the National Science Foundation of the Royal Society, Belgium, Professor and Head of three departments, Agriculture, Soil Sciences and Basic Sciences at the University Center, Dschang, The Republic of Cameroon, and a Senior Professor at the University of Fort Hare, Alice, Republic of South Africa, where he was invited by the late Mr Nelson Mandela, to build a Faculty of Agriculture in his Alma Mater, from where late Mr. Nelson Mandela started his anti-apartheid struggle, are but, a few instances of his global professional standing in the academic world. Between 1997 and 1998, he was invited by ICAR as a Distinguished Visiting Scientist and located at the Indian Institute of Spices Research, Calicut, where he is now settled with his scientist (nematologist) wife Dr. (Mrs.) Pankajam Nair.

Prof. Nair has won a string of awards, both national and international, for developing "**The Nutrient Buffer Power Concept**" and I am very delighted to write this "Foreword" for the book for the betterment of science and society. I have a deep conviction that there ought to be fundamental paradigm shift in soil management in India if Indian agriculture is to move forward. My current professional assignment as Vice Chancellor of the G.B. Pant University of Agriculture and Technology, Pantnagar, Uttarakhand, where both Prof. Nair and Dr. (Mrs) Nair were on the Faculty (Department of Agronomy and Plant Pathology, respectively) at the College of Agriculture, between 1971 and 1980, is an added impetus.

It is a pleasure for me to recommend this book to every student and researcher in agricultural science, not merely in India, but in the world.

Dr. Mangala Rai

Vice-Chancellor, G.B. Pant University
of Agriculture and Technology
Pantnagar, Uttarkhand, India

Former Director General, Indian Council
of Agricultural Research
New Delhi, India

Preface

India's great President, late Dr. A.P.J. Abdul Kalam, launching the book , "ÏSSUES IN NATIONAL AND INTERNATIONAL AGRICULTURE" authored by Professor Kodoth Prabhakaran Nair, in Raj Bhavan, Chennai

Soil has always fascinated me. Unlike many, who look at it as an inert medium, I look at it as the real repository of life for humans, animals and plants. Imagine a world without soil. Will there be life at all? For the plants we grow for food for our sustenance, for the water we drink for our sustenance, for the bricks we need for building our houses, for the material we need for building our roads, for the base we need to grow our beautiful gardens, for all of these, we need soil and without it all this would simply be impossible. In my opinion, soil is that invaluable gift of God to man on planet earth, and the word SOIL, I would expand simply as the "Soul of Infinite Life," substituting each word for the first letter of the word SOIL. This is how I addressed soil in the World Congress on Soil Science in Hamburg, Germany, in 1986, to which I was an invited speaker to the mirth of the several assembled distinguished delegates. Whenever soil degraded, civilizations collapsed. The Roman Empire collapsed when the North African soils desertified, because the North African soils were the granary for the Roman army. Nearer home (India), we have the example of the *Thar* desert, which once was the home for dense forests.

Having graduated in agriculture, and then gone on to do a doctorate in the prestigious Indian Agricultural Research Institute in New Delhi, and later landing in Belgium with a postdoctoral fellowship, I was determined to know more about soil, and, in particular, the thermodynamics of soil reactions fascinated me. It was in 1980, when I received the Senior Fellowship of the world renowned Alexander von Humboldt Foundation of the Federal Republic of Germany (FRG), and was placed at the prestigious Institute of Plant Nutrition, affiliated to Justus von Liebig University, Giessen (FRG), the seat of world chemistry, that I really set my mind on to intensely research the dynamics of soil reactions. There was a reason behind this. By early eighties, the so-called green revolution, which was nothing but a highly intensive sort of chemical farming, on the pattern of industrial agriculture, had run out of steam and fallen on its face in India, and I suspected that it had much to do with the way Indian soil resources were managed. Soils in Punjab, the "cradle" of the green revolution, were heavily degraded, ground water was polluted with nitrogenous residues from unbridled use of urea, aquifers dried, and biodiversity fast disappearing due to the continuous wheat-rice monoculture. Vast stretches of Punjab soils were turning into barren lands. Ideally, it would have been most desirable to work with soils from Punjab, but the German protocol and the understanding I had with the Director of the Institute of Plant Nutrition, the very distinguished Plant Nutritionist, Professor Konrad Mengel, mandated that I concentrate my research on German soils. In September 1984, I presented my concept, which is now globally known as "The Nutrient Buffer Power Concept," in the International Colloquium for the Optimization of Plant Nutrition in Montpellier, France, following which I was named to the National Chair of the Science Foundation, The Royal Society, Belgium. This provided further impetus to my research on the concept, and I tested its validity in Belgian soils. During the tenure of this assignment, I was requested by the Government of The Republic of Cameroon to help set up an agricultural university in the country, patterned along the "Land Grant Pattern" of USA. I was appointed Professor and Head of the Departments of Agriculture, Soil Sciences and Basic Sciences, at The University Centre in Dschang, which was until then following the classical French type of education, unlike the USA's Land Grant Pattern. The changed circumstances in Cameroon also provided an impetus to my research. I assigned one of my very hard working postgraduate students, Ms. CHI Felicitas BIH, a theme related to the phosphorus and potassium nutrition of white clover (*Trifolium repens* L.), a fodder crop with immense potential for the tottering dairy industry in the country, on the basis of "The Nutrient Buffer Power Concept," and she produced an excellent thesis, under my very keen scientific supervision, which was very highly ranked for its quality and also won a cash award of 250000 CFA (Central African Franc), equivalent to US $ 1000, given by the ministry of higher education and scientific research of the Government of The Republic of Cameroon and the event was celebrated nationally. In 1993, I received an invitation from the University of Fort Hare, Alice, Republic of South Africa, to help set up a Faculty of Agriculture. This provided further impetus to my research on the concept in South African soils. At the same time, I received an invitation from late Professor Horst Marschner, Director of the Institute of Plant Nutrition, University of Hohenheim, FRG, to examine the relevance of my concept in zinc nutrition of the problematic Central Asian (Turkey) soils in wheat production. The concept was very successfully tested there as well. Retuning to India, I was requested by the Indian Council of Agricultural Research in New Delhi to join as Distinguished Visiting Scientist at the Indian Institute of Spices Research at Calicut, Kerala. This provided me an opportunity to work on the concept with regard to both zinc nutrition of Black pepper and potassium nutrition of cardamom. Following my research, the Government of Kerala rescheduled its fertilizer recommendations for these two crops, based on "The Nutrient Buffer Power Concept." All of these success stories are catalogued in this book.

By 2010, I had generated so much of research data that I thought the best way to put it across to the scientific world was in the form of a book. Meanwhile, I received several international awards for developing "The Nutrient Buffer Power Concept." I was also getting frequent invitations to speak on the concept in several national and international conferences, colloquia, and workshops, all of which led to a remarkable global scientific recognition of "The Nutrient Buffer Power Concept." I offer this book in the humble hope that the reader will be enriched in his/her knowledge of soils and how they should be intelligently managed. In fact, the fate of the failed green revolution could have been totally different if only those at the helm of scientific affairs with regard to soil management in India had taken note of my pioneering research in Europe and Africa in the 1980s and in Asia in the 1990s. But, it is never too late, even now we can redeem a hopeless situation, and set the record straight, if only those power-that-be get hold of a copy of this book and critically and seriously study the message contained in it.

Calicut, India Kodoth Prabhakaran Nair

Acknowledgments

At the Feet of Sai
With utmost respect, I dedicate this book, written under the most trying circumstances, to the memory of my late parents, Kuniyeri Pookkalam Kannan Nair and Kodoth Padinhareveetil Narayani Amma, both of whom left me an orphan at a very young age, but whose boundless love and blessings made me what I am today. My eldest brother late Dr. Kodoth Padinhareveetil Krishnan Nair and his wife Shrimathi Varikkara Sathy Nair financed my undergraduate studies, when I had no parents to support me. My elder brother, late Kodoth Padinhareveetil Rathnakaran Nair, was the one who had thoughtfully motivated me to study agriculture, at a time, when it was the least sought after subject in India, and partially supported my postgraduate studies in agronomy. My eldest sister late Shrimathi Kodoth Padinhareveetil Karthiayini Amma and elder sister late Dr. (Mrs) Kodoth Padinahreveetil Meenakshi Amma have always been very supportive of me in my studies during my younger days.

Pankajam, my wife, a nematologist trained in Europe, who chose to be a home maker rather than continue as a scientist, our children, Kannan, our son, Ammu, our daughter, and Arvind, our son-in-law, all of whom lent me inspirational advice at testing times during the course of writing this book.

Black and Charlie, our canine fleet, are a source of perpetual joy to me.

It was Professor Konrad Mengel, one of world's leading Plant Nutritionists, Director of the prestigious Institute of Plant Nutrition, at the Justus von Liebig University, Giessen, The Federal Republic of Germany, and Curator of the Liebig Museum, my former colleague, who first prodded me to research a theme on the difficult dynamics of phosphate nutrition in 1980, when I was selected as a Senior Fellow of the world renowned Alexander von Humboldt Foundation, The Federal Republic of Germany. His unstinting academic support led to the development of "The Nutrient Buffer Power Concept" which finally led to the compilation of this book.

Late Professor Dr. Ir. A.H. Cottenie, former Rector of the State University of Ghent, Belgium, and member of The Royal Academy of Science, Letters and Fine Arts, one of the finest men that I have had the very good fortune to interact with, during the late 1960s, with whom I commenced my postdoctoral research in the dynamics of micronutrients, in Belgium, and who, finally, was instrumental in naming me to the National Chair of the Science Foundation of The Royal Society. Late Professor S.S. Bains and late Professor Rajat De, both Heads of the Department of Agronomy, one succeeding the other, at the prestigious Indian Agricultural Research Institute, New Delhi, India, and late Professor A. Mariakulandai, Dean, College of Agriculture, at the Tamil Nadu Agricultural University, Coimbatore, India, and late Professor Rajagopalan, a distinguished geneticist, at the same University, all of whom were my inspirational guides during my postgraduate research.

A genuine word of gratitude to Dr. (Mrs) Saroja Raman, a distinguished soil chemist, who very kindly permitted me to liberally use material from her book on agricultural sustainability, in the context of the discussion of "The Nutrient Buffer Power Concept" *vis-a-vis* global sustainable agriculture.

Last, but never the least, late Mr. Peter Nye at Oxford and late Dr. Stanley Barber at Purdue, eminent soil scientists, were not only my intellectual inspiration, but who through mechanistic-mathematical models proved the importance of the "Nutrient Buffer Power" in nutrient dynamics, and, left it to me, perhaps as a Providential coincidence, to prove the relevance of the laboratory findings in field experimentation. This book also catalogues the field success stories of "The Nutrient Buffer Power Concept," a project on which I have spent more than 35 years of my professional life in three continents: Europe, Africa, and Asia. "The Nutrient Buffer Power Concept" project was shortlisted for the very prestigious US $ 1 Million Rolex Awards for Enterprise 2012 of The Rolex Foundation, Geneva, Switzerland, for its originality from more than 3500 nominations worldwide and is the only project selected for this coveted distinction from the Asian continent. Also, it was nominated, the only nomination from the Indian subcontinent, for the prestigious 2014 Norman Borlaug Award for Crop Nutrition Research, instituted by the International Fertilizer Industry Association, Paris. I have now been nominated for the very prestigious 2018 Volvo Environment Prize, coordinated by The Royal Swedish Academy of Sciences, Stockholm, and the 2018 Right Livelihood Award, for developing "The Nutrient Buffer Power Concept," and bringing succor to millions of poor and marginal farmers across Africa, Asia, and Latin America.

A Word of Appreciation

I am very pleased to gratefully acknowledge the very purposeful discussions that I have had with Ms. Margaret Deignan, Publisher, Springer. She has been a constant source of guidance and encouragement in this book writing project.

And, finally, to you reader, for that leap of faith, in picking up this book, as a validation of the firm belief of a dedicated and untiring scientist, in a refreshingly new idea, which might open up a totally new chapter in soil management.

Addendum

During the final stages of writing this book, we lost Black, on January 4, 2016, noon time, due to a brief illness, leaving behind Charlie, her son, and a void in my heart. The love and joy she gave us is one of the very best gifts of my life, which I shall always endearingly cherish. May her soul rest in peace.

Black's passing away added further grief to my already saddened heart because of the earlier passing away of Cheri, our dear pure bred German shepherd, whom we lost in September 2005, after she lived with us for about 12 years, while I was away in Esfahan, The Republic of Iran, to deliver the keynote address in an international conference on sustainable agriculture. Both Cheri and Black were such a great source of joy to our life.

An Appreciation of the Book

Sustainability of agricultural production depends on health and nutrient status of soil. The large increases in agricultural production could be achieved through the use of fertilizers, and judicious management of moisture supply was well demonstrated through the "Green Revolution" in the past several decades. However, inadequate attention to balanced fertilization, which resulted in nutrient imbalances, and depletion of organic carbon status of the soil, have clearly shown in the recent years that the productivity increases in agriculture cannot be taken for granted. There has been a serious decrease in factor productivity on the lands that brought us "Green Revolution" because of these lacunae in our soil fertility management. Sustaining the increases in the productivity of agriculture is absolutely essential to meet the food, feed, and fiber needs of the growing world population, which is going to exceed 9 billion by 2050, most of whom will be in the developing world. There is therefore an urgent need to pay attention to improving soil fertility and attaining balanced nutrient status in the soil. The strategies to achieve these will greatly depend on understanding the existing soil nutrient status and the operating factors that are creating the imbalances. Once these aspects are well understood, interventions can be easily devised that will ensure sustained balanced nutrient status in the soil for different cropping systems and production environments.

The concept of "Nutrient Buffer Power" is a novel approach for the judicious management of soil fertility and balanced nutrient availability in the soil for sustained agriculture. The concept is well enunciated in the book *Intelligent Soil Management for Sustainable Agriculture: The Nutrient Buffer Power Concept* by an expert who has devoted his lifetime to studying and researching the dynamic processes occurring in the soil which enable it to sustain agricultural production. Ensuring the mineral nutrient availability to plants at the rate at which they need them and are able to use them so that their use efficiency is maximized and their losses are minimized, giving economic benefit to the farmers and preventing environmental pollution, has been the thrust area of research of Professor Prabhakaran Nair. He has done great service to the discipline of soil science and plant nutrition by widely disseminating the results of his research through publications in highly reputed scientific journals (Advances in Agronomy, Soil Science, Journal of Soil

Science, Oxford, Plant and Soil, Journal of Agricultural. Science, Cambridge, European Journal of Agronomy, Journal of Plant Nutrition and Experimental Agriculture, Cambridge) and books. He has created a cadre of young scientists in this field of academic pursuit to further its horizons and to tackle the next generation problems that will crop up because of changing environment.

Dr. Nair has had a most distinguished career working with world renowned plant nutritionists and soil scientists at well-known seats of learning. As a Senior Alexander von Humboldt Fellow, he undertook advanced research at the Institutes of Plant Nutrition, Justus von Liebig University, Giessen, and University of Hohenheim, Germany. Because of his outstanding research accomplishments he was named to the National Chair of the Science Foundation, The Royal Society, Belgium. He served as Professor and Head of the Departments of Agriculture, Soil Sciences and Basic Sciences, at the University Center, Dschang, The Republic of Cameroon. He was a Senior Professor at the University of Fort Hare, Alice, Republic of South Africa, where he was invited by late Mr Nelson Mandela, to build a Faculty of Agriculture. He was designated as an Eminent Scientist by the General Council of The Kerala Agricultural University and as a Distinguished Visiting Scientist of ICAR at the Indian Institute of Spices Research, Calicut, Kerala. He has received several honors and awards for his research accomplishments, particularly for developing *"The Nutrient Buffer Power Concept."* To cite a few, he was the first runner-up for the Norman Borlaug Crop Nutrition Research Award of the International Fertilizer Industry Association, Paris; was shortlisted (only nominee from Asia) for the US $ 1 Million Rolex Awards for Enterprise of the Rolex Foundation, Geneva; received ***Swadeshi Sastra Puraskar*** of the Swadeshi Science Movement; and was awarded "Life Time Achievement Award as a Distinguished Mentor" by the Alumni Almamater Advancement Association of the GB Pant University of Agriculture and Technology, Pantnagar. The publication *Intelligent Soil Management for Sustainable Agriculture: The Nutrient Buffer Power Concept* draws heavily on the research he carried out during the course of his long scientific journey. The concept is presented in a style that could benefit all those involved in the sustainable development of agriculture. I congratulate Dr. Nair for sharing his lifelong experiences in the field of plant nutrition and soil fertility management through this publication.

Mohan C. Saxena, Ph D (IARI)
Dr agr (Hohenheim), Dr Sci (hc)

Former Professor of Agronomy and Director Research, GB Pant University of Agriculture and Technology

Former Assistant Director General, International Center for Agricultural Research in the Dry Areas

Contents

1 Introduction 1
1.1 The Challenge 3
References 4

2 Efficient Plant Nutrient Management – The Key Factor in Sustainable Soil Management 5
2.1 Soil Tests and Nutrient "Availability" 5
2.2 Rating Soil Tests to Define Nutrient Availability and a Fertility Index 6
References 7

3 The Buffer Power and Effect on Nutrient Availability 9
3.1 Basic Concepts 9
3.2 Measuring the Nutrient Buffer Power and Its Importance in Affecting Nutrient Concentration on Root Surfaces 11
References 13

4 Quantifying the Buffer Power of Soils and Testing Its Effect on Nutrient Availability 15
4.1 Phosphorus 15
4.1.1 P Buffer Power Measurement vs Soil Test P Data for Dependability of Availability Prediction 17
4.1.2 P Buffer Power and Q/I Relationship 18
References 21

5 Case Studies with Asian Soils 23
5.1 Potassium 23
5.1.1 The Importance of K Buffer Power Determination in Predicting K Availability to Perennial Crops 23
5.1.2 The Commercial Significance of K Buffer Power Determination in K Fertilizer Management for Perennial Crops 28

5.1.3 The Role of the NH_4^+Ion on K+ Availability 28
5.1.4 K Buffer Power Measurement vs Contemporary Soil Test K Data for Dependability of K Bio Availability Prediction – The Context of MYR (Maximum Yield Research) and MEY (Maximum Economic Yield) Approaches 30
5.1.5 K Buffer Power and Q/I Relationship 32
References.. 35

6 The Role of Electro-Ultrafiltration (EUF) in Measuring P and K Intensity for the Construction of Buffer Power Curves 39
References.. 42

7 Quantifying the Buffer Power for Precise Availability Prediction – Heavy Metals 45
7.1 Zinc 45
7.2 Quantifying Zn Buffer Power 47
References.. 49

8 Case Studies with South Asian Soils 51
8.1 Quantifying the Zn Buffer Power of the Pepper – Growing Soils 52
References.. 55

9 Case Studies with Central Asian Soils......................... 57
9.1 Establishment of the Zn Buffer Power..................... 58
References.. 61

10 Case Studies with African Soils with Regard to P and K 63
10.1 Other Heavy Metals..................................... 67
10.1.1 Molybdenum 67
10.1.2 Iron... 68
10.1.3 Manganese.................................. 68
10.1.4 Boron 69
References.. 72

11 The Changing Face of Global Agriculture 75
References.. 79

12 Sustainable Agricultural Production on a Small Farm............ 81
12.1 Sustainable Agriculture 82
12.1.1 Historical Perspective 82
12.1.2 Malthus and His Prediction 84
12.1.3 Commencement of the Faustian Bargain........... 86
12.2 Socio-Economic Fallout of the Industrial Agriculture......... 88
12.2.1 The Plight of the Developing Countries 88
12.3 Sustainable Development as a Harbinger of Sustainable Agriculture 91

12.4 Elements of Sustainable Agriculture 92
12.5 Key Aspects of Elements of Sustainable Agriculture 95
12.5.1 Productivity 96
12.5.2 Ecological Viability 97
12.5.3 Economic Viability 98
12.5.4 Social Responsibility 100
12.6 Priorities and Trade-offs among Elements of Sustainability 100
12.6.1 Hierarchy of Agricultural Systems and Trade–Offs 103
12.6.2 Hierarchical Linkages and Sustainability 104
12.6.3 Temporal Dimension of Agricultural Sustainability 105
12.6.4 Contextual Nature of Agricultural Sustainability 105
12.6.5 Diversity of Agricultural Systems 106
12.7 Determinants of Agriculture 106
12.7.1 Population Density and Production Processes 107
References 108

13 General Profile of Current Agricultural Systems 111
13.1 Industrialized Countries 111
13.2 Fairly Developed Countries 112
13.3 Least Developed Countries 113
13.4 Impacts of the Changes in the Determinants of Agriculture 113

14 Sustainability Conundrums 115
14.1 Sustainable Agriculture: An Evolving and Adaptive Process . . . 116
14.2 United Principles, but Different Approaches 117
14.3 Conceptual Framework for Sustainability of Agricultural Production Systems 118
14.3.1 General Framework for Sustainable Agriculture 118
14.4 Operationalization of Sustainable Agriculture 121
14.4.1 Resources and their Importance for Sustainable Agriculture 121
14.4.2 Ecosystem Goods and Services 122
14.4.3 Importance of Ecosystem Functions 123
14.4.4 Value of the World's Ecosystem Services and Natural Capital 123
14.4.5 Sustainable Agriculture and Ecosystem Services 124
References 129

15 Land Management for Sustainable Agriculture 131
15.1 Land and Crop Production 133
15.2 Global Distribution of Arable Land 133
15.3 Per Capita Land Availability 134
15.4 Land Quality and Distribution 136

15.5 Soil Degradation and Its Impact on Agriculture 137
15.5.1 Pre-historic Farming and Soil Degradation 137
15.5.2 Extent of Global Soil Degradation. 139
15.5.3 Causes of Soil Degradation 140
15.6 Types of Soil Degradation in Agriculture 140
15.6.1 Water and Wind Erosion 140
15.6.2 Chemical Degradation 142
15.6.3 Physical Degradation 143
15.6.4 Biological Degradation 144
15.7 Soil Degradation and Crop Production 146
15.8 Sustaining Land Productivity. 147
15.9 Challenges to Sustainable Land Management. 148
15.10 Framework for Land Management for Sustainable Agriculture 149
15.10.1 Appropriate Land Use Patterns 150
15.10.2 Agroecological Zones and Land Use. 151
15.11 Soil Management and Sustainable Agriculture 151
15.11.1 Soil Resilience and Sustainable Soil Management . . . 152
15.11.2 Factors Affecting Soil Resilience. 153
15.11.3 Soil Resilience and Degradation 154
15.12 Core Sustainable Soil Management Strategies 155
References 157

16 Erosion Control and Maintenance of Good Soil Tilth 161
16.1 Erosion Management. 162
References 163

17 Soil Fertility and Nutrient Management. 165
17.1 Role of Soil Organic Mater 168
17.1.1 Organic Matter Fractions. 169
17.2 Soil Structure and Agricultural Sustainability 171
17.2.1 Soil Management and Soil Structure 172
17.2.2 Management Constraints in Soils with Poor Soil Structure 173
17.3 Efficient Use of Inputs and Sustainable Agriculture 173
17.3.1 Need-Based Input Application. 176
17.4 Stressed Ecosystems and Context-Based Technologies 176
17.4.1 Climate. 177
17.4.2 Soil Constraints 178
17.5 Socioeconomic Compatibility 179
17.6 Soil Quality and Sustainable Agriculture 180
17.6.1 Components of Soil Quality 181
17.6.2 Importance of Soil Quality 182
17.6.3 Measuring Soil Quality in Agriculture 183
17.6.4 Qualitative or Descriptive Indicators of Soil Quality 183

17.6.5 Quantitative or Analytical Indicators of Soil Quality . . . 184
17.7 Reference Values for Soil Quality Indicators . . . 185
17.7.1 Statistical Limits . . . 185
17.7.2 Scoring Functions . . . 186
17.7.3 Soil Quality Index . . . 186
17.7.4 Analytical or Quantitative Approach to Soil Quality Index . . . 186
References . . . 187

18 How to Manage Water Use for Sustainable Agriculture? . . . 191
18.1 The Supply-Demand Chain of Fresh Water . . . 193
18.2 Ground Water . . . 194
18.3 Availability of Water Region-Wise . . . 195
18.4 Demand for Water . . . 196
18.5 Water Needs of Agriculture . . . 197
18.5.1 Current Picture . . . 197
18.5.2 The Nature of Things to Come . . . 198
18.5.3 What are the Future Constraints for Irrigated Agriculture? . . . 199
18.5.4 The Consequences of Irrigated Agriculture –Pollution . . . 200
18.5.5 Salinization and Waterlogging of Irrigated Lands . . . 200
18.5.6 Overexploitation of Water Resources . . . 202
18.5.7 Unsustainable Use of Ground Water . . . 203
18.5.8 Food Security Concerns Related To Water Shortages . . . 204
18.5.9 Creating New Water Resources for Irrigated Agriculture . . . 206
18.6 Water Supply and Global Warming . . . 207
18.7 Conservation and Sustainable Use of Water in Agriculture . . . 207
18.8 How to Improve Productivity from Water Use in Crop Production? . . . 209
18.8.1 Varietal Improvement . . . 211
18.8.2 Soil and Water Management . . . 211
18.8.3 The Right Agronomic Practices and Crop Choice . . . 213
18.8.4 Need-Based Irrigation . . . 214
18.9 Technical Improvement of Water Delivery Systems . . . 215
18.9.1 Micro Irrigation Techniques . . . 215
18.10 Water Management in Rainfed Agriculture . . . 216
18.11 Effective Use of Water of Marginal Quality . . . 217
18.12 Need-Based Technologies . . . 218
18.12.1 Equitable Distribution and Sharing of Water . . . 220
18.12.2 Getting the Price Right . . . 221
18.12.3 Trading Water . . . 222
18.12.4 Institutions and Regulatory Mechanisms . . . 222

18.13 The Role of Nitrogen in Sustainable Agriculture 223
18.13.1 Nitrogen Use, Crop Growth and Yield. 224
18.13.2 Nitrogen, Photosynthesis and Plant Growth 225
18.13.3 Synchronization of N Supply and N Demand 226
References 228

19 Primary Productivity and Biodiversity 233
19.1 Nitrogen Use at the Farm and Global Levels. 234
19.1.1 Arable Cropping Systems 234
19.1.2 Mixed Farming Systems 236
References 237

20 Environment and Management 239
20.1 Plant-Soil-Atmosphere 240
20.1.1 Scale and Systems 241
References 242

21 Policy Making and Regulations 243
References 244

22 Phosphate Solubilizing Microorganisms and Their Role in Sustainable Agriculture 245
22.1 Imminent Need for Phosphate Solubilizing microorganisms in Phosphate Nutrition of Crop Plants in Sustainable Agriculture 246
22.2 Nature of Phosphatic Fertilizers 247
22.3 Phosphate Solubilizing Microorganisms. 247
22.4 The Phosphorus Solubilizing Microorganisms- Trail of Search for It 248
22.5 Phosphorus Solubilization –How Does it Happen?. 249
22.6 Application of PSM. 252
22.6.1 Factors Which Affect Survival of PSM 253
22.6.2 Crop Response to Composite Inoculants. 253
22.6.3 Interaction Between PSM and Nitrogen Fixing Organisms 254
22.6.4 Symbiosis Between PSM and Arbuscular Mycorrhizal Fungi 256
22.6.5 Tripartite Symbioses Between Nitrogen Fixers, Phosphate Solubilizers and Arbuscular Mycorrhizal Fungi 258
22.6.6 The Reasons for Failure of Phosphate Solubilizing Microorganism. 259
22.6.7 Application of Genetic Engineering in developing Super phosphate Solubilizing Microbial Inoculants. 260
References 262

23 Energy Management in Sustainable Agriculture 267
23.1 Energy and Agriculture 268
23.1.1 Concern about the Quality and Content of Energy 268
23.1.2 Pattern in Energy Use 269
23.1.3 Energy Intensification in Agriculture 269
23.2 Sustainable Energy Management in High Input Agriculture 274
23.3 Energy Audit of High Energy Systems 275
23.3.1 Sustainable Energy Management in Low – Input Systems 276
23.3.2 Responding to Low Energy Problems 276
23.4 Food-Energy Nexus and Integration 277
23.5 Agriculture as Energy Provider 277
23.6 Biomass Energy and Rural Livelihoods 278
23.6.1 Biomass Energy in Commercial and Industrial Production 279
23.6.2 Examples of Benefits to the Agricultural Sector and Livelihoods 280
23.6.3 Opportunities for Expanding Bioenergy Initiatives . . . 281
23.6.4 Other Renewable Sources of Energy 281
References 283

24 Measurement of Agricultural Sustainability 285
24.1 Dimensions of Agricultural Sustainability 286
24.2 Partial Indices of Agricultural Sustainability 286
24.3 Productivity/Biophysical Indicator 287
24.4 Total Factor Productivity (TFP) 287
24.5 Economic Index 288
24.6 Total Natural Resource Productivity (TNRP) or Social Factor Productivity (SFP) 288
24.7 Soil Quality as an Indicator of Sustainability 289
24.8 The Dynamic Assessment Approach 290
24.9 Future Prospects 291
24.10 Framework for the Evaluation of Sustainable Land Management 292
24.11 Challenges to Global Agricultural Sustainability 292
24.11.1 Current Strengths and Concerns 293
24.11.2 Prognosis for the Future 294
24.12 Proposed Models and Pathways for Sustainability 295
24.12.1 Proposed Pathways for Achieving Production Targets 295
24.12.2 How Realistic Are the Models? 296
24.12.3 How Effective Are the Proposed Pathways for Sustainability? 299

24.13 Future Challenges for Agricultural Sustainability 300
24.13.1 Physical Constraints. 301
24.13.2 Natural Resource Constraints 302
24.14 Conflicting Trends 309
24.15 Technological Constraints 310
References 313

25 Climate Change and Agricultural Production 315
25.1 Challenges to Agriculture from Climate Change Parameters. 317
References 318

26 Achieving Agricultural Sustainability – The Future Challenge 319
26.1 Focus on Differential Capabilities of Resources 320
26.2 Sustainable Intensification of High Potential Lands 321
26.2.1 Differential Intensification of Low-Potential Lands and the People Who Inhabit Them 321
26.3 Expanding the Concept of Sustainable Intensification of High Potential Lands 323
26.3.1 Natural Ecosystems as Models for Agricultural Systems 323
26.4 Agroecosystems. 324
References 325

27 Holistic Technologies. 327
27.1 Conservation Tillage 328
27.2 The Comparative Merits/Demerits of Conservation Tillage 328
References 330

28 Integrated Plant Nutrient Management 331
28.1 Sources of Plant Nutrients and Their Impact. 332
28.2 Merits of Organic Fertilizers 333
28.3 Additional Advantages of Integrated Plant Nutrient Systems . . . 333
28.4 Crop Rotation in Sustainable Agriculture 334
28.4.1 Improvement in Crop Yield 335
28.4.2 Soil Quality 335
28.4.3 The Role of Crop Rotations in Pest and Disease Control. 336
28.4.4 Biological Diversity. 337
28.4.5 Crop Rotation Effect on Reduction of Greenhouse Gases 337
28.4.6 Constraints on Implementing Crop Rotations 338
28.5 Frontier Technologies for Sustainable Agriculture 338
28.5.1 Information Technology and Precision Farming 338
28.6 Sustainable Intensification of Low-Potential Areas. 345
28.7 Agroecological Principles and Protocols. 346
28.8 Biodiversity and Sustainable Agriculture 346

28.8.1 Biodiversity and Its Importance. 346
28.8.2 What Is Biodiversity and What Is Its Importance? . . . 348
28.8.3 Agroecosystems and Agricultural Biodiversity 349
28.8.4 Importance of Agricultural Biodiversity 349
28.8.5 Agricultural Intensification and Biodiversity. 350
28.8.6 Agricultural Pollution and Off-Farm Loss of Biodiversity 351
28.8.7 Conservation and Sustainable Use of Agricultural Biodiversity 352
28.8.8 Current Lacunae. 352
References 353

29 The Salient Conclusions 355

Photos 357

References 365

About the Author

K. P. Prabhakaran Nair is an eminent international agricultural scientist having worked for over three decades in Europe, Africa and Asia, holding some of the most prestigious academic positions, such as, Professor, National Science Foundation, The Royal Society, Belgium; Professor and Head, Departments of Agriculture, Soil Sciences and Basic Sciences, The University Center, The Republic of Cameroon; Senior Professor, University of Fort Hare, Alice, Republic of South Africa. He is globally known for developing "The Nutrient Buffer Power Concept," a revolutionary soil management technique, which is changing the face of modern farming in the developing world. He has also authored nine books, three of which were published by the world's number one science publisher Elsevier and one launched by late and former President of India Dr Kalam in Raj Bhavan, Chennai. Prabhakaran Nair has won a string of national and international awards and now lives with his wife, Dr Pankajam Nair, a trained (from Belgium) nematologist, in Calicut, Kerala, India and is actively engaged in global agricultural research. He is the only agricultural scientist in the world, living or dead, to have been invited six times to write invitational chapters for *Advances in Agronomy,* the *magnum opus* of agricultural science.

Chapter 1
Introduction

Abstract The chapter discusses the centrality of soil in the entire food chain, and, expands the author's perception of Soil as follows: "SOIL: S-Soul O- Of I-Infinite L-Life".

Keywords Concept of soil · Soul of infinite life

Many years ago, in one of the early editions of Advances in Agronomy, Roy W. Simonson writing a chapter entitled "Concept of Soil," noted "Someone has said that the fabric of human life is woven on earthen looms – it everywhere smells of the clay". More than five decades later, we Agronomists and Soil Scientists have come very far in our understanding of "the fabric of human life" which "everywhere smells of the clay." That "the fabric of human life" which is so very intimately linked to soil has dramatically changed is beyond dispute. Yet, there is no denying the fact that this "fabric of human life" will always be linked to the soil which is "the pragmatic, the reality, the entity that dictates much of what societies can do" (Boul 1994). Soil, in my opinion, is that invaluable gift of God to life on planet Earth and can aptly be termed "The Soul of Infinite Life." Though the basic concept of soil, since its early description as a "thin mantle over the land surface" has vastly changed over the years, this thin mantle has always been the focal point since it is the medium of plant growth. For early man it was nothing more than a physical support for his predation. Quite likely, some areas were known to provide better footing than others, and some were to be avoided if possible. It is amazing that even after decades of research in soil science which has provided such invaluable information on this "thin mantle over the land surface" so crucial to the existence of life, human, plant and animal, on planet earth, this basic instinct of predation has remained unchanged. How else can we explain the disdain and callousness so often witnessed in modern societies, propelled by an insatiable greed to acquire unlimited wealth, which leads to the abuse of soil, this invaluable gift of God to man? Undoubtedly, the earliest shift in attitude toward soils must have originated at a time when man began to grow food, rather than gather it as his ancestors did. In many ways, this shift in attitude was the precursor to modern day soil science. Though this shift must have occurred in pre-Christian times, about 9000 years ago, and focused on the inevitability of a

K. P. Nair, *Intelligent Soil Management for Sustainable Agriculture*,
https://doi.org/10.1007/978-3-030-15530-8_1

proportionally smaller land surface supporting a larger human population, it is only in recent times that we have witnessed the magnitude of the impact of this shift in attitude on human existence. Much land has degraded and become unsuitable for agriculture since a century ago. The 1992–1993 World Resources Report (Stammer 1992) from the United Nations on the status of the world soils contains very alarming conclusions. For example, nearly 10 million ha of the best farm lands of the world have been so ruined by human activity since World War II that it is impossible to reclaim them. Over 1.2 billion ha that have been seriously damaged and can be restored only at a great cost. This loss in soil capability could mean that there will be enormous food shortages in the next 20–30 years and, as is but natural, the people of disadvantaged nations will suffer the most. Many factors have contributed to this alarming state of affairs, one of the prime factors being the "high input agriculture," or more specifically the chemical agriculture, euphemistically known as the "green revolution," where unbridled use of chemical fertilizers led to soil ruination. Punjab, the "cradle of green revolution" in India is a testimony to this sad state of affairs, where unbridled use of chemical fertilizers to boost the yield of dwarf wheat and rice varieties led to soil degradation, the ground water is loaded with high amounts of fertilizer residues (especially nitrate from urea hydrolysis in soil) which has made water no more potable, led to soil salinity, dried aquifers and also led to vanishing bio diversity due to continuous monoculture of wheat and rice. There are hundreds of acres of land stretches which have become barren, where once stood lush wheat and rice fields. Crop yields have plateaued or drastically declined. Two-thirds of the seriously eroded land is in Asia and Africa. About 25% of the cropped land in Central America is moderately to severely damaged. In North America, this is a small percentage – only 4.4%. Since the time of this "green revolution," food production has declined dramatically in 80 developing countries in the past decade. Soil degradation is the major factor. Nearly 40% of the world's farming is done on very small parcels of 1 ha or less (Robison et al. 1981). Ignorance and poverty characterize this situation. Yet, emphasis on agriculture has been confined mostly to large-scale farming. Large-scale farming, grand projects at huge costs and huge profits, have been the order of the day for many decades. In a lighter vein, it can be said that even the "lebens raum" concept of Adolf Hitler had an echo in the inevitability of this modern day fact. What else can justify the ruthless conquest of vast territories of land by this master strategist who set out to conquer the world – or, more appropriately, the world's soils?

Despite the complexity of soil science and the emergent soil management practices, the basic concept of soil as a medium of plant growth can be expected to persist for an indefinite length of time. But it is becoming increasingly clear that the earlier views on soil as merely the "supportive medium" for plant growth is giving place to newer ones on "managerial concepts" of this supportive medium. This is amply illustrated by the shift in focus from the green revolution phase of the 1960s to mid-1970s where application of increasing quantities of soil inputs such as fertilizers and pesticides was emphasized, to the "sustainable agriculture" phase from the early 1980s to the present (probably to continue?); sustainable agriculture places more reliance on biological processes by adopting genotypes to adverse soil conditions, enhancing soil biological activity and optimizing nutrient cycling to minimize external inputs, such as fertilizers, and maximize their efficiency of use. In fact, the paradigm of the earlier phase has given way to the emergent new paradigm (Sanchez 1994) and this is clearly reflected in the dialogue of the world leaders during the first Earth Summit in 1991 in Rio de Janeiro, Brazil, where Agenda 21 has incorporated six chapters on soil management issues (Keating 1993). The focus of this book will be on the second paradigm inasmuch as prescriptive soil management is concerned with regard to understanding soil nutrient bio availability and its efficient management in crop production.

1.1 The Challenge

Soils are regarded by the International Policy Community as increasingly important in world development issues, such as, food security, poverty alleviation, land degradation, and the provision of environmental services (Wood et al. 2000). Soils are a crucial component of terrestrial ecosystems and a determinant of their capacity to produce goods and services. Soils exert production, buffering, filtering and biological functions. Solar energy, carbon dioxide and nitrogen from the air, and nutrients from the soil are converted into plant products that provide animals and humans with food, fiber and biofuels. Soils hold water from episodic rainfall or irrigation as well as nutrients applied as organic inputs or mineral fertilizers, releasing them at rates plants can utilize for longer periods of time. Soil biota decomposes organic minerals, cycle nutrients, and regulates gas fluxes to and from the atmosphere. Soils filter non-hazardous and toxic compounds through surface adsorption and precipitation reactions and largely determine the quality of terrestrial waters. Soils, therefore, deliver many of our basic needs and play a central role in determining the quality of the environment.

Though we all, as soil scientists, will not dispute what has been written above, how is it that soil science has been pushed to the backyard, while others have made spectacular advances, or at least, others are noticed by people, at large, the world over, while we continue to be ignored by the world? In fact, the most classic example, of late, is the science of genetically modified organisms (GMOs), be they of plant or animal origin. Notwithstanding the controversies surrounding them, they have captured the attention of people all over the world. Ironically, none stops to think that, after all, even a genetically modified plant, say for instance a Bt cotton plant, cannot grow in outer space, but, needs a fertile soil to grow on. Tragically, the science of soil, is not even recognized as a "science," in the sense others are, as for example, plant science or medicine, or even economics. More often than not, my wife and two grown up children, an elder son and a younger daughter, chide me as to where I have reached in life during the last more than forty years of professional commitments, dirtying my hands with soil, and thinking of the science of soil as a "science." Their difficulty in trying to understand what I have done, and continue to do, is what public awareness people call a "brand failure." Our lack of visibility is related to our culture as "reductionist scientists," operating largely within our limited scientific circles and with land users who utilize our knowledge of soil science and our extension services. The take home message is that we soil scientists are the problem. But, we can become the solution by undertaking the kinds of synthesis research that are of direct use to policy makers and communicating to them in a way they readily accept it and put it to use. A classic example of global success is that of the Bt technology, where policy makers, starting from the United States of America to India (where the science of the Bt technology was totally unknown even up to as late as 2002), got the total involvement of the governments, and, so the policy makers who run them. Soil science has been brilliantly informed by reductionist physics and chemistry, poorly informed by biology, ecology and geography and largely uninformed by the social sciences (Swift 1999). In a survey of the global environment, The Economist magazine (6 July 2002) reported that leading experts could not reach a firm conclusion about the state of the environment, because, much of the information they needed was incomplete or missing altogether. This article went on to say that businessmen always say "What matters gets measured." While soil scientists cannot be accused of not measuring soil properties, we are perhaps guilty in the lack of synthesis, integration, and interpretation of those measurements as they relate to environmental goods and services. We should certainly take on this challenge. The central focus of this book would be how "The Nutrient Buffer Power Concept" could be made to be an acceptable idea for the policy makers, so that it attains the status of a "brand."

References

Boul, S. W. (1994). Soil, society's yoke to the earth. In *Proceeding XV of the ISSS symposium* (Vol. 1, pp. 89–104), Acapulco, Mexico, July 10–16, 1994.

Keating, M. (1993). *The Earth Summit's Agenda for Change* (A Plain Language Version of Agenda 21 and the Other Rio Agreements). Geneva: Centre for our Common Future.

Robison, L. R., Johnston, N. P., & Hill, J. M. (1981). Small-scale agriculture, an untapped giant. *Benson Institute, 4*(2), 4–7.

Sanchez, P. A. (1994). Tropical soil fertility research: Towards the second paradigm. In *Proceeding XV ISSS symposium, Acapulco, Mexico, July 10–16, 1994*," Vol. 1, pp. 89–104).

Stammer, L. B. (1992, January). Study finds serious harm to 10 percent of world's best soil. *Los Angeles Times* (quotations from World Resources Institute of U.N. Environmental Program).

Swift, M. J. (1999). Integraing soils, systems and society. *Nature & Resources, 35*, 12–20.

Wood, S., Sebastian, K., & Scherr, S. J. (2000). *Pilot analysis of global ecosystems: Agroecosystems*. Washington, DC: International Food Policy Research Institute/World Resources Institute.

Chapter 2
Efficient Plant Nutrient Management – The Key Factor in Sustainable Soil Management

Abstract The chapter discusses key aspects of efficient plant nutrient management – the key factor in sustainable soil management. The supporting aspects, such as, soil tests and nutrient availability with reference to rating of soil tests to define nutrient bio availability and fertility index are also discussed.

Keywords Soil tests · Nutrient availability · Sustainable soil management

Agricultural systems differ from natural systems in one fundamental aspect: while there is a net outflow of nutrients by crop harvests from soils in the first, there is no such thing in the second (Sanchez 1994). This is because nutrient losses due to physical effects of soil and water erosion are continually replenished by weathering of primary minerals or atmospheric deposition. Hence, the crucial element of sustainability of crop production is the nutrient factor. But, of all the factors, the nutrient factor is the least resilient (Fresco and Kroonenberg 1992). The thrust of high input technology, the hallmark of the "green revolution," in retrospect, or the moderation by low input technology, the foundation stone of "sustainable agriculture," in prospect, both dwell on this least-resilient nutrient factor. If the pool of nutrients in the soil, both native and added, could be considered as the "capital," efficient nutrient management might be analogous to raising the "interest" accrued from this capital in such a way that there is no great danger of the erosion of this capital. Hence, sound prescriptive soil management should aim at understanding the actual link between the "capital" and the "interest" so that meaningful management practices can be prescribed.

2.1 Soil Tests and Nutrient "Availability"

It is universal knowledge that soil tests are the basis for predicting nutrient "availability." There are, perhaps, as many soil tests for each nutrient as there are nutrients. This book will not dwell on the merits or demerits of any single soil test or group of them. Suffice to say that fertilizer recommendations traditionally are made at the point where marginal revenues equal marginal costs which involve some positive

K. P. Nair, *Intelligent Soil Management for Sustainable Agriculture*,
https://doi.org/10.1007/978-3-030-15530-8_2

Table 2.1 Nutrient balance (kg/ha/year) in intensively managed Arable soils

	N	P	K
Inputs			
Fertilizers	156	39	119
Other	32	–	9
Total	188	39	128
Outputs			
Harvest	103	16	91
Removal	24	–	–
Other			
Total	127	16	91
Balance	61	23	37

After Frissel (1978)

synergism (DeWit 1992). The most common result of this approach is the vast build-up in the soil nutrient pool in intensively cultivated soils (Whitmore and van Noordwijk 1994). Data in Table 2.1 indicate positive balances (in kg/ha/year) for N(61), P(23), and K(37) in intensive crop production systems (Frissel 1978).

Over several decades, such positive balances can lead to a huge build-up of the nutrient capital, especially in the case of high-input, intensive agricultural systems as in the case of many European, North American, and Scandinavian countries. A dangerous consequence of such huge nutrient build-up is nutrient contamination of groundwater to such extremes that "environmental soil tests" become necessary to assess critical limits of nutrient pollution (Sharpley et al. 1993). Nitrogen is a prime candidate for this scenario especially in the temperate zone. At the other end of the spectrum are the marginal areas of the tropical zone where inadequate replenishment of nutrient removal by crop growth and also nutrient loss by soil and water erosion has left that capital "in the red." Initially fertile Alfisols of much of Africa with subsequent severe depletion of N and P (Yates and Kiss 1992) are an example of this nutrient "bankruptcy." Either way, contemporary soil tests are the basis on which prescriptive management practices are formulated.

2.2 Rating Soil Tests to Define Nutrient Availability and a Fertility Index

Most soil test laboratories around the world use some kind of "rating system" to evaluate soil test values. These rating systems invariably use qualitative terms such as "low," "medium," or "high" to describe the availability of a specific plant nutrient. Admittedly, these terms denote different meanings in the context of availability of a particular plant nutrient and, at best, are empirical terminologies. This problem has been recognized by researchers over the years. Morgan (1935) suggested a scale of 1–10 with 8 equal to the point of no response. Bray and Kurtz (1945) used relative yield or percentage sufficiency to describe the degree of deficiency, with 100 defined as the point of no response. The index below 100 follows the curvilinear relationship between soil-test values and yield without addition of the element. Above 100, the index displays a straight-line relationship indicating the relative margin of adequacy or the proximity to an excessive level. To eliminate the need for a percent sign, the values are referred to as "Fertility Indices" and they are reported to the nearest multiple of 10 from 0 to 9990 (Cope and Evans 1985). In addition to ratings, most laboratories use some method of reporting results more precisely, mainly for use by farmers in record keeping and monitoring soil fertility. Some report kg/ha, lb/a, or ppm extracted, but these would be confusing to farmers, because each element has a different level for a specific degree of adequacy (Cope and Evans 1985). For instance, the adequate or critical level for one soil may be 25 ppm P, 120 ppm K, 200 ppm Ca, and 30 ppm Mg. Adequate levels in other soils and from other extracting procedures would be different for each element (Cope and Evans 1985).

Despite the fact that a number of soil tests and others such as Diagnosis and Recommendation Integrated Systems (DRIS) are in vogue, to predict nutrient availability, it must be said that a universal picture is yet to emerge in this field with regard to precise availability prediction. This is primarily because a soil test in the laboratory can never simulate plant root absorption of a nutrient in a field soil, though most of the time the assumption is that it does. In the final analysis, it is the plant and plant alone which will decide whether or not the nutrient is available. This book will examine the question whether a better and more reliable alternative exists.

References

Bray, R. H., & Kurtz, L. T. (1945). Determination of total organic and available form of phosphorus in soils. *Soil Science, 59*, 39–45.

Cope, J. T., & Evans, C. E. (1985). Soil testing. *Advances in Soil Science, 1*, 201–228.

DeWit, C. T. (1992). Resource use efficiency in agriculture. *Agricultural Systems, 40*, 125–151.

Fresco, L. O., & Kroonenberg, S. B. (1992, July). Time and spatial scales in ecological sustainability. *Land use policy, 9*, 155–167.

Frissel, M. J. (Ed.). (1978). *Cycling of mineral nutrients in agricultural ecosystems*. Amsterdam: Elsevier.

Morgan, M. F. (1935). *A universal soil testing system* (Vol. 372). New Haven: Connecticut Agricultural Experiment Station.

Sanchez, P. A. (1994). Tropical soil fertility research: Towards the second paradigm. In *Proceeding XV ISSS symposium, Acapulco, Mexico, July 10–16, 1994*," Vol. 1, pp. 89–104).

Sharpley, A. N. I., Pierzynki, G., & Sims, J. T. (1993). Innovative soil phosphorus availability indices: Assessing inorganic phosphorus. In *Agronomy Abstracts*. Madison: American Society of Agronomy.

Whitmore, A. P., & van Noordwijk, M. (1994). Bridging the gaps between environmentally acceptable and agronomically desirable nutrient supply. In D. Glen (Ed.), *Proceedings of the Long Ashton Symposium: Agriculture in the 21st century*. in press.

Yates, R. A., & Kiss, A. (1992). Using and sustaining Africa's soils. In *Agricultural and rural development series* (Vol. 6). Washington: World Bank.

Chapter 3
The Buffer Power and Effect on Nutrient Availability

Abstract The chapter discusses at length the basic concepts pertaining to nutrient bioavailability. The focus of the chapter is on the thermodynamics of soil nutrient bioavailability enmeshing diffusive process within the soil and in the plant cell, based on mechanistic-mathematical models, and lays down the fundamentals of "The Nutrient Buffer Power Concept", and, how precise quantification of the "Buffer Power" of each of the investigated nutrients, namely, Phosphorus, Potassium, and Zinc, lead to a clear understanding of plant nutrient bioavailability.

Keywords Basic concepts · Nutrient concentration · Nutrient buffer power

3.1 Basic Concepts

In any nutrient management approach that is sound and reproducible, one must start with a basic understanding of the chemical environment of plant roots. When we consider this, the first term that we come across is the "soil solution," because the plant root is bathed in it and is most affected by its chemical properties. The Soil Science Society of America (1965) defines soil solution as "the aqueous liquid phase of the soil and its solutes consisting of ions dissociated from the surfaces of the soil particles and of other soluble materials." Adams (1974) has given a simple definition: "The soil solution is the aqueous component of a soil at field-moisture contents." Perhaps it is important to emphasize here that much of contemporary soil testing has considered a soil extract as synonymous with the soil solution. Since soil extraction is supposed to simulate plant root extraction, it is pertinent to consider the chemical environment of the root, though briefly, from this angle. It is worth noting that the chemical environment of roots in natural soil systems is so obviously complex that both soil scientists and plant physiologists have been unable to provide a precise definition. If this complex chemical system is to be accurately quantified, thermodynamic principles will need to be used to evaluate experimental data. Even then, the limitations are obvious, as in the case of K where the thermodynamic investigations are quite often inapplicable under field conditions. This is because, although a quasi-equilibrium in K exchange can be achieved in the laboratory, these

K. P. Nair, *Intelligent Soil Management for Sustainable Agriculture*,
https://doi.org/10.1007/978-3-030-15530-8_3

conditions are seldom, if ever, attained under field conditions (Sparks 1987). Agricultural soils are, for the most part, in a state of disequilibrium owing to both fertilizer input and nutrient uptake by plant root. It thus appears that a universal and accurate definition of a root's chemical environment awaits the proper application of thermodynamics for the root's ambient solution (Adams 1974) or even kinetics, as in the case of K (Sparks 1987), where thermodynamics has been found inadequate.

Soil extractions with different extractants provide a second approach in defining the root's chemical environment. This approach has been particularly successful in understanding cases like P insolubility, soil acidity, and K fixation. However, this approach also fails to define precisely the root's chemical environment. Though this approach also suffers from deficiencies, such as the extractants removing arbitrary and undetermined amounts of solid-phase electrolytes and ions (or the extractants causing precipitation of salts or ions from the soil solution) and the soil–plant interrelationship defined in terms of the solid phase-component of the soil, even though the solid phase is essentially inert except as it maintains thermodynamic equilibria with the solution phase (Adams 1974), the latter part could be researched more to understand how the solid phase–solution phase equilibria can be interpreted to give a newer meaning to quantifying nutrient bio availability. It is in this context that the role of the plant nutrients' "buffer power" assumes crucial importance.

The close, almost linear, relationship in a low concentration range of <0.5 mM for $NO_3^-N, NH_4^+-N, H_2PO_4^-, HPO_4^{2-}$, which has been established by numerous solution culture experiments, can be quantitatively described by the equation

$$U = 2\pi r \alpha C_r$$

where U = uptake of a 1-m root segment, r = root radius, C_r = concentration of the ion at the root surface and, α = root absorbing power (Mengel 1985). The metabolic rate of the root determines its absorbing power. A high root absorbing power would imply that a relatively high proportion of nutrient ions coming in contact with the root surface is absorbed and vice versa. The nutrient ion concentration at the root surface (C_r) depends on a since a high root absorbing power tends to decrease C_r; it also depends on the rate of movement from bulk soil toward the root (Mengel 1985). Diffusion and/or mass flow controls this movement. But it is now established that nearly 95% of this movement for nutrients such as P, K, and Zn (among heavy metals) and, possibly NH_4^+-N, is by way of diffusion. When root uptake of an ion species is less than its movement toward it, accumulation of the ion species on the root surface is bound to occur, as has been shown to be the case with Ca^{2+} where mass flow contributes to this accumulation (Barber 1974). The diffusive path for ions, such as P and K, which plant roots take up at high rates but which are in low concentration in the soil solution near the root, is the concentration gradient. In a sense, the effective diffusion coefficient which quantifies the diffusive path and the buffer power are analogous because the diffusive flux across the root surface is integrally related to the nutrient buffer power. This has been shown to be true in the case of P where a highly significant positive correlation between the two was found to exist in 33 soil samples obtained from experimental sites in the U.S.A. and Canada (Kovar and Barber 1988). However, in a routine laboratory set-up, it is far easier to measure the buffer power than the effective diffusion coefficient and this book will further focus on the question of how buffer power can be quantified without recourse to cumbersome analytical techniques and how its integration into routine soil test data will considerably improve predictability of nutrient uptake, and, consequently lead to dependable fertilizer recommendations for the nutrients discussed above.

3.2 Measuring the Nutrient Buffer Power and Its Importance in Affecting Nutrient Concentration on Root Surfaces

The ability to predict the mobility of dissolved chemicals, such as fertilizers, in the soil is of considerable value in managing fertilizer applications. Soil testing, in its essence, aims to achieve this. While modeling transport and retention of ions from thermodynamic (Selim 1992), kinetic (Sparks 1989), and mechanistic (Barber 1984) angles could be very informative, the importance of translating this information into practically feasible procedures in crop production calls for an understanding not only of the basic concepts but of their intelligent application as well. In a dynamic state of plant growth, the concentration of any nutrient on the root surface is nearly impossible to measure since both the nutrient in the plant tissue and the root absorbing power, which directly affects it, change quickly due to root metabolic processes. The inability of even mechanical mathematical models to accurately predict nutrient influx rates has recently been highlighted (Lu and Miller 1994). Hence, if an effective soil testing procedure is to be devised for a nutrient, which is an alternative to defining the plant root's chemical environment, one must resolve the problem of quantifying the nutrient concentration on the root surface indirectly, even if it is impossible to resolve it directly, for the reasons mentioned above.

Using Fick's first law,

$$F = -D(dC/dx)$$

where F = the flux, dC/dx = concentration gradient across a particular section, and D = the diffusion coefficient, Nye (1972) has suggested that the formula can be applied to both ions and molecules. The negative sign for D implies net movement from high to low concentration. Although for molecules in simple systems like dilute solutions D may be nearly constant over a range of concentrations, for ions in complex systems like soils and clays D will usually depend on the concentration of the ion, and on that of other ions as well (Nye 1972). Nye (1972) has further suggested that though Fick's first law may be derived from thermodynamic principles in ideal systems, in a complex medium such as the soil the above equation may be regarded as giving an operational definition of the diffusion coefficient. Thus, Nye (1972) defines the diffusion coefficient as

$$D = D_l \Theta f_l \left(dC_l / dC\right) + D_E$$

where D_l = diffusion coefficient of the solute in free solution, Θ = the fraction of the soil volume occupied by solution and gives the cross section for diffusion, f_l = an impedance factor, C_l = concentration of solute in the soil solution, D = an excess term which is zero when the ions or molecules on the solid have no surface mobility, but represents their extra contribution to the diffusion coefficient when they are mobile. D_E can generally be neglected since only in rare instances will it play any role in diffusion of plant nutrient ions in soil (Mengel 1985). From the point of view of nutrient availability, dC_l/dC, which represents the concentration gradient, assumes crucial importance, as we shall see below.

The term dC_l/dC, where C_l = concentration of the nutrient ion in the soil solution and C = concentration of the same ion species in the entire soil mass, assumes considerable significance in lending a practical meaning to nutrient bio availability. If we ascribe the term "capacity" or "quantity" to C and "intensity" to C_l, we have in this term an integral relationship between two parameters that may crucially affect nutrient bio availability. Since the concentration gradient of the depletion profile of the nutrient in the zone of nutrient uptake depends on the concentration of the ion species in the entire soil mass (represented by "capacity" or "quantity") in relation to the rate at which this is lowered on the plant root surface by uptake (represented by "intensity"), it could be argued that a quantitative relationship between the two should represent the rate at which nutrient depletion and/or replenishment in the rooting zone should occur (Nair 1984a, b). This relationship has been functionally quantified by Nair and Mengel (1984) for P in eight widely differing central European soils (Table 3.1) and the term dC_l/dC has been referred to as the "nutrient

Table 3.1 Comparison of P buffer power of eight widely differing central European soils (Determined by two different techniques)

	Regression		r	
Soil	(1)	(2)	(1)	(2)
Benzheimer Hof	Y = 18.8x + 7.94	Y = 0.23x + 8.98	0.912	0.995
Hungen	Y = 38.2x −1.03	Y = 4.32 + 0.25x	0.967	0.997
Oldenburg B6	Y = 49.8x + 0.52	Y = 0.72 + 0.26x	0.994	0.999
Wolfersheim	Y = 70.3x + 0.03	Y = 0.11 + 0.27x	0.998	0.983
Obertshausen	Y = 70.5x + 2.66	Y = 2.89 + 0.30x	0.966	0.998
Oldenburg B3	Y = 73.6x + 2.07	Y = 0.61 + 0.31x	0.994	0.997
Klein–Linden	Y = 75.0x + 0.38	Y = 1.81 + 0.32x	0.999	0.991
Gruningen	Y = 75.4x + 0.89	Y = 3.62 + 0.36x	1.000	0.996

Note. The b values in the regression functions represent the P buffer power of each soil. In regression function (1) (after Nair and Mengel 1984) Y = CAL-P (Schüller's method) and in regression function (2) (after Nair 1992a, b) Y = the author's method. x in both refers to electro-ultrafiltrable P Note the very high r values in all the cases. The soils are arranged in their sequential increase in P buffer power

buffer power." Nair and Mengel (1984) used electro-ultrafiltration to quantify C_l while using an incubation and extraction technique to quantify C. For P, C was found to closely approximate isotopically exchangeable P (Keerthisinghe and Mengel 1979), but in the experiments conducted by Nair and Mengel (1984), it was estimated by the extraction of incubated soil with an extractant which was a mixture consisting of 0.1 M Ca lactate + 0.1 M Ca acetate + 0.3 M acetic acid at pH 4.1. The extractant exchanges adsorbed phosphate and dissolves Ca phosphates except apatites; the method known as the "CAL-method," developed by Schüller (1969), is now widely used in central Europe. In the case of K+ and $NH_4^+ - N$, C denotes the concentration of exchangeable, and to some extent nonexchangeable, fractions (Mengel 1985). Since very low concentrations in the range of 2.0 μM may be attained on the root surface for both P and K (Claassen and Barber 1976; Claassen et al. 1981; Hendriks et al. 1981), Nair and Mengel (1984) had to use electro-ultrafiltration to quantify C_l. Thus, the nutrient depletion around the roots which is caused by the diffusive flux of the nutrients toward the root surface is related to both the quantity and the intensity parameters, and a quantifiable relationship between both represents the buffer power specific to the nutrient and the soil. A growing root will at first encounter a relatively high concentration of P which is in the range of the concentration of the bulk soil solution (Nair and Mengel 1984). As uptake continues, a depletion will occur at the root surface. This depletion profile gets flatter with enhanced nutrient uptake (Lewis and Quirk 1967; Claassen et al. 1981; Hendriks et al. 1981). But it is the capacity of the soil to replenish this depletion which ensures an optimal supply of nutrient ions to the plant root without greatly depressing its average concentration on the root surface. It is the nutrient's buffer power that decides these depletion and/or replenishment rates. A soil with a high P buffer power implies that the P absorbed from the soil solution is rapidly replenished. In such a case, P concentration at the root surface decreases only slowly and mean P concentration at the root surface remains relatively high. In soils with a low P buffer power, the reverse is true, and P concentration at the root surface is rapidly diminished and remains relatively low. This has been proved to be true experimentally for P (Nair and Mengel 1984; Nair 1992a, b). This phenomenon holds true for Zn^{2+} (Nair 1984a, b), K^+ (Nair et al. 1997), and (Mengel 1985) as well.

References

Adams, F. (1974). Soil solution. In E. W. Carson (Ed.), *The plant root and its environment* (pp. 441–481). Charlottesville: University Press of Virginia.

Barber, S. A. (1974). Influence of the plant root on ion movement in soils. In E. W. Carson (Ed.), *The plant root and its environment* (pp. 525–564). Charlottesville: University Press of Virginia.

Barber, S. A. (1984). *Soil nutrient bioavailability: A mechanistic approach*. New York: Wiley.

Claassen, N., & Barber, S. A. (1976). Simulation model for nutrient uptake from soil by growing plant root system. *Agronomy Journal, 68*, 961–964.

Claassen, N., Hendriks, K., & Jungk, A. (1981). Rubidium-Verarmung des wurzelnahen Bodens durch Maispflanzen. *Zeitschrift für Pflanzenernährung und Bodenkunde, 144*, 533–545.

Hendriks, L., Claassen, N., & Jungk, A. (1981). Phosphatverarmung des wurzelnahen Bodens und Phosphataufnahme von Mais und Raps. *Zeitschrift für Pflanzenernährung und Bodenkunde, 144*, 486–499.

Keerthisinghe, G., & Mengel, K. (1979). Phosphatpufferung verschiedener Böden und ihre Veränderung infolge Phosphatalterung. *Mitteilungen der Deutschen Gesellschaft, 29*, 217–230.

Kovar, J. L., & Barber, S. A. (1988). Phosphorus supply characteristics of 33 soils as influenced by seven rates of phosphorus addition. *Soil Science Society of America Journal, 52*, 160–165.

Lewis, D. G., & Quirk, J. P. (1967). Phosphate diffusion in soil and uptake by plants III. 31P-movement and uptake by plants as indicated by 32P-autoradiography. *Plant and Soil, 26*, 445–453.

Lu, S., & Miller, M. H. (1994). Prediction of phosphorus uptake by field-grown maize with the Barber-Cushman model. *Soil Science Society of America Journal, 58*, 852–857.

Mengel, K. (1985). Dynamics and availability of major nutrients in soils. *Advances in Soil Sciences, 2*, 65–131.

Nair, K. P. P. (1984a). Towards a better approach to soil testing based on the buffer power concept. In *Proceeding of 6th International colloquium for the optimization of plant nutrition* (Vol. 4, pp. 1221–1228), 2–8 September, Pierre-Martin Prevel, Montpellier, France.

Nair, K. P. P. (1984b). Zinc buffer power as an important criterion for a dependable assessment of plant uptake. *Plant and Soil, 81*, 209–215.

Nair, K. P. P. (1992a). Measuring P buffer power to improve routine soil testing for phosphate. *European Journal of Agnon, 1*(2), 79–84.

Nair, K. P. P. (1992b). Soil testing procedure to economise on fertilizer use in land management. In *The spirit of enterprise — The 1993 rolex awards* (pp. 323–325).

Nair, K. P. P., & Mengel, K. (1984). Importance of phosphate buffer power for phosphate uptake by rye. *Soil Science Society of America Journal, 48*, 92–95.

Nair, K. P. P., Sadanandan, A. K., Hamza, S., & Abraham, J. (1997). The importance of potassium buffer power in the growth and yield of cardamom. *Journal of Plant Nutrition, 20*(7 & 8), 987–997

Nye, P. H. (1972). Diffusion of ions and uncharged solutes in soils and clays. *Advances in Agronomy, 31*, 225–272.

Schüller, H. (1969). Die CAL-Methode, eine neue Methode zur Bestimmung des pflanzen verfügbaren Phosphates in Böden. *Zeitschrift für Pflanzenernährung und Bodenkunde, 123*, 48–63.

Selim, H. M. (1992). Modeling the transport and retention of inorganics in soils. *Advances in Agronomy, 47*, 331–384.

Sparks, D. L. (1987). Potassium dynamics in soils. *Advances in Soil Sciences, 6*, 2–63.

Sparks, D. L. (1989). *Kinetics of soil chemical processes*. San Diego: Academic.

Chapter 4
Quantifying the Buffer Power of Soils and Testing Its Effect on Nutrient Availability

Abstract The chapter elaborately discusses how the "Nutrient Buffer Power" can be precisely quantified in the laboratory and its significance in test crop nutrient uptake can be measured by appropriate statistical computations. The first instance is with regard to Phosphorus, in Central European soils, also in African soils, followed by Potassium in African and South Asian soils, and Zinc, in South and Central Asian soils. All the three are the most important diffusible nutrients.

Keywords Quantifying buffer power · Nutrient bio availability · Phosphorus buffer power

4.1 Phosphorus

When the exchangeable or desorbable nutrient ions are plotted on the y axis and the corresponding concentration of the same ions in the equilibrated solution on the x axis, we obtain a slope, the steepness of which represents the buffer power for that ion. The P buffer powers of eight central European soils widely differing in their physicochemical properties are reported in Table 3.1.

The "quantity" factor (C) in the Eq. (24.3) was determined by two methods: (i) by an incubation and extraction technique (Nair and Mengel 1984) and (ii) by an absorption–desorption equilibrium technique (Nair 1992a, b) while the "intensity" factor (C_1), was determined by electro-ultrafiltration (Nemeth 1979). It is highly remarkable that the correlation coefficients, in the case of all the soils investigated, between C (represented by y) and C_1 (represented by x) are very highly statistically significant as indicated by the r values, which in almost all cases were close to unity [including one soil (Gruningen) in which it was absolute unity]. This clearly establishes the fact that the techniques employed will accurately quantify the concentration gradient for P in the soils investigated when plants are grown in those soils. If this hypothesis is true, the impact of the P buffer power will clearly be reflected in the P uptake by the test plant since the P buffer power becomes the most crucial factor in predicting P bio availability. This can be tested by correlating P uptake with the routine soil test data without integrating the P buffer power and also by

K. P. Nair, *Intelligent Soil Management for Sustainable Agriculture*,
https://doi.org/10.1007/978-3-030-15530-8_4

Table 4.1 Correlation coefficients (r) for the relationship of P uptake vs routine P Soil tests without (A) and with (B) integration of the P buffer power

	A	B
P uptake vs EUF-P	0.393*	0.8870**
P uptake vs CAL-P	0.714**	0.9415**

After Nair and Mengel (1984)
Note. Electro-ultrafiltrable P (EUF-P) and Ca lactate + Ca acetate + acetic acid mixture extractable P (CAL-P) (Schüller 1969), respectively, used routinely to characterize P availability in central European soils
*, ** Significant at 5 and 1% confidence levels, respectively

integrating the P buffer power. Nair and Mengel (1984) used the multiple linear regression eq. $Y = a + bx + cz$, where Y = P uptake by the test plant, x = routine soil test data, z = P buffer power, and a and b = constants, to test the hypothesis. The impact of the P buffer power on P uptake *vis -a -vis* P bio availability is clearly shown by the data in Table 4.1.

In both cases of routine soil test data, the predictability of uptake was remarkably increased by integrating the P buffer power. The coefficient of determination, which is a measure of the precision of these computations, increased from 15 to 79% in the case of EUP-P and from 51 to 89% in the case of CAL-P This in effect means that by integrating the P buffer power, the precision of P uptake predictability increased by 427% in the case of EUF-P and 74% in the case of CAL-P. Obviously, the increase has been very remarkable in the case of EUF-P. This also shows that the quantification of either the intensity factor alone (as in the case of EUP soil test data) or the quantity factor alone (as in the case of CAL soil test data) is insufficient to precisely predict P availability. A substantial proportion of the variance in P uptake is contributed by the P buffer power. From the above arguments, it follows that soils with a low buffer power for a particular nutrient requires a high initial concentration of the ions and vice versa. Mengel and Busch (1982) provided experimental evidence to substantiate this in the case of K^+ uptake of rye grass growing in nine soils of widely differing physicochemical properties. This would also point to the crucial fact that the relationship between buffer power for the nutrient and the "critical" or "cut off" limit where plant yield is perceptibly reduced have an inverse relationship—a soil with a low buffer power for the nutrient resulting in a higher critical concentration in the plant tissue as compared to a soil with a relatively high buffer power having a lower critical concentration in the plant tissue for the same nutrient. This has been found in the case of K (Mengel and Busch 1982) and P (Nair 1988 unpublished data). Nair (1988, unpublished data) found a very highly significant negative correlation between the plant top critical concentration for P and the P replenishment rate in the soil substrate which the P buffer power represents in summer rye (*Secale cereal* L.). In the case of dry matter yield this relationship was significant, but positive. This suggests that the P buffer power controls the P depletion and/or replenishment in the rooting zone and its accurate measurement would be the key to a precise quantification of P bio availability, consequently, how much P fertilizer is required for the crop. Remarkably, the P buffer power of the soils had no direct relationship to clay or humus content, or soil pH (Nair and Mengel 1984).

4.1.1 P Buffer Power Measurement vs Soil Test P Data for Dependability of Availability Prediction

The important question on which the choice between P buffer power and/or contemporary or routine soil tests has to be made depends on what one aims to achieve inasmuch as prescriptive soil management practices, with reference to fertilizer application, are concerned. Contemporary soil tests are based on philosophies and procedures which were developed several decades ago, with essentially no change in the general approach. In order to be accurate, a soil test must measure both solution concentration of nutrients and their rates of movement through the soil. The "buffer power approach" seems to accomplish this. Virtually no soil test provides information about the most important factors which govern phosphate availability, which are phosphate concentration in the soil solution and the phosphate buffer power. In the widely used bicarbonate method (Olsen et al. 1954) or using water extraction (van der Paauw 1971), the P quantities recovered may be 10–100 times more than that found in the soil solution (Mengel 1985). Thus, all the methods which use an acid extraction, extract much more phosphate from the "labile pool" (Larsen 1967). Acid extractions, in certain instances, can also extract phosphate that under normal crop growth will be found unavailable to plant roots. The discrepancy between contemporary soil tests and crop response to P application has recently been highlighted (Nair 1992a, b). In an extended study comparing eight central European soils with extremely differing physicochemical properties, Nair (1992a, b) found that the response to a wide range of P application to summer rye grass (***Secale cereale*** L.) has no relevance at all to routine soil test data. Three routinely used soil tests, namely, the 0.01 M $CaCl_2$ test (Mattingly et al. 1963), the CAL-P test (Shcüller 1969), and the EUF-P test (Nemeth 1979), were considered. The fluctuations in shoot P concentration with regard to P application were not at all reflected by the soil test values. Two of the soils which were at either extremes of "availability" (as characterized by the soil tests) showed totally divergent responses to P application, while the response was found to be very consistent considering the soil's P buffer power. This proves that the accuracy of routine soil tests can be vastly enhanced by integrating the corresponding P buffer power of the soil.

Buffer curves are extremely important in determining phosphate bio availability. The equilibration of a soluble fertilizer incorporated into a soil is not merely a question of adsorption and/or precipitation, but also of diffusion. Even if the fertilizer source is soluble and well mixed with the soil, the phosphate distribution will not be homogenous, resulting in areas of high and low phosphate concentration. Since phosphate diffusion is a slow process, an even distribution brought about by diffusion takes time. The contact period between phosphate fertilizer and the soil has an impact on P availability, which has practical importance, and this can be depicted by buffer curves as shown in Fig. 4.1 (Barekzai 1984). This figure represents on the y axis P quantity (Ca lactate soluble) and on the x axis P intensity (electro-ultrafiltrable) and shows that the solubility of the applied P fertilizer decreased considerably during the 6 weeks of incubation as compared to the time immediately after application.

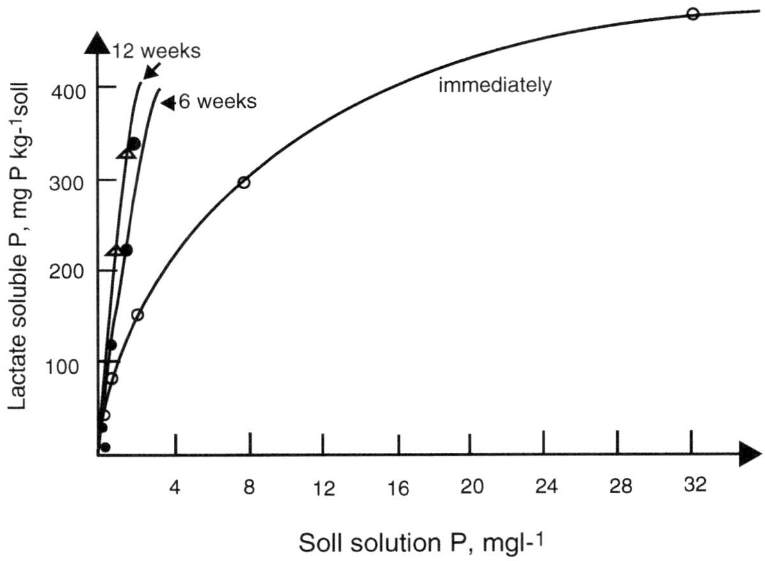

Fig. 4.1 Phosphate buffer curves of an acid soil obtained immediately after phosphate application and 6 and 12 weeks after phosphate application. (After Barekzai 1984)

In fact these curves show the P aging process and reflect the phosphate concentration of the bulk soil solution. The steepness of the curve at a given phosphate concentration reflects the buffer power and this steepness for eight widely differing central European soils was quantified by computing a regression function between P quantity (Ca lactate soluble P) and P intensity (electro-ultrafiltrable P); the b values in the regression functions were calculated to represent the P buffer power (Nair and Mengel 1984). Considerable differences can be found among soils in their P buffer power; acid tropical soils invariably have a high buffer power, meaning that at a given quantity of available P, the P concentration of the soil solution is relatively low (Pagel and Huay 1976). With soils widely differing in their P buffer power, at an identical quantity of available P, the P concentration of the soil solution can differ by a factor of 100 (Barekzai 1984).

4.1.2 P Buffer Power and Q/I Relationship

Almost half a century ago Schofield (1955), while introducing the Q/I (Quantity/Intensity) concept to soil P, argued that the amount of P available to plants is not necessarily dependent on the size of the labile soil P pool and suggested that the work required to remove P from the labile pool might be crucial in determining the uptake of this element by plant roots. In fact, the P buffer power might be an indirect measure of the work required to remove P from the labile pool to affect plant uptake.

There has been research on both sorption and desorption Q/I relationships (Raven and Hossner 1994) and their theoretical representations are shown in Figs 4.2 and 4.3. As soil P release is a key feature in P supply to plants, the study of P desorption Q/I characteristics might be more adequate than P sorption Q/I characteristics when evaluating plant available P and P buffer power of soils. P buffer power quantification by the sorption method would only be valid if both sorption and desorption curves overlap and if the proportion of solution P in the total desorbed P is negligible and all of the sorbed P is also desorbed (Raven and Hossner 1994). The success of the adsorption–desorption equilibrium approach in quantifying exactly the P buffer power of eight widely varying central European soils (Nair 1992a, b) lies in the above fact. However, P sorption and desorption Q/I curves generally do not coincide and sorption–desorption hysterisis is commonly observed (Fox and Kamprath 1970; Okajima et al. 1983; Barrow 1983). However, P buffer power values derived from sorption curves appear to be adequate capacity indexes in terms of P uptake by plants (Bowman and Olsen 1985a, b; Holford 1988). The latter author, using results from 15 soils of diverse P characteristics, reported that the buffer capacities derived from P adsorption isotherms are satisfactory for characterizing P desorption, as well as adsorption, in the context of P diffusion in the soil and P uptake by plants. He further stated that during the later stages of P uptake, buffer capacities were always greater than indicated by adsorption measurements, suggesting that non labile P was making a significant contribution to the replenishment of the soil solution. Recent

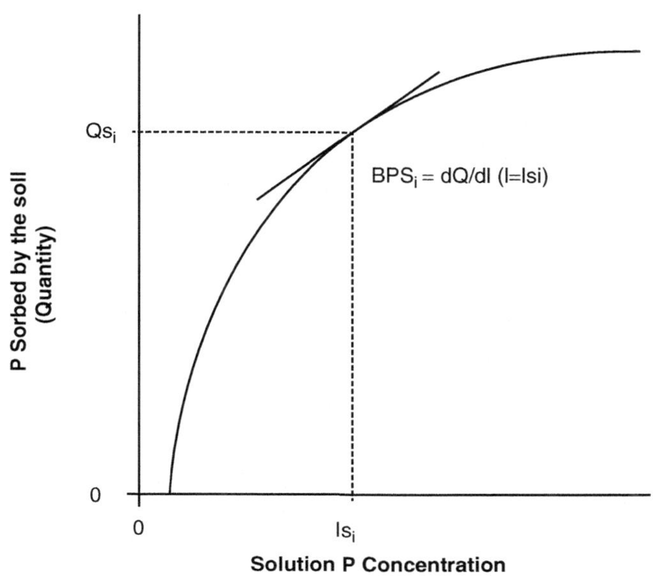

Fig. 4.2 Theoretical soil P sorption curve and related parameters (Qsi, amount of P sorbed by the soil in situation i; Isi, solution P concentration in situation i; and Bpsi, P buffering power at ISi). (After Raven and Hossner 1994)

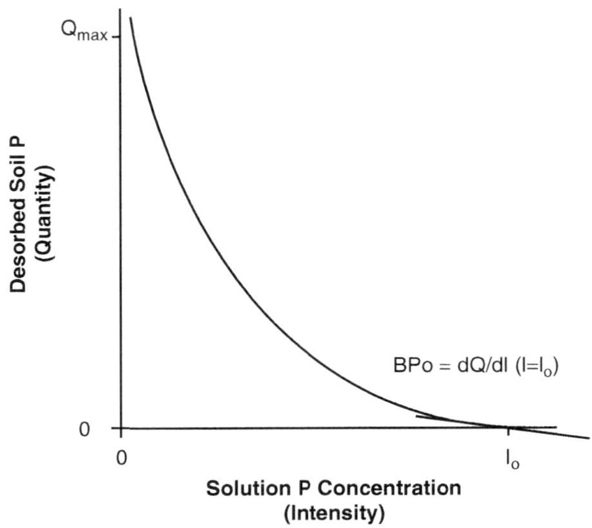

Fig. 4.3 Theoretical soil P desorption Q/I curve and related parameters (Qmax, maximum desorbed soil P, labile soil P, or resin-extractable P at a 4/3 soil/resin weight ratio; 10, solution P concentration when no P is desorbed or Q = 0; and BPO, P buffering power at 10). (After Raven and Hossner 1994)

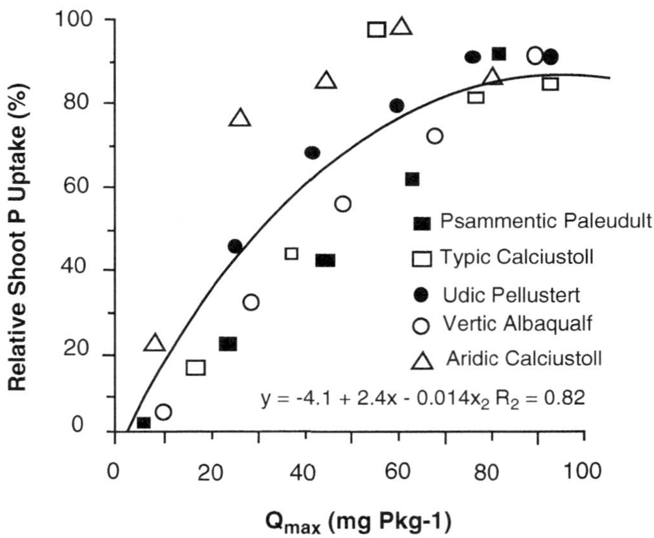

Fig. 4.4 Relationship between the relative shoot P uptake of corn plants and the maximum desorbed soil P (Qmax). (After Raven and Hossner 1994)

investigations (Raven and Hossner 1994) indicate that P buffer power values derived from sorption Q/I curves differ from those of desorption Q/I curves. They have reported that test plant P uptake (in this case a single corn plant in a short-term container culture) was better related to the quantity parameter derived from the desorption Q/I relationship (Q_{max} as shown in Fig. 4.4. Fox and Kamprath (1970) developed a standard procedure to determine sorption curves that allow computation of the quantity of P fertilizer required to establish a desired P concentration in the soil solution. Though Beckwith (1965) tentatively suggested that adjustment to 0.2 mg P liter^{-1} would be adequate for successful growth of most plants, these critical values would depend on plant species and nutritional factors (Fox 1981). Further, recourse to merely a critical solution P concentration value would neglect P buffering effect. Procedures evaluating buffering capacities of soils are often time consuming, but the question as to the need for precise documentation of soil parameters quantifying P uptake with precision for field crops, where other important factors may come into play, cannot be underestimated. Empirical correlations based on Mitcherlich-type equations have been fairly successful in extension work, but these equations are restricted to areas with similar soil, climate, and management. Since these equations cannot accommodate soils of divergent phosphate buffering capacities, the need to include a numerical value for P buffering capacity can hardly be overlooked (Bowman and Olsen 1985b). The usefulness, for field application, of Q/I relationships obtained in the laboratory is often questioned. Q/I relationships can give an idea of the relative ease with which the "readily extractable P" will be released (buffered)—something that other conventional soil testing procedures do not give (Bowman and Olsen 1985b). Estimates can be made of the number of years the soil can supply P before deficiency levels set in. Hence, the interpretation of soil tests for soils of identical Q but with a higher P buffer power will need to be different. Another important question is whether it is critical to employ laboratory desorption isotherm procedures merely because the procedure more realistically simulates plant uptake (Bowman and Olsen 1985b). The scientific understanding of the uptake process seems more appropriately explained by the desorption isotherm process since plant uptake is basically a desorption process. However, Bowman and Olsen (1985b) suggest that where field responses are concerned, adsorption isotherms are more appropriate to standardize P buffer power of soils. The work of Nair and Mengel (1984) and Nair (1992a, b) suggests that both adsorptive and desorptive procedures are appropriate, the crucial question being how accurately these processes are determined in the laboratory and by plant culture techniques.

References

Barekzai, A. (1984). *Alterung von wasserlöslichem Phosphat-untersucht in Giefäß und Modellversuchen*. Ph.D. thesis, FB 1g, Justus Liebig University, Giessen.

Barrow, N. J. (1983). On the reversibility of phosphate sorption by soils. *Journal of Soil Science, 34*, 751–758.

Beckwith, R. S. (1965). Sorbed phosphate at standard supernatant concentration as an estimate of the phosphate needs of soils. *Australian Journal of Experimental Agriculture and Animal Husbandry, 5*, 52–58.

Bowman, R. A., & Olsen, S. R. (1985a). Assessment of phosphate buffering capacity. 1. Laboratory methods. *Soil Science, 140*, 287–291.

Bowman, R. A., & Olsen, S. R. (1985b). Assessment of phosphate buffering capacity. 2. Greenhouse methods. *Soil Science, 140*, 387–392.

Fox, R. L. (1981). External phosphorus requirements of crops. In R. H. Dowdy et al. (Eds.), *Chemistry in the soil environment* (ASA Special Publication 40) (pp. 223–239). Madison: ASA and SSSA.
Fox, R. L., & Kamprath, E. J. (1970). Phosphate sorption isotherms for evaluating the phosphate requirement of soils. *Soil Science Society of America Proceedings, 34*, 902–907.
Holford, I. C. R. (1988). Buffering of phosphate in the soil solution during growth of ryegrass compared with buffering by sorption. *Plant and Soil, 111*, 3–9.
Larsen, S. (1967). Soil phosphorus. *Advances in Agronomy, 19*, 151–210.
Mattingly, C. E. G., Russell, R. D., & Jephcott, B. M. (1963). Experiments on cumulative dressings of fertilizers on calcareous soils in Southwest England. II. Phosphorus uptake by ryegrass in the green house. *Journal of the Science of Food and Agriculture, 14*, 629–637.
Mengel, K. (1985). Dynamics and availability of major nutrients in soils. *Advances in Soil Sciences, 2*, 65–131.
Mengel, K., & Busch, R. (1982). The importance of the potassium buffer power on the critical potassium level in soils. *Soil Science, 133*, 27–32.
Nair, K. P. P. (1984). Zinc buffer power as an important criterion for a dependable assessment of plant uptake. *Plant and Soil, 81*, 209–215.
Nair, K. P. P. (1992a). Measuring P buffer power to improve routine soil testing for phosphate. *European Journal of Agnon, 1*(2), 79–84.
Nair, K. P. P. (1992b). Soil testing procedure to economise on fertilizer use in land management. In *The spirit of enterprise — The 1993 rolex awards* (pp. 323–325).
Nair, K. P. P., & Mengel, K. (1984). Importance of phosphate buffer power for phosphate uptake by rye. *Soil Science Society of America Journal, 48*, 92–95.
Nemeth, K. (1979). The availability of nutrients in the soil as determined by electroultrafiltration (EUF). *Advances in Agronomy, 31*, 155–188.
Okajima, H., Kubota, H., & Sakuma, T. (1983). Hysteresis in the phosphorus sorption and desorption process of soils. *Soil Science & Plant Nutrition, 29*, 271–283.
Olsen, S. R., Cole, C. V., Watanabe, F. S., & Dean, L. A. (1954). Estimation of available phosphorus in soils by extraction with sodium bicarbonate. USDA Cir. No. 939.
Pagel, H., & van Huay, H. (1976). Wichtige Parameter der Phosphat-Sorptionskurven einiger Böden der Tropen und Subtropen und ihre Zeitliche Veränderung durch P—Düngung. *Archiv für Acker- und Pflanzenbau und Bodenkunde, 20*, 765–778.
Raven, K. P., & Hossner, L. R. (1994). Sorption and desorption—Quantity–intensity parameters related to plant-available soil phosphorus. *Soil Science Society of America Journal, 58*, 405–410.
Schofield, R. K. (1955). Can a precise meaning be given to "available" soil phosphorus? *Soil Fertility, 18*, 373–375.
Schüller, H. (1969). Die CAL-Methode, eine neue Methode zur Bestimmung des pflanzen verfügbaren Phosphates in Böden. *Zeitschrift für Pflanzenernährung und Bodenkunde, 123*, 48–63.
van der Paauw, F. (1971). An effective water extraction method for the determination of plant-available soil phosphorus. *Plant and Soil, 34*, 467–481.

Chapter 5
Case Studies with Asian Soils

Abstract This chapter discusses the determination of Potassium buffer power and its integration into the computations to enhance the precision of the predictability of nutrient uptake of Potassium by perennial crops, such as Cardamom, growing widely in Southern India, in the States of Kerala and Karnataka. Also, the influence of K buffer in determining cardamom capsule yield.

Keywords Potassium buffer power · Perennial crops

5.1 Potassium

As discussed earlier with regard to P, the dynamics of K availability follows the same pattern as that of P, especially in the range of low concentrations. Beckett (1971, 1972) has used the activity ratio for K^+ and Ca^{2+} to determine K bio availability. Since interlayer K would play an important role in K bio availability, it would be more logical to consider K buffer power in determining K bio availability. Routinely, it is the ammonium acetate extractant that is widely used to characterize K bio availability. The reason that this may not be suitable to characterize exchangeable K is that in the routine extraction, only the top layer is extracted, while interlayer K from which deep-rooted plants can feed is ignored (Nair et al. 1997). There is extensive evidence to substantiate this (Mengel 1985). The importance of interlayer K in the nutrition of deep rooted and perennial crops, such as cardamom, (***Elettaria cardamomum*** M.), the world's most valuable spice crop, has been highlighted recently by Nair et al. (1997).

5.1.1 The Importance of K Buffer Power Determination in Predicting K Availability to Perennial Crops

As in the case of P, the K buffer power assumes great importance in predicting K bio availability, especially with regard to deep-rooted and perennial crops. With annual or biennial crops, K bio availability has been studied with reference to the exchangeable K. However, with perennial and deep-rooted crops non exchangeable and

K. P. Nair, *Intelligent Soil Management for Sustainable Agriculture*,
https://doi.org/10.1007/978-3-030-15530-8_5

interlayer K play a crucial role in K bio availability. Three soil parameters that control the rate of K supply to plant roots which have been used for predicting K absorption by plants are the K intensity in the soil solution, the K buffer power, and the effective diffusion coefficient (Beckett 1971; Mengel and Kirby 1980; Claassen et al. 1986). K buffer power can be directly obtained from the K quantity–intensity relationship. The effective diffusion coefficient depends on, among other factors, the buffer power. Plants feed not only from exchangeable K but also from nonexchangeable K^+, which mainly consists of K^+ trapped in the interlayers of nonexpanded 2:1 clay minerals (Hoagland and Martin 1933; Schachtschabel 1937).

Exchangeable K comprises that which can be exchanged with the NH_4^+ ion, and is primarily planar K^+, the interlayer K^+ of expanded 2:1 clay minerals and some of the K^+ at the interlayer edges of non-expanded 2:1 clay minerals. Interlayer K^+ of non-expanded clay minerals such as illites and interlayer and lattice K^+ micas (present in feldspars) constitutes the non-exchangeable K^+. The interlayer K^+ is of particular importance in the nutrition of deep-rooted and perennial crops such as cardamom, as demonstrated by the recent work of Nair et al. (1997), and also for annual crops such as rye grass (*Lolium perenne.* cv. Taptoe) (Mengel and Uhlenbecker 1993).

In most of the soil tests for K^+ bio availability, non-exchangeable K is not considered. For cereals such as wheat (*Triticum aestivum*) 80% of the K^+ extracted by the crop came from the non-exchangeable K pool. This is one of the most important reasons for the poor soil test crop response relationship with regard to K fertilizer application based on such tests (Kuhlmann and Wehrmann 1984). The contribution of nonexchangeable K to plant bio availability was assessed by 1 M HCl extraction by Schachtschabel (1961), similar to the 1 M HNO_3 extraction proposed by Pratt (1965) and McLean and Watson (1985). However, the efficiency of 1 M HCl extraction to quantify plant available K from the non-exchangeable pool has been disputed (Boguslawski and Lach 1971; Grimme 1974; Kuhlmann and Wehrmann 1984). Soils containing primarily 2:1 clay minerals such as vermiculite and illite have interlayer K in excess of crop demand. However, the bio availability of interlayer K of non-expanded minerals is independent of the quantity of interlayer K as such, but dependent on its release rate depending on the type of K^+ bearing minerals (Sparks 1987). Release of K from interlayer positions is an exchange and diffusion process (von Reichenbach 1972). While exchange depends on the cation species and their concentration near the surface of the mineral, diffusion depends largely on the expansion of the mineral and therefore on soil moisture. Net release of K will only occur if the K concentration of the adjacent solution is low (Mengel 1985). Martin and Sparks (1983), while studying the release of non-exchangeable K^+ from sandy loam and loamy sand extracted with a H^+ charged ion exchanger resin, found large release of K^+ with a K^+ concentration of about 1–2 μM in the contact solution. This concentration may approximate the rhizosphere concentration level. Under submerged conditions as in rice, there can be a depletion zone for K in the rhizosphere (Xu and Liu 1983). Plant roots act as a sink for K and maintain the K solution concentration at very low levels (Kuchenbuch and Jungk 1984); this would cause further release of interlayer K (Mengel 1985). These considerations point to the important fact that a precise quantification of K bio availability, where nonexchangeable interlayer K is concerned, hinges primarily on its release rate, which the K buffer power attempts to quantify, as we shall see in the following discussion.

Nair et al. (1997) selected cardamom (***E. cardamomum*** M.) to demonstrate the importance of non-exchangeable and interlayer K on K availability *vis-a-vis* the K buffer power. The K buffer power curves were constructed by a two-step extraction in which 1 N HNO_3 was used to determine the quantity of K (Wood and De Turk 1941) and 1 N NH_4 OAc was used to determine K intensity. The NH_4 OAc extractant is universally used to determine exchangeable K. The contribution of non-exchangeable K to plant availability has been assessed both by extraction with 1 M HCl (Schachtschabel 1961) and 1 M HNO_3 (Pratt 1965; McLean and Watson 1985). Nair et al. (1997) regressed 1 N HNO_3 extractable K (y) over 1 N NH_4 OAc extractable K (x) to obtain the K buffer power (Table 5.1).

Data in Table 5.1 clearly indicate that the Coorg soils, which had a much higher K buffer power, produced cardamom yield which was twice that obtained in the Idukki soils. The higher K buffer power of the Coorg soils was clearly reflected in the cardamom yield. By comparison, the 1 N NH_4 OAc extractable K had no significant relationship with leaf K (Table 5.2) and further the integration of K buffer power with the routine soil test K data (NH_4 OAc extractable K) remarkably improved this relationship.

Table 5.1 K buffer power of cardamom-growing soils from two regions of southern India extensively growing this crop

Region	Regression function (Y = a + bx)	R	Crop yield (kg ha^{-1})
Coorg	142.38 + 1.4443x	0.8561**	155
Idukki	592.46 + 0.9172x	0.5799*	80

After Nair et al. (1997)

Note. b values in the regression function represent the K buffer power of the soil. The K buffer power refers to pooled values of soil samples obtained from 94 locations covering an area of more than 20,000 ha in the two cardamom-growing regions of southern India, namely, **Coorg** and **Idukki**. Yield data refer to the same locations

*, ** Significant at a confidence level of 5 and 1%, respectively

Table 5.2 Correlation coefficients and regression functions for the relationship leaf K (Y) vs exchangeable K (x, NH4OAc extraction)

Details	Regression function (Y = a + bx)	Correlation coefficient (r)
Leaf vs exchangeable K	**Coorg**	0.2064
	Y = 1.2701 + 0.0004	
Leaf vs exchangeable K	**Idukki**	−0.006
	Y = 1.6448 + 0.000006	

After Nair et al. (1997)

Note. The correlation coefficients refer to the leaf samples from 94 locations from which soil samples were also obtained to calculate the K buffer power. In cardamom the fifth pair of leaves from the top of each panicle bearing tillers are sampled for K analysis

Table 5.3 Correlation coefficients (r) for the Relationship between leaf K (Y) and exchangeable K (x, NH_4 OAc extraction) for the pooled data (94 locations) from two regions (Coorg and Idukki) without (A) and with (B) K buffer power integration

	Correlation coefficient	
Details	A	B
Leaf vs exchangeable K	0.2510	0.4367**

After Nair et al. (1997)
**Significant at 1% confidence level

Cardamom is a heavy feeder of K, and in India, which grows most of this valuable spice crop, and in other countries on the Asian and African continents where also this crop is grown, its K fertilizer needs are almost always based on the exchangeable K determined by 1 N NH_4 OAc. Data in Table 5.2 unequivocally show the inability of this extraction to precisely predict K bio availability; further, data in Table 5.3 show how the situation is remarkably improved by the integration of the K buffer power into the computations. A substantial variation (302.7%) in leaf K is attributable to the K buffer power. These results have been obtained from a very extensive area (more than 20,000 ha), which demonstrates their significance.

The K buffer power in this instance integrates both exchangeable (NH_4 OAc extraction) and non-exchangeable and/or interlayer K (HNO_3 extraction) and this gives an accurate measurement of K depletion around the plant roots. In an investigation (Mengel and Uhlenbecker 1993) on K availability from interlayer K to rye grass (*L. perenne* L. cv. Taptoe) it was observed that the rate constant (b values) obtained by correlating K released (from the interlayers of clay minerals) and time periods by a modified electro-ultrafiltration (EUF) technique was closely related to K uptake and represented the K bio availability index from non-exchangeable K. These rate constants, according to the authors, are of the utmost importance because they provide information on the bio availability of non-exchangeable K in attaining maximum yield; and a set of "critical b" values toward attaining this objective have been reported. It appears that the rate constants of Mengel and Uhlenbecker (1993) are analogous to the K buffer power values reported by Nair et al. (1997), because, though the techniques differ in their details, they have accomplished the same objective of precisely predicting K bio availability from the nonexchangeable pool and/ or the interlayer K. The capability of tapping interlayer K varies among plant species. For instance, Steffens and Mengel (1979) found that rye grass (***Lolium perenne***) could feed from interlayer K^+ for a longer period without yield depression, while red clover (Trifolium pratense) could not. These authors reported that since ***L. perenne*** had a longer and deeper root system compared to ***Trifolium pratense,*** the former could grow satisfactorily while relatively low K^+ concentration at which the latter would already suffer from K deficiency (Steffens and Mengel 1981). The differences in root mass, root length and root morphology between monocots and dicots explain the better K^+ feeding capacity from interlayer K^+ of the former compared to the latter (Mengel 1985).

Cotton, ***Gossypium hirsutum*** L., is another deep-rooted long-duration crop, on which the K buffer power exerts considerable influence on K acquisition. Brouder and Cassman (1994) evaluated K uptake by cotton in a vermiculitic soil using mechanistic models and observed that initial model output produced both substantial under- and over predictions of whole-plant K accumulation. Model predictions were greatly enhanced by estimating the K buffer power. They further concluded that the contribution of the fixed K pool to the plant available K pool was likely to be substantial and that this influence must be captured in estimates of the soil K buffer power. These studies were conducted after observing in a San Joaquin Valley cotton field in California that cotton exhibited late season K deficiency while other crop species remained unaffected.

In such cases, the precise estimation of K buffer power will lead to far more dependable K fertilizer recommendations than estimations by the routine NH_4 OAc extraction.

Though it has long been recognized (Schachtschabel 1937) that the soil K fraction which is not exchangeable by NH_4^+ ions (non exchangeable K) may be important for the supply of K to plants, it is only of late that researchers have paid more attention to this aspect. The work of Sparks and Huang (1985) has very carefully examined the release mechanism from the nonexchangeable source and the factors controlling it. Considerable portions of initially non exchangeable K can be utilized by plant roots even within a few days (Kuchenbuch and Jungk 1984). The depletion zone, however, extends into the ambient soil for 2 mm only. Hinsinger et al. (1992) embedded phlogopite in agar and observed that the interlayer K of this mineral was entirely lost in the close vicinity of ryegrass roots within 4 days. Since the process limiting the rate of K uptake in the rhizosphere may be K transport through the soil rather than the release from minerals as such, some researchers have focused their attention on this aspect. One such example is the mechanistic mathematical model of Claassen and Barber (1976). Claassen et al. (1986) and Claassen (1990) have successfully applied the model referred to above to predict K depletion profiles in soil around plant roots. Meyer and Jungk (1993) have used these models to predict K uptake by test plants from exchangeable and nonexchangeable K sources. They reported that 64–79% of the K taken up by wheat (***Triticum aestivum*** L.) and sugar beet (***Beta vulgaris*** L.) was derived from the rapidly released exchangeable and 21–36% from the non exchangeable or less mobile soil K fraction.

The buffer power describes the relationship between adsorbed K and the K concentration of the ambient solution. In simulation models it is assumed that this relationship is linear and hence independent of soil solution concentration. However, in desorption studies with soil, a sharply curved buffer relationship could be found and Meyer and Jungk (1993) have referred to it. Very near the plant roots the soil can be subjected to a curved buffer function since plant roots strongly reduce the soil solution concentration.

An important aspect to be considered in the utilization of nonexchangeable K is the role of plant roots. Plant species differ in their ability to utilize non exchangeable K and this has been attributed to the differences in root length (Mengel and

Steffens 1985). Radial distance between two single roots decreases, consequently increasing root density and this would result in the overlapping of the depleted soil volumes between these roots. This would also lead to a decrease in the rate of K uptake per unit root. In the case of the rapidly diffusing K fraction, which has a higher mobility, the competition effect between roots could be very intense. There is evidence to support this view, as shown by the work of Mitsios and Rowell (1987), who observed that the contribution of non-exchangeable K increased with a corresponding increase in root density. Additionally, the differences in root hair length and density among plant species (Föhse et al. 1991) affect their ability to acquire soil K. Accordingly, the work of Meyer and Jungk (1993) has shown that K uptake was higher when they included root hairs as well in their model calculations. Since root hairs contribute to an increase in root absorbing surface, a reduction in the distance of diffusion from the site of K release to the site of K uptake, and an increase in the K concentration gradient, they can be expected to exert a pronounced effect on K availability from the less mobile K fraction.

5.1.2 The Commercial Significance of K Buffer Power Determination in K Fertilizer Management for Perennial Crops

The commercial significance of K buffer power determination for dependable K fertilizer recommendations assumes great importance in those countries which are faced with importing these fertilizers at a huge cost to the national exchequer. India is a case in point. The recent decontrol of prices of both phosphate and potassium fertilizers by the government of India resulted in an overnight escalation of their market prices. In a situation like that, the farmers become extremely wary of their field use and unless the fertilizer application is cost effective, faith in their use, especially those mentioned above, would be shattered.

The K fertilizer recommendation for cardamom has been based exclusively on NH_4 OAc extraction. The investigation of Nair et al. (1997) showed its ineffectiveness. Though the importance of K buffer power in predicting K availability has been reported earlier, these research reports related mainly to annual crops such as white clover (During and Duganzich 1979) and rye grass (Mengel and Busch 1982); the work of Nair et al. (1997) was the first of its kind in perennial crops, in the world.

5.1.3 The Role of the NH_4^+ Ion on K+ Availability

One of the frequent assumptions made in predicting K availability in soils is that results from binary (two-ion) exchange systems can be extrapolated to ternary (three-ion) systems by using appropriate equations. The K–Ca exchange reactions in soils are often investigated in laboratory studies. Most of the research carried out on soil clay minerals and soils as exchanger surfaces (Vanselow 1932; Gapon 1933; Argersinger et al. 1950; Sposito 1981a, b; Sposito et al. 1981, 1983; Jardine and Sparks 1984) are binary exchange systems. However, field soils are at least ternary systems (Adams 1971; Curtin and Smillie 1983). The evaluation of soils as binary systems implies that these reactions can be used to predict results in ternary systems such as field soils. For this assumption to be valid, the binary exchange selectivity coefficients need to be independent of exchanger-phase composition (Lumbanraja and Evangelou 1992). But, the work of Shu-Yan and Sposito (1981) showed that it is impossible to predict exchange phase–solution phase interactions in a ternary system such as the field soil from a binary system such as the laboratory sample. This focuses the importance of the ternary systems. As far as K availability is concerned, it would be important to include the NH_4^+ ion as well. The work of Lumbanraja and Evangelou (1992) has shown that K^+ adsorption to soil surfaces is suppressed in the presence of added NH_4^+ ion, while the adsorption of NH_4^+ ion to the same surface is enhanced in the presence of added K^+. These observations point to the influence of added NH_4^+ ion on the desorption potential (chemical potential) of adsorbed K or vice versa (Lumbanraja and Evangelou 1992) and would be relevant to the determination of K buffer power especially when agents containing NH^+ ions, such as NH_4OAc, are used in determining K buffer power (Nair et al. 1997). The work of Lumbanraja and Evangelou (1992), though, clearly demonstrates the effect of the NH_4^+ ion on K^+ desorption, with an increase in K^+ desorption in the presence of added NH_4^+ ion. In its absence, it might be safe to conclude that the shape of the K buffer power curve will not appreciably change even if larger quantities of K are removed due to cropping and therefore can be considered as a relatively constant property of soils. There is evidence to support this view. These authors, while investigating the Q/I relationship on K uptake by wheat (***T. aestivum***) in calcareous vertisols and inceptisols of southwestern Spain, found that 80% of the K extracted came from the nonexchangeable K pool.

These observations, coupled with the ones discussed earlier, suggest that the precision of predicting K bio availability can be substantially enhanced by first quantifying the K buffer power of the soil in which the crop is intended to be grown. Admittedly, the rate-limiting steps involved in K dynamics are not entirely understood (Sparks 1987). Notwithstanding this limitation, if one must move forward in devising better management of K fertilizers in crop production, a starting point has to be made with regard to precisely quantifying K bio availability. Quantifying the K buffer power of soils and basing K fertilizer recommendations on this seems to be the best starting point.

5.1.4 K Buffer Power Measurement vs Contemporary Soil Test K Data for Dependability of K Bio Availability Prediction – The Context of MYR (Maximum Yield Research) and MEY (Maximum Economic Yield) Approaches

The importance of K buffer power determination in devising effective K fertilizer recommendations must be viewed against the twin objectives of maximizing fertilizer use efficiency to attain high crop productivity—the crucial concern of developing countries—and minimizing nutrient loss to the environment—the crucial concern of industrialized countries. Potassium losses are primarily due to leaching and occur on coarse-textured soils under humid conditions (Mengel 1985). Where such leaching losses are acute, split application of K fertilizer adjusted to the crop demands and to the fluctuating weather conditions are considered a useful measure to enhance efficiency of K use by crops. Despite the fact that predictive kinetic models of K dynamics are absent and the K-rate limiting steps are still inadequately understood (Sparks 1987) a better understanding would doubtless result in more precise prediction of the fate of applied K, and, consequently, assist in enhancing crop yields and K uptake efficiency, through more effective K fertilizer recommendations. The K fertilizer recommendations still continue to revolve around quantifying available K, mostly by standard extractions with NH_4 OAc or acids such as 1 N or 1 M HNO_3 or 1 M HCl. A case in point is the MYR (Maximum Yield Research) or MEY (Maximum Economic Yield) approach. The recent findings of Narang et al. (1994) in India indicate that in a rice–wheat rotation of five crop seasons, the decline in available K was 12.3 μg g^{-1} at 120–30–30/120–60–30 (N–P–K rates) and 15.6 μg g^{-1} at 180–30–30/180–60–30 (N–P–K rates) over their respective initial levels, pointing out the anomalous and inconsistent fact that at identical K application rates the reduction in available K is variable. Obviously, such a conclusion is subject to scrutiny from the viewpoint of both quantifying available K (in this case the standard NH_4 OAc extraction was employed) and of the varying N and P rates where interaction between the three nutrients (N, P, and K) is bound to play a role in their individual bio availability. However, one cannot ignore the inadequacy of such approaches, because these contemporary soil tests fail to detect precisely the differences in soil K bio availability. Contradictory results have been obtained where higher K fertilizer application resulted in a corresponding higher available K (Singh and Nambiar 1986; Narang and Bhandari 1990; Roy et al. 1990) while the same practice resulted in a decline of extractable K (Swarup and Singh 1987; Hegde and Dwivedi 1992). Obviously, these inconsistent results cast a deep shadow of doubt on the dependability of contemporary soil tests for K availability in soils. A similar situation prevails in the case of rice production for MEY in Costa Rica. Using a modified Olsen's extractant, a mixture of $NaHCO_3$ + EDTA, Cordero and Espinosa (1994) classified soil K critical levels as low (<0.1 $cmol_c$ $liter^{-1}$), medium (0.11–0.2 $cmol_c$ $liter^{-1}$), and high (>0.2 $cmol_c$ $liter^{-1}$). The K critical level was set at

0.1 cmol$_c$) liter^{-1} and the K fertilizer recommendations varied from 30 to 60 kg K_2O ha^{-1}. But responses to K fertilizer up to 80 kg KO ha^{-1} were obtained even in soils with a K critical level of 0.15 cmol$_c$ liter^{-1} in Vertisols and Mollisols with high Ca + Mg contents (Cordero and Espinosa 1994) indicating that the K critical level based on the soil extraction is not a dependable index of K bio availability and the situation can vary from one soil to another. The above situation would call for different K fertilizer recommendations for different soils based on extractable K and the Ca + Mg/K ratio. This disallows K fertilizer recommendations in a universal manner. These are not the limitations which the K buffer power approach suffer from. Since the K buffer power quantifies the average K concentration on the plant's root surface, though indirectly (Nair 1984a, b), it would provide greater precision in K bio availability prediction and, hence, result in more dependable and uniform K fertilizer recommendations. The supportive evidence provided by During and Duganzich (1979), Mengel and Busch (1982), Mengel and Uhlenbecker (1993), and Nair et al. (1997) reinforces this assumption. The MYR approach focuses on the BMP (Best Management Practice) and aims to maximize crop yields wherever possible. Yet some of the key inputs, such as decisions on fertilizer rates, are still based on contemporary soil tests based on philosophies and procedures developed several decades ago with essentially no change in the general approach. In an interesting paper entitled "Maximum Yield Research: Friend of the Environment," presented at the 15th World Congress of Soil Science in Acapulco, Mexico, Ludwick makes a strong case for this approach in maximizing crop production. He cites Norman Borlaug (1990) the Nobel Peace Prize Winner, and Borlaug and Dowswell (1993) who said, "The only way for agriculture to produce sufficient food to keep pace with population and to alleviate the hunger of the world's poor is to increase the intensity of agricultural production under those ecological conditions which lend themselves to intensification while decreasing the intensity of production in the more fragile ecologies. Most of the yield increase in food production needed over the next several generations must be achieved through yield increases on land now under cultivation. Moreover, these yield increases must be achieved through the application of technology already available or well advanced in the research pipeline. This will not only lead to economic development but it will also do much to solve the serious environmental problems that come as a consequence of trying to cultivate lands that are not suited for crop production. Fortunately many of the more-favored agricultural lands currently under cultivation are still producing food at yield levels far below their potential."

He further cites Cassman (1990) who mentions that the greatest challenge to global food output is to understand how soil fertility management affects soil quality, and to apply this knowledge to integrate all aspects of crop management to optimize yield and input utilization efficiency over the long term. Soil fertility management governs plant nutrition, which in turn influences susceptibility to disease, plant growth and yield.

In a nutshell, the MYR approach lays great stress on nutrition, yet the nutrient management aspects move along the beaten track. Unless imaginative attempts are made to revise the basic concepts of quantifying nutrient availability, even the MYR approach is bound to falter as has been clearly shown in the case of K recommendations for rice–wheat rotations in India and rice production in Costa Rica. These are but a few examples which have been reported and are reviewed here, and yet there could be many more which are not reported and which are outside the purview of this review. These facts call for a new direction in soil testing for nutrient availability. The buffer power concept appears to provide this. The field experience of this new approach is detailed in the book "The Spirit of Enterprise—The 1993 Rolex Awards" which summarizes the work of Nair (1992a, b) in the farmers' fields in the Republic of Cameroon.

5.1.5 K Buffer Power and Q/I Relationship

The use of the Q/I relationship to describe the K status of soils is based on Woodruff's (1955a, b) observations that K availability to plants can be characterized by the free energy of K–Ca exchange with regard to solid-solution reactions. Beckett (1971) listed a number of assumptions as prerequisites for Woodruff's observation to be valid. An important assumption is that the rate of K uptake by plant root must be regulated by "the difference in the equivalent free energies of K and Ca as offered to the uptake sites." Maas (1969) has provided experimental evidence to substantiate this from uptake data of excised maize roots. These data demonstrate that $Ca2^+$ uptake in the presence of K^+ is a competitive process. The rate of K uptake was maximized in the ΔG range predicted by Woodruff (1955b). Such experimental evidence of K uptake by excised maize roots provides supportive proof to the validity of Woodruff's observation that K availability could be described by the free energy of K–Ca exchange as long as all other soil factors related to plant growth are not limiting and are held constant (Beckett 1972). The use of Q/I in predicting K bio availability to plants in soils has been widely tested (Beckett 1972; Bertsch and Thomas 1985). However, it has been shown that the Q/I relationship does not enjoy universal acceptance since a single relationship for all soils between K uptake by a given crop and AR_K does not exist, perhaps due to the nature of the soil components regulating AR_K (Evangelou et al. 1994) (Fig. 5.1). The reader is advised to read the excellent review of Evangelou et al. (1994) for a better appreciation of the new developments and perspectives on soil K Q/I relationships. The reasons for the failure of the Q/I concept to precisely predict soil K bio availability in all soils are varied and an important aspect that has to be recognized is that Q/I concept does not deal with physiological effects, but appears to be influenced by them (Evangelou et al. 1994). Mengel and Busch (1982) conducted an interesting investigation dealing with some of these physiological effects. They demonstrated that a very close correlation coefficient ($r^2 = 0.908$), reflected by a parabola ($y = c/x$; Figs. 5.2 and 5.3), existed between K buffer power and the critical K^+ concentration of the soil solution. They estimated the K buffer power by correlating exchangeable K (0.2 N $BaCl_2$ exchangeable) as y (K, "quantity") with soil solution K (electro-ultrafiltrable) as x (K, "intensity"), and the b values in the regression functions were taken as a

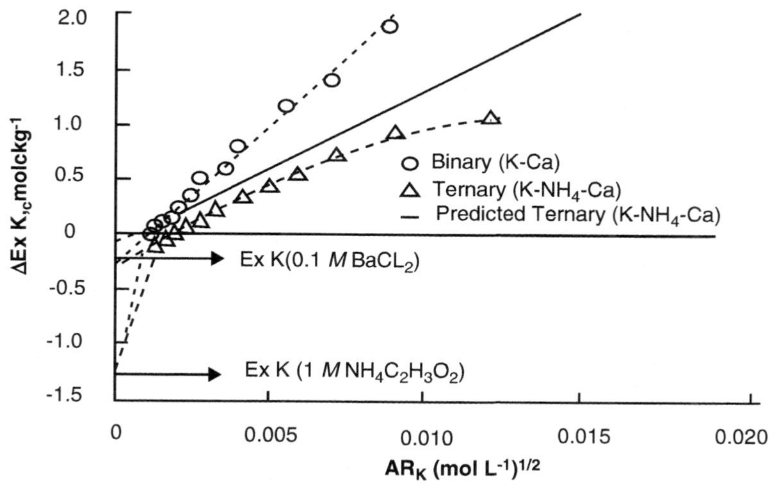

Fig. 5.1 Potassium quantity/intensity (Q/I) plots of Eden soil in the absence (binary) and presence (ternary) of NH4 [ternary prediction is from experimental binary Vanselow exchange selectivity coefficients for K–Ca and NH4–Ca exchange, and from experimental ternary activity ratios (AR) for K and NH4; ExK is exchangeable K]. (After Lumbanraja and Evangelou 1990)

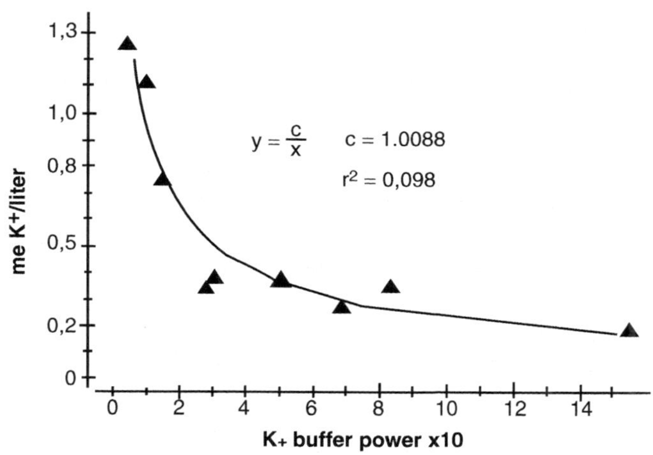

Fig. 5.2 Relationship between the critical K+ concentration of the soil solution and the K+ buffer power: y = critical K+ concentration; x = K+ buffer power. (After Mengel and Busch 1982)

measure of the K buffer power. Nine central European soils of extremely differing nature were tested for K availability to a test crop of ***L.multiflorum*** (cv. Avance). These authors reported that as the K buffer power increased, the critical K concentration (the critical K concentration defined as that level below which a yield increase is obtained when K fertilizer is applied) decreased.

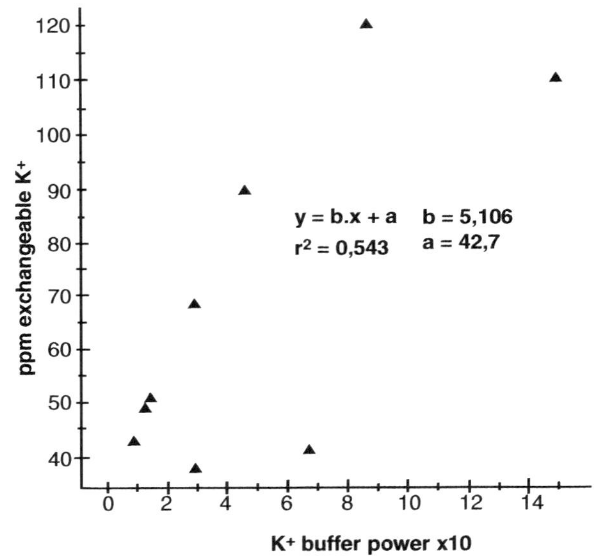

Fig. 5.3 Relationship between the critical level of exchangeable K+ and the K+ buffer power: y = critical level of exchangeable K+; x = K+ buffer power. (After Mengel and Busch 1982)

Though the Q/I concept does not deal with physiological effects, the latter influences it. For a better appreciation of the Q/I concept vis-a-vis the K buffering effect on prediction of K uptake, one should be quite familiar with plant nutrition concepts. Even if nutritional effects are assumed to play only a limited role in K uptake under a well-defined experimental set-up, the ability of a soil to replenish K^+ in the soil solution must be considered a highly complex process and not solely dependent on K–(Ca + Mg) exchange alone (Evangelou et al. 1994). K bio availability is also controlled by additional factors, such as hysteresis or exchange irreversibility effects, anion effects, multiplication effects, potential-determining ion effects, and kinetic effects (Evangelou et al. 1994). To predict K bio availability with precision, these factors need to be further scrutinized. But, most importantly, it is the soil/soil solution–plant root surface interactions which merit closest scrutiny in predicting K bio availability. The importance of K buffer power has to be examined in this context. Since the ultimate objective in perfecting prescriptive soil management practices with regard to K fertilizer application, *vis-a-vis* soil testing for plant available K, hinges on the soil/soil solution–plant root interactive effects, and also since plant root surface K concentration during the dynamic state of plant growth is nearly impossible to measure in a field soil, Nair (1984a, b) has argued that a precise quantification of K buffer is an important prerequisite to accurately predict K bio availability to growing plants and must be considered as such as an indirect measure of the plant root surface K concentration. The experimental evidence provided by Mengel and Busch (1982) and Nair et al. (1997) substantiates this view.

References

Adams, F. (1971). Ionic concentrations and activities in soil solutions. *Soil Science Society of America Proceedings, 35*, 420–426. Anderson.

Argersinger, W. J., Jr., Davidson, A. W., & Bonner, O. D. (1950). Thermodynamics and ion exchange phenomena. *Transactions of the Kansas Academy of Science, 53*, 404–410.

Beckett, P. H. T. (1971). Potassium potentials–A review. In *Potash Review* (pp. 1–41). Bern: International Potash Institute.

Beckett, P. H. T. (1972). Critical cation activity ratios. *Advances in Agronomy, 24*, 376–412.

Bertsch, P. M., & Thomas, G. W. (1985). Potassium status of temperate region soils. In R. D. Munson (Ed.), *Potassium in Agriculture* (pp. 131–162). Madison: Am. Soc. Agron.

Boguslawski, E. V., & Lach, G. (1971). Die K-Nachlieferung des Bodens im Pflanzen experiment im Vergleich mit dem austauschbaren Kalium. *Zeitschrift für Acker- und Pflanzenbau, 134*, 135–164.

Borlaug, N. E. (1990). The challenge of feeding 8 billion people. *Farm Chemical International*, summer issue, 1990.

Borlaug, N. E., & Dowswell, C. R. (1993). Fertilizer: To nourish infertile soil that feeds a fertile population that crowds a fragile world. *Better Crops With Plant Food, 77*(3), 6–7.

Brouder, S. M., & Cassman, K. G. (1994). Evaluation of a mechanistic model of potassium uptake by cotton in vermiculitic soil. *Soil Science Society of America Journal, 58*, 1174–1183.

Cassman, K. G. (1990). The role of soil fertility research in developing sustainable food production systems. *Better Crops With Plant Food, 74*(4), 16–19.

Claassen, N. (1990). *Die Aufnahme von Nährstoffen aus dem Boden durch die höhere Pflanze als Ergebnis von Verfügbarkeit und Aneignungsvermögen*. Gottingen: Severin-Verlag.

Claassen, N., & Barber, S. A. (1976). Simulation model for nutrient uptake from soil by growing plant root system. *Agronomy Journal, 68*, 961–964.

Claassen, N., Syring, K. M., & Jungk, A. (1986). Verification of a mathematical model by simulating potassium uptake from soil. *Plant and Soil, 95*, 209–220.

Cordero, A., & Espinosa, J. (1994). Maximum economic rice yield in Costa Rica. Research for maximum yield in harmony with nature. In *XV world congress of soil science* (pp. 29–39), Acapulco, Mexico, July 10–16.

Curtin, D., & Smillie, G. W. (1983). Soil solution composition as affected by timing and incubation. *Soil Science Society of America Journal, 47*, 701–707.

During, C., & Duganzich, D. M. (1979). Simple empirical intensity and buffering capacity measurements to predict potassium uptake by white clover. *Plant and Soil, 51*, 167–176.

Evangelou, V. P., Wang, J., & Phillipis, R. E. (1994). New developments and perspectives on soil potassium quantity/intensity relationships. *Advances in Agronomy, 52*, 173–227.

Föhse, D., Claassen, N., & Jungk, A. (1991). Phosphorus efficiency of plants. II. Significance of root radius, root hairs, and cation–anion balance for phosphorus influx in seven plant species. *Plant and Soil, 132*, 261–272.

Gapon, E. N. (1933). Theory of exchangeable adsorption in soil. *Journal of General Chemistry (USSR), 3*, 144–163. [Abstract in Chem Abs. 28, 41–49, 1934.].

Grimme, H. (1974). Potassium release in relation to crop production. In International Potash Institute (Ed.), *Potassium research and agricultural production* (pp. 131–136). Berne: Int. Potash Inst.

Hegde, D. M., & Dwivedi, S. S. (1992). Nutrient management in rice-wheat cropping system in India. *Fertility News, 37*, 27–42.

Hinsinger, P., Jaillard, B., & Dufey, J. E. (1992). Rapid weathering of a trioctahedral mica by roots of ryegrass. *Soil Science Society of America Journal, 56*, 977–982.

Hoagland, R., & Martin, J. C. (1933). Absorption of potassium by plants in relation to replaceable, non-replaceable and soil solution potassium. *Soil Science, 36*, 1–34.

Jardine, P. M., & Sparks, D. L. (1984). Potassium-calcium exchange in a multi-reactive soil system II. Thermodynamics. *Soil Science Society of America Journal, 48*, 45–50.

Kuchenbuch, R., & Jungk, A. (1984). Wirkung der Kaliumdüngung auf die Kalium verfügbarkeit in der Rhizosphäre von Raps. *Zeitschrift für Pflanzenernährung und Bodenkdunde, 147*, 435–448.
Kuhlmann, H., & Wehrmann, J. (1984). Kali-Düngeempfehlung auf der Grundlage von 81K-Düngeversuchen zu Gertreide und Zuckerrüben. *Zeitschrift für Pflanzenernährung und Bodenkunde, 147*, 349–360.
Lumbanraja, J., & Evangelou, V. P. (1990). Binary and ternary exchange behavior of potassium and ammonium at low exchangeable fractional loads. *Soil Science Society of America Journal, 54*, 698–705.
Lumbanraja, J., & Evangelou, V. P. (1992). Potassium quantity–intensity relationships in the presence and absence of NH4 for three Kentucky soils. *Soil Science, 154*, 366–376.
Maas, E. V. (1969). Calcium uptake by excised maize roots and interactions with alkali cations. *Plant Physiology, 44*, 985–989.
Martin, H. W., & Sparks, D. L. (1983). Kinetics of nonexchangeable potassium release from two coastal plain soils. *Soil Science Society of America Journal, 47*, 883–887.
McLean, E. O., & Watson, M. E. (1985). Soil measurement of plant-available potassium. In R. D. Munson (Ed.), *Potassium in agriculture* (pp. 277–308). Madison: ASA, CSSA, and SSA.
Mengel, K. (1985). Dynamics and availability of major nutrients in soils. *Advances in Soil Sciences, 2*, 65–131.
Mengel, K., & Busch, R. (1982). The importance of the potassium buffer power on the critical potassium level in soils. *Soil Science, 133*, 27–32.
Mengel, K., & Kirby, E. A. (1980). Potassium in crop production. *Advances in Agronomy, 33*, 59–110.
Mengel, K., & Steffens, D. (1985). Potassium uptake of ryegrass (Lolium perenne) and red clover (Trifolium pratense) as related to root parameters. *Biology and Fertility of Soils, 1*, 53–58.
Mengel, K., & Uhlenbecker, K. (1993). Determination of available interlayer potassium and its uptake by ryegrass. *Soil Science Society of America Journal, 57*, 761–766.
Meyer, D., & Jungk, A. (1993). A new approach to quantify the utilization of nonexchangeable soil potassium by plants. *Plant and Soil, 149*, 235–243.
Mitsios, I. K., & Rowell, D. L. (1987). Plant uptake from exchangeable and non-exchangeable potassium. I. Measuring and modelling for onion roots in a chalky boulder clay soil. *Journal of Soil Science, 38*, 53–63.
Nair, K. P. P. (1984a). Towards a better approach to soil testing based on the buffer power concept. In *Proceeding of 6th International colloquium for the optimization of plant nutrition* (Vol. 4, pp. 1221–1228), 2–8 September, Pierre-Martin Prevel, Montpellier, France.
Nair, K. P. P. (1984b). Zinc buffer power as an important criterion for a dependable assessment of plant uptake. *Plant and Soil, 81*, 209–215.
Nair, K. P. P. (1992a). Measuring P buffer power to improve routine soil testing for phosphate. *European Journal of Agnon, 1*(2), 79–84.
Nair, K. P. P. (1992b). Soil testing procedure to economise on fertilizer use in land management. In *The spirit of enterprise — The 1993 rolex awards* (pp. 323–325).
Nair, K. P. P., Sadanandan, A. K., Hamza, S., & Abraham, J. (1997). The importance of Potassium buffer power in the growth and yield of Cardamom. *Journal of Plant Nutrition, 20*, 987–997.
Narang, R. S., & Bhandari, A. L. (1990). Integrated nitrogen management in rice–wheat system. Paper presented at the rice-wheat workshop. Directorate for Cropping System in Punjab, India. In *Proceedings of symposium on maximum yield research*, November 16–18, 1990, New Delhi, India.
Narang, R. S., Tiwanria, U. S., and Dev, G. (1994). Maximum yield research studies in rice-wheat system and soil productivity—The Indian experience. Research for maximum yield in harmony with nature. Satellite symposium sponsored by PPI (pp. 46–55). XV World Congress of Soil Science, Acapulco, July 10–16.
Pratt, P. F. (1965). Potassium. In C. A. Black et al. (Eds.), *Methods of soil analysis, part 2. Agron. Monogr* (Vol. 9, pp. 1023–1031). Madison: ASA.

Roy, H. K., Kumar, A., & Sinha, K. (1990). Response of wheat to K in red loam (Alfisols) soils of Ranchi. *Journal of Potassium Research, 6*, 23–28.

Schachtschabel, P. (1937). Aufnahme von nicht austauschbarem Kalium durch die Pflanze. *Zeitschrift für Pflanzenernährung und Bodenkunde, 48*, 107–133.

Schachtschabel, P. (1961). Fixierung und Nachlieferung von Kalium-und Ammonium-Ionen. Beurteilung und Bestimmung des Kaliumversorgungs-grades von Böden. *Landwirtschaftliche Forschung, 14*, 29–47.

Shu-Yan, C., & Sposito, G. (1981). The thermodynamics of ternary cation exchange systems and the subregular model. *Soil Science Society of America Journal, 45*, 1084–1089.

Singh, G. B., & Nambiar, K. K. M. (1986). Crop productivity and fertility under intensive use of chemical fertilizers in long-term field experiments. *Indian Journal of Agronomy, 31*, 115–127.

Sparks, D. L. (1987). Potassium dynamics in soils. *Advances in Soil Sciences, 6*, 2–63.

Sparks, D. L., & Huang, P. M. (1985). Physical chemistry of soil potassium. In R. D. Munson (Ed.), *Potassium in agriculture* (American Society of Agronomy) (pp. 201–276). WI: Madison.

Sposito, G. (1981a). Cation exchange in soils. A historical and theoretical perspective. In R. H. Dowday (Ed.), *Chemistry in the soil environment*. Madison: Soil Science Society of America.

Sposito, G. (1981b). *The Thermodynamics of the soil solution*. Oxford: Clarendon Press.

Sposito, G., Holtzelaw, K. M., Johnston, C. T., & Le Vesque, C. S. (1981). Thermodynamics of sodium-copper exchange on Wyoming bentonite at 198°K. *Soil Science Society of America Journal, 45*, 1079–1084.

Sposito, G., Holtzelaw, K. M., Charlet, L., Jonany, C., & Page, A. L. (1983). Sodium-calcium and sodium magnesium exchange on Wyoming bentonite in perchlorate and chloride background ionic media. *Soil Science Society of America Journal, 47*, 51–56.

Steffens, D., & Mengel, K. (1979). Das Aneignungsvermögen von Lolium perenne im Vergleich zu Trifolium pratense für Zwischenschilht-Kalium der Touminerale. *Landw. Forsch. SH., 36*, 120–127.

Steffens, D., & Mengel, K. (1981). Vergleichende Untersuchungen zwischen Lolium perenne und Trifolium pratense über das Aneignungsvermögen von Kalium. *Mitt. Dtsch. Bodenk. Ges., 32*, 375–386.

Swarup, A., & Singh, K. N. (1987). Movement of potassium in a sodic soil profile as influenced by long-term use of inorganic fertilizers under rice-wheat rotation. *Journal of the Indian Society of Soil Science, 35*, 253–256.

Vanselow, A. P. (1932). Equilibria of the base-exchange reactions of bentonites, permutities, soil colloids, and zeolites. *Soil Science, 33*, 95–113.

von Reichenbach, H. (1972). Factors of mica transformation. In *Potassium in Soil* (pp. 33–42). Berne, Switzerland: Int Potash Institute.

Wood, L. K., & De Turk, E. E. (1941). The adsorption of potassium in soils in replaceable forms. *Soil Science Society of America Proceedings, 5*, 152–161.

Woodruff, C. M. (1955a). Ionic equilibria between clay and dilute salt solutions. *Soil Science Society of America Proceedings, 19*, 36–40.

Woodruff, C. M. (1955b). The energies of replacement of calcium by potassium in soils. *Soil Science Society of America Proceedings, 19*, 167–171.

Xu, M. L., & Liu, Z. Y. (1983). The nutrient status of soil-root interface. II. Potassium accumulation and depletion in rhizosphere soils. *Acta Pedologica Sinica, 20*, 295–302.

Chapter 6
The Role of Electro-Ultrafiltration (EUF) in Measuring P and K Intensity for the Construction of Buffer Power Curves

Abstract The chapter discusses the importance of constructing "Buffer Power Curves" before embarking on how these determinations in the laboratory can be mathematically employed to precisely measure plant uptake. Details of Electro Ultrafiltration Technique are given in the chapter.

Keywords Electro ultra filtration · Buffer power curves · Phosphorus · Potassium intensity

In the construction of the buffer power curves it is imperative that both the "quantity" and the "intensity" factors are measured accurately. While the former can be quantified by an incubation and extraction technique (Nair and Mengel 1984) or an adsorption–desorption equilibrium technique (Nair 1992a, b), the precise determination of the latter poses far greater problems. Here, the role of electro-ultrafiltration in quantifying the intensity factor is briefly discussed. Electro-ultrafiltration is a flow and batch technique combined to determine the kinetics of K bio availability (Nemeth 1971, 1979). It combines electrodialysis and ultrafiltration. Originally electro-ultrafiltration was used to purify colloidal suspensions (Kottgen and Diehl 1929) but it was later modified for soil testing purposes (Grimme 1979). These authors attempted to extract the labile fraction of the plant nutrients without a chemical treatment such as chemical extraction. The EUF method eliminates accumulation of reaction products in the electrode compartments, and also charge on membranes separating the compartments as well as the effect of electroosmosis (Nemeth 1972; Grimme 1979). Nemeth (1971, 1972) modified the procedure to extract soil at three field strengths instead of one as done originally. Thus, an extraction (desorption) time of 10 min at 50 V and 20 °C was followed by two sequential extractions of 10 min each at 200 V and 20 °C and five sequential extractions of 5 min each at 400 V and 80 °C. The extraction period can be modified for different time sequences. For instance, Nair and Mengel (1984) collected three fractions, EUF-I + EUF-II, extracted at 200 V and 20 °C for 30 min and EUF-III extracted at 400 V and 80 °C between 30 and 35 min, in the case of P, while Mengel and Uhlenbecker (1993) collected EUF-I extracted at 200 V and 20 °C for 30 min followed by EUF-II extracted at 400 V and 80 °C for another 30 min in the case of K

K. P. Nair, *Intelligent Soil Management for Sustainable Agriculture*,
https://doi.org/10.1007/978-3-030-15530-8_6

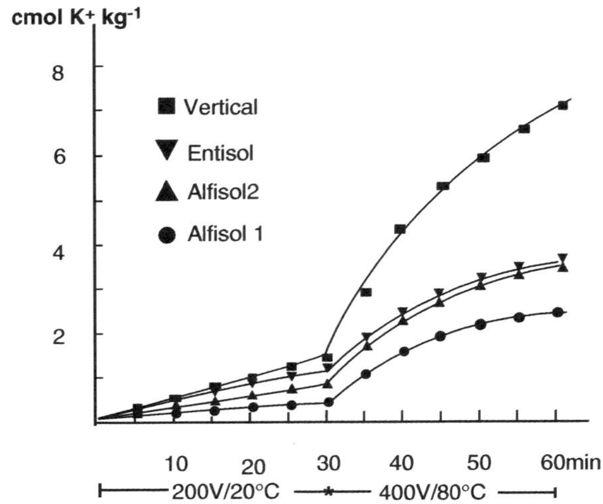

Fig. 6.1 Cumulative K + extracted by electro-ultrafiltration (EUF). Extraction at 200 V, 20 °C, 30 min (EUF0–30fraction); extraction at 400 V, 80 °C, 30 min (EUF30–60 fraction). (After Mengel and Uhlenbecker 1993)

(Fig. 6.1). Each extraction period was 5 min. In the case of P, Nair and Mengel (1984) considered all the three fractions as a measure of P "intensity," while in the case of K, the first two fractions (5–10 min) were considered a measure of K "intensity" (Sparks 1987). Since the stepwise extractions will provide an assessment of the changes in the mobility of the nutrient, the slopes of the desorption curves will indicate the rate at which the available quantity of the nutrient will be exhausted. For K, the ratio of the seventh and second fraction has been taken as a measure of K buffer power of the soil in question (Sparks 1987). Nair and Mengel (1984) and Nair (1992a, b) correlated the EUF fractions with CAL-extractable P (Schüller 1969) and desorbable P (Nair 1992a, b), respectively, and the b values in the regression functions were used as a measure of P buffer power. The very high r values obtained in both of the cases discussed earlier indicate that these b values will indeed provide a very dependable measure of P replenishment rates in the rooting zone which the P buffer power, in fact, attempts to measure. While EUF can provide dependable information on both "intensity" and "quantity" factors of P and K (Simrad and Tran 1993; Simrad et al. 1991), it must be pointed out that this facility may not be universally available and is, of course, quite expensive to provide. Thus, in routine soil testing procedures, there is a need to examine alternative procedures to estimate the P and K "intensity" factors. Nair (1994, unpublished data) has used extractions with 0.1 M $CaCl_2$ in tropical African and Asian soils to quantity P and K "intensity." Thus, it is relevant to mention that wherever an EUF facility exists, it could be used to quantify the "quantity" and "intensity" factors, and where it is not available, alternate procedures such as extraction with 0.1 M $CaCl_2$ can be used to measure K and P "intensity" factors. But in both cases, the essential point is that quantifying the P and K buffer power of soils and integrating the buffer power into the computations would provide a much better picture of the soil bio availability of these nutrients than contemporary soil tests alone. Additionally, EUF has the possibility to simultaneously provide information on "quantity" and "intensity" factors of other plant nutrients, besides P and K, at the same extraction cycles, as for Zn (Nair 1984a, b).

In a recent study (Steffens 1994), EUF has been used to study P kinetics. He quantified P kinetics by the Elovich b values [similar to that used by Mengel and Uhlenbecker (1993) in the case of K] by stepwise extraction using electro-ultrafiltration. In all, 11 fractions were used to study the P kinetics. The first two fractions were obtained, respectively, at 20 °C, 200 V, and <15 mA during an extraction period of 30 min and subsequently at 80 °C, 400 V, and <150 mA during an extraction period of 5 min, that is, a total of 30–35 min. This procedure was similar to that adopted by Nair and Mengel (1984) and Nair (1992a, b). After these two fractions were extracted, nine other fractions were extracted at 5-min intervals at 80 °C, 400 V, and <150 mA. The Elovich b values were computed using mg P kg^{-1} soil extracted vs time. P uptake was studied during a 4-year period using soils which were treated with basic slag phosphate, super phosphate, partially acidulated phosphate rock, and phosphate rock (Fig. 6.2). The test crops which were successively grown were rye grass (***L. perenne*** L., cv. Taptoe) in the first year, rape (***Brassica napus*** L., cv. Akella) in the second year, corn (cv. Badischer Landmais) and summer wheat (***T.aestivumL.,*** cv. Selpek) in the third year, and summer rye (***Secale cereale*** L., cv. Sorom) in the fourth year. The Elovich b values were more closely correlated to cumulative plant P uptake than water soluble P (van der Paauw 1971), CAL-Calcium acetate + calcium lactate + acetic acid mixture extractable P (Schüller 1969), or Mehlich 3 extractable P (Mehlich 1984). The cumulative plant P uptake from the 15 test soils was very well predicted (r = 0.99) when the Elovich b values were also integrated into the multiple regression model y = a + bx + cz used earlier

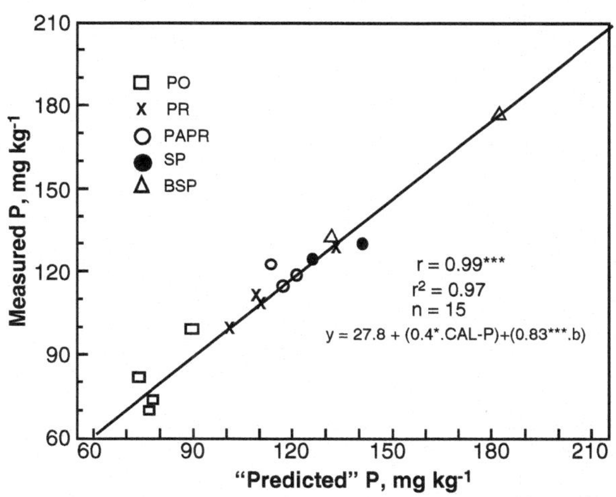

Fig. 6.2 Relationship of measured and predicted plant P uptake from pot experiments during a 4-year period (where CAL-P denotes the amount of calcium acetate–calcium lactate–acetic acid–extractable P, and b denotes the slope term of the electro-ultrafiltration extraction curves as described by the Elovich equation). PO, without P fertilization (control); *PR* phosphate rock, *PAPR* partially acidulated phosphate rock, *BSP* basic slag phosphate, *SP* superphosphate. *, ***Significant at the 0.05 and 0.001 probability levels, respectively. (After Steffens 1994)

by Nair and Mengel (1984) and Nair (1992a, b), where y denotes the plant P uptake, x denotes the CAL-P, and z denotes the Elovich b value. This indicates that considering the Elovich b value as a parameter for P release and the quantity of CAL extractable P for determining P bio availability following different P fertilizer applications, provides a much more reliable estimation of plant P uptake than consideration of only a single parameter of P bio availability such as the water extractable P, CAL-P, or Mehlich P. The Elovich b values used in this investigation are analogous to the P buffer power values of Nair and Mengel (1984) for P, who, for the first time, showed that precise prediction of P bio availability can only be done when both the labile P and the P "intensity" are integrated within one framework. Larsen (1967) had postulated such a view much earlier, but experimental evidence utilizing the EUF facilities to measure P "intensity" as shown by the work of Nair and Mengel (1984) and Nair (1992a, b) for P and by Nair (1984a, b) for Zn came much later. All of these studies underscore the great importance of EUF in modern-day soil testing methods.

References

Grimme, H. (1979). Factors controlling potassium availability to plants. In A. Saurat, & M. M. El-Fouly (Eds.), *Procedding of International Potassium Workshop, Cairo, Egypt* (pp. 33–47).

Kottgen, P., & Diehl, R. (1929). Lieber die anwendung der dialyse und electroultrafiltration zur bestimmung des nahvstoff bedurfnisses des bodens. *Zeitschrift für Pflanzenernährung und Bodenkdunde, 12*, 65–105.

Larsen, S. (1967). Soil phosphorus. *Advances in Agronomy, 19*, 151–210.

Mehlich, A. (1984). Mehlich 3 soil test extractant. A modification of the Mehlich 2 extractant. *Communications in Soil Science and Plant Analysis, 15*, 1409–1416.

Mengel, K., & Uhlenbecker, K. (1993). Determination of available interlayer potassium and its uptake by ryegrass. *Soil Science Society of America Journal, 57*, 761–766.

Nair, K. P. P. (1984a). Towards a better approach to soil testing based on the buffer power concept. In *Proceeding of 6th International colloquium for the optimization of plant nutrition* (Vol. 4, pp. 1221–1228), 2–8 September, Pierre-Martin Prevel, Montpellier, France.

Nair, K. P. P. (1984b). Zinc buffer power as an important criterion for a dependable assessment of plant uptake. *Plant and Soil, 81*, 209–215.

Nair, K. P. P. (1992a). Measuring P buffer power to improve routine soil testing for phosphate. *European Journal of Agnon, 1*(2), 79–84.

Nair, K. P. P. (1992b). Soil testing procedure to economise on fertilizer use in land management. In *The spirit of enterprise — The 1993 rolex awards* (pp. 323–325).

Nair, K. P. P., & Mengel, K. (1984). Importance of phosphate buffer power for phosphate uptake by rye. *Soil Science Society of America Journal, 48*, 92–95.

Nemeth, K. (1971). Die characteriesierung des K-haushaltes von boden mittels K-desorptionskurven. *Geoderma, 5*, 99–101.

Nemeth, K. (1972). The determination of desorption and solubility rates of nutrients in the soil by means of electroultrafiltration (EUF). *Proceedings of the ... Colloquium of the International Potash Institute, 9*, 171–180.

Nemeth, K. (1979). The availability of nutrients in the soil as determined by electroultrafiltration (EUF). *Advances in Agronomy, 31*, 155–188.

Schüller, H. (1969). Die CAL-Methode, eine neue Methode zur Bestimmung des pflanzen verfügbaren Phosphates in Böden. *Zeitschrift für Pflanzenernährung und Bodenkunde, 123*, 48–63.

Simrad, R. R., & Tran, T. S. (1993). Evaluating plant available phosphorus with the electroultrafiltration technique. *Soil Science Society of America Journal, 57*, 404–409.
Simrad, R. R., Tran, T. S., & Zizka, J. (1991). Evaluation of the electro-ultrafiltration technique (EUF) as a measure of the K supplying power of Quebec soils. *Plant and Soil, 132*, 91–101.
Sparks, D. L. (1987). Potassium dynamics in soils. *Advances in Soil Sciences, 6*, 2–63.
Steffens, D. (1994). Phosphorus release kinetics and extractable phosphorus after long-term fertilization. *Soil Science Society of America Journal, 58*, 1702–1708.
van der Paauw, F. (1971). An effective water extraction method for the determination of plant-available soil phosphorus. *Plant and Soil, 34*, 467–481.

Chapter 7
Quantifying the Buffer Power for Precise Availability Prediction – Heavy Metals

Abstract The chapter discusses how to quantify the "Buffer Power" for precise prediction of nutrient bio availability for heavy metals. This chapter discusses elaborately details of experimental findings in the case of the heavy metal Zinc.

Keywords Buffer power · Zinc buffer power

7.1 Zinc

Since Zn, among the heavy metals, has become the most problematic micronutrient on a global scale, the scope of this review and the thrust my research was confined to this very important plant nutrient. There is a great paucity of published material on the effect of buffer power on bio availability of heavy metals.

Plants obtain most of their fertilizer Zn from reaction products and not applied sources as such, implying that any source of Zn added to soil has to necessarily conform to a chain reaction involving adsorptive, desorptive, and resorptive processes that govern the maintenance of an equilibrium between adequate Zn concentration in the soil solution nearest to the zone of Zn depletion on the one hand and plant uptake on the other. The Zn buffer power defines this. As Zn concentration in soil solution is normally very low, the supply to plant roots by mass flow can only account for a very small fraction of plant demand. For instance, with a transpiration coefficient of 300 liters kg^{-1} dry matter and a corresponding Zn concentration of 10^{-7} M in the soil solution, approximately 2 mg of Zn can be supplied by mass flow against a demand of 10–30 mg Zn kg^{-1} dry weight of plant tissue. In calcareous soils, as the Zn concentration is of a much lower order of approximately 10^{-8} M, the supply by mass flow could be very much lower (Marschner 1994) indicating that mass flow can only contribute very negligibly to meet plant needs of Zn. Hence, Zn movement to the plant root surface is principally by diffusion and is essentially confined to a zone around the plant root which hardly extends beyond the root hair cylinder (Marschner 1994).

Though published papers on the relevance of Zn buffer power in soils to Zn availability are very limited, there is no dearth on other aspects. Genotypic differences to

K. P. Nair, *Intelligent Soil Management for Sustainable Agriculture*,
https://doi.org/10.1007/978-3-030-15530-8_7

Zn nutrition in a very important crop such as maize, with regard to its hybrids, inbreds, and composites, was reported almost four decades ago in India (Nair and Prabhat 1977); this study focused on the importance of genetic engineering for tailoring maize varieties for better efficiency of Zn utilization, while the complicated mechanism of Zn–P–Fe interaction was researched even prior to this (Nair and Babu 1975). In a recent review on the mechanism of Zn uptake, Marschner (1994) indicated that flow culture experiments with various species showed adequate ranges of Zn concentration in the range of 6×10^{-8} to 8×10^{-6} M which are concentrations greater than those that would be expected in the soil solution of most soils. He further pointed out that although work using chelate-buffered solutions has indicated adequate Zn concentrations between 10^{-10} and 10^{-11} M, extremely low adequate Zn concentrations required a concomitant excess of about 100 μM Zn-chelate as buffer at the plasma membrane of the root cells. This implies a need for an unlimited Zn pool for replenishment of Zn^{2+}at the plasma membrane. When plants grow in soil, it is impossible to expect a Zn buffer of this size to exist, and free $Zn^{2}{}_{+}$ and chelated-Zn concentrations will be at least threefold lower. Hence, critical deficiency or sufficiency concentrations obtained through research employing chelate-buffered solutions cannot be applied to soil-grown plants.

Most of the work on Zn bio availability to plants in soil is based on chemical extractions among which DTPA is the most frequently used. The DTPA extraction quantifies a labile fraction of soil Zn comprising water soluble, exchangeable, adsorbed, chelated, and some occluded Zn. The critical soil level of DTPA extractable Zn can vary from 0.3 to 1.4 mg kg^{-1} soil which equates to about 900–4200 g ha^{-1} of Zn in heavy soils and about 600–2800 g ha^{-1} Zn in light soils in the plough layer (0–20 cm). The crop requirements, on the other hand, are quite small, in the range of 100–300 g ha^{-1} for a total dry matter production of about 10 t ha^{-1} (Marschner 1994). The inadequacy of DTPA extraction to reflect plant Zn demand shows that other important factors, such as replenishment of soil solution Zn (Nair 1984a, b), mobility, and transport to the root surface (Wilkinson et al. 1968; Nair et al. 1984), and also the activity of the roots themselves (Wilkinson et al. 1968; Marschner 1994), are involved. Since the Zn buffer power is intricately involved in all three factors, the focus of this book is mainly on that attribute.

As early as almost five decades ago, it was suggested that colloidal Zn was released by some specific process associated with root activity (Wilkinson et al. 1968). Conditions in the rhizosphere and particularly root-induced changes markedly affect Zn availability. A difference in rhizosphere pH of as much as 2, higher or lower compared to bulk soil, can be expected to occur as a result of imbalance in ionic uptake. For instance, any acidifying fertilizer such as $(NH_4)_2SO_4$ can result in a net excretion of H^+ ions and others such as NH_4NO_3 can result in a net excretion of HCO_3^- H or OH^- ions. Additionally, secretion of organic acids and enhanced CO_2 production, as well, will affect rhizosphere pH, and all of the above-mentioned changes will markedly affect Zn bio availability. However, the scope of this book is confined to the kinetic/dynamic aspects of the changes occurring in the rooting zone mirrored in the Zn buffer power rather than changes in soil reaction in the rhizosphere *per se* on Zn bio availability.

The distribution of Zn between the solid and solution phases can be described by the buffer power. The bio availability of soil Zn to the plant depends on the initial Zn concentration, Zn buffer power and effective diffusion coefficient (Barber 1984). The Langmuir equation gives the relation between B and C1 as

$$\frac{C_1}{(x/m)} = \frac{1}{aB} + \frac{C_1}{B}$$

where C_1 = Zn concentration in the soil solution, x/m = the amount of Zn adsorbed per unit of soil, B = adsorption maximum, and a = a constant related to the soil's bonding energy for Zn. A straight line is obtained when C_1 (/(x/m) is plotted against C_1 with a slope of 1/B and intercept of 1/aB. The inverse of C_1/(x/m) = b, the Zn buffer power, where C_1 and x/m are both expressed in volume units (Barber 1984). Using this approach, Shuman (1975) estimated the buffer power values varying from 5 to 100 for four soils representing different major physiographic regions of Georgia. Based on the diffusion model of Drew et al. (1969), Nair (1984a, b) has argued that the $\bar{c}$ in the equation $U = 2\pi\alpha a\bar{c}t$ (Drew et al. 1969) where U = quantity of Zn absorbed per centimeter root length, a = root radius in cm, α = root absorbing power, c = average Zn concentration on the root surface, and t = duration of the absorption period, in fact represents an indirect measure of the Zn buffer power. As we already know, the bulk of Zn uptake is by diffusion (Wilkinson et al. 1968; Elgawhary et al. 1970; Barber 1984). This diffusive process will maintain a concentration gradient in the rooting zone. This concentration gradient will directly affect Zn uptake because of its effect on the average Zn concentration on the root surface. The Zn buffer power will affect this concentration gradient, because the rate of Zn depletion and/or replenishment is mirrored by it. In a sense the effective diffusion coefficient and the buffer power are analogous for nutrients which are principally absorbed by the plant root through diffusive processes (Nair 1989). Hence, the crucial question to examine would be the role of Zn buffer power in influencing Zn bio availability for plant uptake.

7.2 Quantifying Zn Buffer Power

Nair (1984a, b) used a novel method to quantify Zn buffer power. Heavy metal pollution is a major environmental hazard in Europe, especially on the highways due to automobile exhausts; as a consequence, a number of soils are very highly contaminated. Such soils are unsuitable for a precise study on the impact of Zn buffer power on plant Zn uptake, and the choice of soil was narrowed to a heavy clay in the northern part of Belgium containing 1.1 ppm total Zn, 5.5 ppm DTPA extractable Zn, and 0.13 ppm electro-ultrafiltrable Zn. The soil was acidic (pH 6.6, 1 N KCl), low in organic carbon (1.8%), high in $CaCO_3$ (16.5%), containing beidellite as the predominant clay mineral. Zn buffer power was quantified by regressing DTPA extractable Zn (y) on electro-ultrafiltrable Zn (x) by a stepwise extraction of incubated soil, and the b value in the regression function represented the Zn buffer power. Electroultrafiltration has been employed for a rapid estimation of the intensity of a range of heavy metals, especially in central, western, and eastern European soils (Nemeth 1979; Nemeth and Recke 1982). However, earlier investigations of these authors did not give satisfactory results: analysis of only the alkaline cathode filtrate is insufficient to give precise information about availability indices, primarily due to the fact that the heavy metals are collected by the cathode filter as hydroxides and hydrated oxides, which are hardly soluble in the alkaline cathode filtrate. However, as these oxides comprise soluble and desorbable ions of the heavy metal in question, they are of importance to plant nutrition; Nair (1984a, b) used a modified procedure to bring the oxides into an acid medium for subsequent analysis. The studies indicated that Zn “intensity” alone is insufficient to precisely predict Zn uptake by maize used as a test plant in the investigation. The relevant data are given in Table 7.1.

Data in Table 7.1 show that the precision of predicting Zn concentration in the maize top increased remarkably, as substantiated by the increase in the coefficient of determination which works out to 65.9%. With regard to total Zn uptake by the maize plant, the increase is less marked, yet substantial. The data show that the major factor responsible for the variance in Zn concentration in the maize top and total Zn uptake by the maize top is the Zn buffer power, and not Zn “intensity” as quantified by electro-ultrafiltration. In a separate experimental set-up, Nair et al. (1984) monitored Zn transport by simulation experiments in a laboratory set-up using the same field soil and obtained proof to support the hypothesis that it is essentially the buffered Zn that contributes to the major uptake process by the test plant. Dissolved salts in a field soil must move with the liquid, the net movement from a given reference plane being dependent on the activity gradient of salt and water. This naturally brings us to the important question of solute movement within the soil matrix which is dependent on the soil conductivity. Buffered Zn would be in the bulk soil solution to be further transported to plant roots. This transport, though dependent on the soil conductivity, would also be linked to the quantum of Zn *per se*

Table 7.1 Correlation coefficients (r) between Zn intensity and maize top Zn concentration and total Zn uptake by plant top at harvest without (A) and with (B) Zn buffer power integration

	Correlation coefficient	
Details	A	B
Zn intensity vs maize top Zn concentration	0.290	0.812***
Zn intensity vs total Zn uptake by maize top	0.356	0.618**

After Nair (1984a, b)

**, ***Significant at 1 and 0.1% confidence levels, respectively

in the bulk soil solution, which, in turn, would be dependent on the Zn buffer power of the soil in question. Nair et al. (1984) quantified these relationships. In a field soil, it is the buffered Zn which maintains a concentration gradient between the bulk soil solution and the root surface sufficiently conducive for optimal plant uptake. The investigation also showed the ineffectiveness of DTPA extraction as a reliable soil test for Zn availability. Electro-ultrafiltration is being put to use increasingly for commercial soil testing in many countries in Europe, notably, for sugarbeet cultivation in Austria. Determination of Zn intensity singly may not provide reliable information regarding plant bio available Zn. Besides Zn intensity and Zn buffer power, other soil factors have only a minor influence on Zn absorption. Electro-ultrafiltration has the advantage of a rapid assessment of the Zn "intensity" factor. Combining the advantage of rapid assessment of the "intensity" factor of the nutrient with its corresponding buffer power might greatly enhance the reliability of fertilizer recommendations not only for Zn, but, perhaps, for other heavy metals as well, making electro-ultrafiltration a reliable tool in soil testing for heavy metals.

References

Barber, S. A. (1984). *Soil nutrient bioavailability: A mechanistic approach*. New York: Wiley.

Drew, M. C., Nye, P. H., & Vaidyanathan, L. V. (1969). The supply of nutrient ions by diffusion to plant roots in soil. I. Absorption of potassium by cylindrical roots of onion and leek. *Plant and Soil, 30*, 252–270.

Elgawhary, S. M., Lindsay, W. L., & Kemper, W. D. (1970). Effect of EDTA on the self-diffusion of zinc in aqueous solution and in soil. *Soil Science Society of America Proceedings, 34*, 66–70.

Marschner, H. (1994). Micronutr. *News Information, 14*, 3–5.

Nair, K. P. P. (1984a). Towards a better approach to soil testing based on the buffer power concept. In *Proceeding of 6th International colloquium for the optimization of plant nutrition* (Vol. 4, pp. 1221–1228), 2–8 September, Pierre-Martin Prevel, Montpellier, France.

Nair, K. P. P. (1984b). Zinc buffer power as an important criterion for a dependable assessment of plant uptake. *Plant and Soil, 81*, 209–215.

Nair, K. P. P. (1989). Comments on "Phosphorus supply characteristics of 33 soils as influenced by seven rates of phosphorus addition". *Soil Science Society of America Journal, 53*, 984.

Nair, K. P. P., & Babu, G. R. (1975). Zinc–phosphorus–iron interaction studies in maize. *Plant and Soil, 42*, 517–536.

Nair, K. P. P., & Prabhat, G. (1977). Differential response of tropical maize genotypes to zinc and manganese nutrition. *Plant and Soil, 47*, 149–159.

Nair, K. P. P., Ram, N., & Sharma, P. K. (1984). Quantitative relationship between zinc transport and plant uptake. *Plant and Soil, 81*, 217–220.

Nemeth, K. (1979). The availability of nutrients in the soil as determined by electroultrafiltration (EUF). *Advances in Agronomy, 31*, 155–188.

Nemeth, K., & Recke, H. (1982). Reproducibility of Ca, K, Na, and P contents in the different EUF fractions. *Plant and Soil, 64*, 25–33.

Shuman, L. M. (1975). The effect of soil properties on zinc adsorption by soils. *Soil Science Society of America Proceedings, 39*, 454–458.

Wilkinson, H. F., Loneragan, J. F., & Quirk, J. P. (1968). The movement of zinc to plant roots. *Soil Science Society of America Proceedings, 32*, 831–833.

Chapter 8
Case Studies with South Asian Soils

Abstract The chapter discusses, at length, case studies with buffer power estimation for Zinc in South Asian soils, especially India, and its relevance in the field cultivation of Kerala State's economic mainstay, Black pepper (*Piper nigrum*).

Keywords Zinc buffer power · India · Black pepper

The 3rd International Zinc Symposium held in Hyderabad, India, where 195 delegates from 30 countries participated, focused on the cruciality of Zn in both plant and human nutrition. Zn is assuming the role of a major constraint in crop production, in the developing world, in particular, in India. There is widespread Zn deficiency in tropical soils. This section of the book is divided into two parts, where the focus will be first on (A) South Asian soils, India being the classical example and (B) Central Asian soils, Turkey, where severe Zn deficiency in the main wheat growing belt of the country (Anatolia) is threatening wheat production and adversely affecting human nutrition.

Investigations in Indian soils: The lateritic belt of Kerala State, in southern India, was chosen as the test soil. The economic mainstay of the State is Black pepper (***Piper nigrum*** L.). The crop is very susceptible to Zn deficiency and among the major diseases causing havoc in Black pepper plantations, the onset of the "Quick Wilt" caused by ***Phytophthora*** spp. (***Phytophthora capsici***) has been attributed to a severe Zn deficiency in the soil. The "official" recommendation for supplemental Zn fertilizer to control the dreaded disease is based on the classical DTPA extraction, and a "blanket" recommendation for the entire State is made, where, the farmers are advised to apply Zn fertilizer at the rate of 25 kg/ha. for the entire State. This author has, for long, suspected the advisability of making such a "blanket" recommendation, and, had suggested, in its place, investigation of the Zn buffering mechanism of the soils involved and then calibrate the appropriate Zn fertilizer recommendations based on this. Zn fertilizer is quite expensive, and, at prevalent rate of approximately Rs 300/kg (Approx. US $ 5/kg) could turn out to be prohibitively expensive for a poor Black pepper farmer of Kerala State. In the following paragraphs, the salient results are summarized.

K. P. Nair, *Intelligent Soil Management for Sustainable Agriculture*,
https://doi.org/10.1007/978-3-030-15530-8_8

8.1 Quantifying the Zn Buffer Power of the Pepper – Growing Soils

Nair (1984) has used electro-ultra-filtration to quantify Zn "Intensity" in measuring Zn buffer power of European soils. Because the procedure is very highly sophisticated, and due to the non-availability of such expensive instrumentation in a country like India, including many other developing countries, a simple adsorption-desorption equilibrium technique was developed by the author (Nair 2002) to quantify Zn buffer power of Black pepper-growing soils. For this, 200 g of representative soil samples from the principal Black pepper-growing regions were selected from the State, incubated at 60% of the maximum water holding–capacity with graded rates of Zn, over a fortnight, by maintaining the water regime to constancy. At the end of the period, the soil samples were extracted with 0.01 M $CaCl_2$ over a 24-h period and the extract tested for Zn using atomic absorption spectrophotometry. This represented Zn "Intensity." Separate extractions were made with DTPA, which represented the Zn "Quantity." Data in the following Table (Table 8.1) indicate that the Zn buffer power values of the Black pepper-growing soils varied widely.

Data in Table 8.1 bring to light two very important facts. First, the very highly significant correlation coefficients ("r" values) prove that the technique employed allows a precise determination of the Zn buffer power of the soils investigated. Second, the soils varied widely in their Zn buffer power. It is this fact which has to be critically viewed against the existing Zn fertilizer recommendation, emanating

Table 8.1 Zinc buffer power of Black-pepper growing soils

Soil	"r" value	"b" value
Peruvannamuzhi	0.8337***	0.7824
Thamarasseri	0.9304***	1.5786
Ambalavayal	0.9604***	3.0358

Note: All the "r" values are very highly significant, at a confidence level of 99.9% (***) implying that the adsorption-desorption equilibrium technique developed by the author provides an accurate measure of the Zn buffer power of the experimental soils. Further, the soils showed widely varying Zn buffer power, as shown by the "b" values, with the soils from *Peruvannamuzhi* region least buffered for Zn, while the soils from *Ambalavayal* region most buffered, with the soils from *Thamarasseri* region intermediate in their Zn buffer power. These findings have remarkable impact, as we shall see from the results incorporated in Table 8.3

from routine soil tests using the universally employed DTPA extraction, to understand the cruciality of the Zn buffer power for precise formulations of Zn fertilizer application for Black pepper.

Data in Table 8.2, below, clearly indicate that integrating the Zn buffer power has improved the inter-relationship between DTPA routine soil test versus both Zn concentration in plant top and Zn uptake. Most remarkably, the overall relationship between DTPA soil test and dry matter production not only improved very significantly when Zn buffer power was integrated into the computations, but from a negative value to a positive one. This implies two very important conclusions. First, the universally adopted DTPA soil test for Zn is unsuitable for the soils investigated. Second, it is the Zn buffer power that is the decisive factor which determines dry matter production, implying, higher the Zn buffer power, more the yield. Results in Table 8.3 (below) substantiate this very important conclusion.

Data in Tables 8.1, 8.2 and 8.3 have to be reviewed in conjunction to precisely understand the cruciality of the "Buffer Power Concept." Black pepper is the economic mainstay of the Kerala State. Of late, Widespread Zn deficiency has been observed in the State, and the onset of the most devastating "Quick Wilt" disease and "Foot Rot" disease, both of which can exterminate a Black pepper plantation in a matter of a week's time, has been attributed to the infestation by ***Phytophthora*** fungus, which thrives best in Zn deficient soils. The scientists at the Indian Institute of Spices Research in Calicut, Kerala State, India, under the administrative control of the Indian Council of Agricultural Research in New Delhi, have made a "blanket"

Table 8.2 Correlation coefficients ("r" values) for the inter-relationship between routine DTPA soil test versus Zn concentration, Zn uptake and dry matter production without (1) and with (2) Zn buffer power integration in Black pepper

Details	1	2
Zn concentration versus DTPA test	0.88***	0.92***
Zn uptake versus DTPA test	0.78	0.86***
Dry matter production versus DTPA test	−0.74	0.78***

Note: 1 and 2, without (1) and with (2) Zn buffer power integration in the computations, respectively, ***Significant at 99.9% confidence level

Table 8.3 Black pepper yields from farmers' fields weighted against the Zn buffer power

Region	Yield (kg/vine)		
	Targeted	Actual	Deviation (%)
Peruvannamuzhi	0.241	0.401	+ 66
Thamarasseri	0.490	0.487	+ 0.6

Note: Target weighting was done against the highest vine yield obtained from the soils from the *Ambalavayal* region. Note the remarkable closeness between targeted and actual yields in *Thamarasseri* region, whereas in *Peruvannamuzhi* region, there is a huge deviation. Soils in the *Peruvannamuzhi* region are atypical Black pepper soils. Ironically, the "official" recommendation for Zn fertilizer application in Kerala State emanate from the field experiments conducted in *Peruvannamuzhi* soils, since the Indian Institute of Spices Research, which conduct field experiments on Black pepper, affiliated to the Indian Council of Agricultural Research, New Delhi, is located in *Peruvannamuzhi,* which will totally mislead Black pepper farmers in the Kerala State recommendation of soil application of 23 kg $ZnSO_4$/ha, when DTPA extractable Zn is less than 1.6 ppm. In terms of monetary input, this would translate to more than US $ 115 ha^{-1} in terms of farmer investment. This is a substantial investment, given the prevalent economic conditions of the poor and marginal Black pepper farmers of Kerala State. Because the soils are variously Zn buffered (Table 8.1), a single "blanket" dose as the Kerala Government is propagating is totally unscientific, because, where the farmer needs to apply only lesser quantities of Zn, he would end up applying more and thus waste his money, if he were to go by the routine DTPA soil test values alone. To elaborate, the soils from the ***Ambalavayal*** region have a Zn buffer power of 3.0358 as compared soils from the ***Peruvannamuzhi*** region which have a buffer power of only 0.7824. In other words, the ***Amabalavayal*** soils have the potential, which is nearly more than four-fold, to supply Zn as compared ***to Peruvannamuzhi*** soils. In practical terms, this would imply that the ***Ambalavayal*** soils would only require 25% of the total quantum of Zn applied to ***Pervannamuzhi*** soils. Translated into the recommendation, it would imply that the ***Ambalavayal*** soils would only need about 5 kg Zn fertilizer/ha when the ***Peruvannamuzhi*** soils require 23 kg/ha. In the case of ***Thamarasseri*** soils the quantity required would be approximately 50% of the recommended rate. The pecuniary benefits to the poor and marginal Black pepper farmer from this finding can easily be gauged from this sound and practical reasoning, and need, hardly, any underscoring. Translated on a large scale covering the entire State of Kerala, the monetary saving would be huge.

In fact, the scientists at the Indian Institute of Spices Research have made a gross underestimation of the Zn-supplying power of the soils in the State, because, their results and recommendations are all based on the soils of the ***Peruvannamuzhi*** region, where the Black pepper research station is located. The soils here are acutely Zn impoverished, but, that is not the case in the entire State of Kerala. This also shows that, more often than not, practical recommendations from experiment stations, when extrapolated on a large-scale, can run into real problems, especially when such recommendations are based on outdated routine soil testing procedures and the poor farmers could be at the receiving end of losing money. These results (Nair 2004) substantiate the earlier ones (Nair 2002) on the importance of precisely quantifying the Zn buffer power, rather than routinely estimate DTPA-extractable Zn. This approach is further discussed in the case of the author's investigations in Turkey (Central Anatolia), where severe Zn deficiency is causing great adverse effects on wheat production. Farmers have been advised to apply as much as 100 kg $ZnSO_4^{-1}$ ha as a blanket recommendation, without obtaining any tangible and consistent response in wheat yield. Hence, the great economic importance of the "Buffer Power Concept," as against text book based routine soil testing procedures, needs hardly any underscoring. Nair et al. (1997) have conclusively demonstrated the importance of the buffer power concept in the case of potassium nutrition of cardamom (***Elettaria cardamom*** M.), another very important spice crop, besides Black pepper, in the State of Kerala.

References

Nair, K. P. P. (1984). Towards a better approach to soil testing based on the buffer power concept. In *Proceeding of 6th International colloquium for the optimization of plant nutrition* (Vol. 4, pp. 1221–1228), 2–8 September, Pierre-Martin Prevel, Montpellier, France.

Nair, K. P. P. (2002). *"The Nutrient Buffer Power Concept" – A revolutionary soil management technique for sustainable agriculture*. Invitational research paper presented in the 17th World Soil Science Congress, Bangkok, Thailand, August 2002.

Nair, K. P. P. (2004). Sustaining crop production in the developing world through "the nutrient buffer power concept" – A case study with Black pepper (Piper nigrum) growing at low pH in the laterite soils of India. In *Proceeding of 16th international symposium on plant-soil interactions at low pH* (p. 392), Sendai, Japan, August 1–5.

Nair, K. P. P., Sadanandan, A. K., Hamza, S., & Abraham, J. (1997). The importance of Potassium buffer power in the growth and yield of Cardamom. *Journal of Plant Nutrition, 20*, 987–997.

Chapter 9
Case Studies with Central Asian Soils

Abstract The chapter discusses, at length, the importance of the precise determination of Zinc buffer power in wheat cultivation in Central Asia (Turkey, in the wheat belt, Anatolia). The results clearly demonstrate the scientific inadequacy of the commonly used soil test for Zinc bio availability – the DTPA test – in the Anatolia region, and, unequivocally demonstrate how the buffer power concept has been able to enormously help the wheat farmers of the region with regard to Zinc fertilizer application, who have been wasting their monetary resources on unscientific fertilizer application, for example, as much as 100 kg ZnSO4/ha.

Keywords Zinc buffer power · Turkey · Wheat

In a global study initiated by FAO and carried out in 30 countries, it was estimated that about 30% of the agricultural soils of the world are Zn-deficient (Sillanpaa 1982). Based on the analysis of 298 soil samples, Turkish soils were classified as Zn-deficient (Sillanpaa and Vlek 1985). Acute Zn deficiency triggering severe decreases in yields of cereals have been noted in many countries of the tropical belt. Cereal species show large differential sensitivity to Zn deficiency. Large differences in sensitivity to Zn deficiency is also seen among genotypes within a specific cereal species (Graham and Rengel 1993). This indicates the existence of genetic potential to select and breed for Zn deficiency resistant genotypes. Among the cereals, wheat is particularly sensitive to Zn deficiency. Growing wheat on Zn-deficient soils leads to not only decrease in plant growth and yield, but, would lead to low Zn accumulation in grains. This will have a deleterious effect on human nutrition, because, wheat grains are rich in phytate and fiber, which limit absorption and/ or utilization of Zn in human cells (Welch 1993). Approximately 40% of world's population suffer from micronutrients,' including Zn deficiency, (Graham and Welch 1994). High consumption of cereal-based foods is thought to be one of the major reasons for Zn deficiency in humans. The proportion of wheat, maize and rice consumed in diets of people in developing countries comprises about 90% of the average diet (Graham and Welch 1994). In Turkey, in Central Asia, it has been observed that Zn deficiency in humans is widespread where cereal-based foods dominate (Cavdar et al. 1983; Prasad 1982). Turkey is one of the major wheat producing countries in the world.

K. P. Nair, *Intelligent Soil Management for Sustainable Agriculture*,
https://doi.org/10.1007/978-3-030-15530-8_9

Nearly 45% (4.5 million ha) of the wheat producing area in Turkey is located in the highly calcareous region of Central Anatolia. Central Anatolia is semi arid, being the driest region of Turkey. Scanty rainfall (about 325 mm) and low water supply greatly limit wheat yields. The author initiated a project to critically examine the role of the buffer power concept in the Zn nutrition of wheat, as inconsistent results have been observed even with huge quantities (as much as 100 kg $ZnSO_4$/ha) of Zn application to the wheat crop.

Nair (unpublished) employed an adsorption-desorption equilibrium and a soil incubation technique to quantify the Zn buffer power of some highly Zn deficient Turkish soils which have failed to show consistent response to Zn application based on DTPA extraction in wheat grown in these soils. Despite quite comparable DTPA extractable Zn, the Zn buffer power of these soils varied largely from a low value of 27 to a very high value as high as 200. It is being hypothesized that the Zn buffer power could, possibly, hold the key to explaining the differential response of wheat to Zn application in these problematic soils. The technique developed by Nair (unpublished) dispenses with the need to employ highly cost intensive electro-ultrafiltration, which will certainly not be suitable to practical situations in many developing countries constrained by economic considerations, and in its place, employs a far simpler, and, simultaneously, far less costlier, technique of quantifying Zn intensity by an adsorption –desorption equilibrium technique using 0.01 M $CaCl_2$ solution or soil incubation technique.

9.1 Establishment of the Zn Buffer Power

The Zn buffer power curves are represented by the plot between the Zn "Quantity" and Zn "Intensity" parameters (Nair 1996). The "Q" and "I" parameters were established by laboratory procedures, one where an adsorption -desorption equilibrium technique was employed and another where an incubation technique was used, as described below.

Adsorption – Desorption Equilibrium technique to measure Zn "Intensity" 5 g soil were weighed into PVC tubes saturated with 25 ml Zn solution (Zn SO_4 $7H_2O$) in 0.01 M $CaCl_2$ solution in amounts ranging from 0–500 mg Zn/kg soil. The soil suspensions were shaken in an end over end shaker at ambient temperature (25 + – °C) –at 125 cycles per minute. After 24 h the suspensions were filtered through "Blue Ribbon" 589 cube (3) –150 mm-filter paper (Schleicher and Schuell –German make). The soil on the filter paper was thoroughly washed with double distilled water and subsequently with 60% ethanol to remove any Zn ions not bound to the soil matrix and the entire mass of soil was transferred to another set of PVC containers. The filtrate was immediately analyzed for Zn using atomic absorption spectrophotometry.

Estimation of Zn "Intensity": To the new set of PVC containers with the soil mass, 25 ml of 0.01 M $CaCl_2$ solution were added and they were shaken for 24 h under identical conditions, as explained above. After 24 h, the suspensions were filtered, as before, and the filtrate immediately analyzed for Zn. This represents Zn "Intensity."

Estimation of Zn "Quantity": The soil in the PVC containers was extracted with DTPA extractant, which is a mixture of 0.005 M DTPA, 0.01 M $CaCl_2$, and 0.1 M TEA, adjusted to a pH of 7.3 using 6 N HCl. The pH of the extractant was monitored using a portable pH meter. 25 ml of the extractant were pipetted into the PVC containers and the soil suspensions were shaken as before, but, only for 2 h. The extraction period is crucial as originally suggested by the authors, since extraction is not complete after 2 h, the labile and non-labile quantities of Zn continuing to dissolve. The filtrate was immediately analyzed for Zn by atomic absorption spectrophotometry and the Zn concentration in the filtrate was expressed as mg Zn/kg soil (ppm).

Each of the soil series had 6 Zn treatments, including a "No Zn" control treatment, and each treatment was replicated thrice to conform to a Completely Randomized Block Design. Thus, in all, in each cycle of operation, there were 54 samples and each was analyzed, at every stage, in duplicate.

The most important conclusions of great practical significance, emanating from these investigations, are the following. Both the experimental soils from Turkey were critically Zn deficient (<1 mg Zn/kg soil, that is 1 ppm), while the German soil had 3.6 mg Zn/kg soil (3.6 ppm). And a uniform rate of 100 kg $ZnSO_4$/ha as $ZnSO_4.7H_20$, was applied to both soils. Yet, there was a huge margin of difference in the mean wheat grain yield (Table 9.2). The percentage increase in mean wheat grain yield in Konya soils, which are very highly Zn buffered, is 69 while those from Kuruklu, where the Zn power had a very low value, showed but negligent increase in wheat yield. The percentage increase in coefficient of determination for the interrelationship between Zn buffer power and mean increase in wheat grain yield is 74. This simply illustrates that nearly 75% of yield variation can be explained by the Zn buffer power. The soil Zn dynamics which explains this phenomenon is as follows. Photo 1 clearly illustrates the importance of the "Zn Buffer Power" in correcting Zn deficiency of highly Zn-deficient soils, in wheat nutrition.

That Zn is principally transported through the diffusive mechanism is now an established fact and Nair (1984a, b) has provided incontrovertible experimental evidence where Zn buffer power is shown to be the most important criterion affecting the uptake process. Flux rates thus depend on Zn concentration gradients between root surfaces and the bulk soil solution (Nair 1992). Current results have to be viewed against this background. According to the model of Zn uptake proposed by Nair (1984a, b), the average concentration on the root surface is analogous to Zn "Intensity" parameter, and so would critically affect the Zn uptake process. A root growing into the soil would, at first, encounter a relatively high Zn concentration at its surface, which would be more or less equal to the Zn concentration in the bulk soil solution. Due to root uptake of Zn, its concentration on the root surface would diminish. This decrease would not only depend on the uptake rate, but, also on the Zn buffer power, because, the Zn buffering mechanism would govern the rate of replenishment in the zone of Zn depletion around the roots. A soil with a high Zn buffer power would replenish the depletion process much faster than a soil with low Zn buffer power. This differential mechanism is reflected in the Zn buffer power values (Table 9.1) of the soils from ***Konya*** and ***Koruklu.*** Thus, it is logical to conclude that the uptake model proposed by Nair (1984a, b) clearly explains the yield

Table 9.1 Linear regression functions (Y = a + bx) and correlation coefficients ("r" values) for the relationship between DTPA extractable Zn (Zn "Quantity" Y) and 0.01 M $CaCl_2$ extractable Zn (Zn "Intensity" x)

Regression function					
Soil	Adsorption-Desorption Equilibrium	Soil incubation	"r" value	A	B
Konya (Central Anatolia)	Q = 201bI + 5.3	Q = 679bI-5.9		0.951***	0.978***
Koruklu (Gap region)	Q = 17bI + 17.9	Q = 375bI-8.7		0.015	0.784***
German soil (reference soil)	Q = 8bI + 6.5	Q = 11bI + 7.8		0.972***	0.942***

Note: Q = Zn "Quantity," I = Zn "Intensity," b = Zn buffer power. A without Zn buffer power integration in the computations, and B with the Zn buffer power integration. The experimental technique employed has resulted in very highly significant correlation coefficients ("r" values), except in the case of soils from the Gap region (Koruklu) in the case of the soil incubation experiments, which suggests the high accuracy of the adsorption-desorption equilibrium technique to determine Zn "Intensity," which is a true reflection of the Zn buffer power. Note the remarkable improvements in the "r" values (B compared to A), in all cases, including that of the soils from the Koruklu region, which unequivocally shows that the integration of the Zn buffer power remarkably improves the relationship. For both the Turkish soils, inclusion of the Zn buffer power in the computations resulted in increasing the coefficient of determination (CD) from 90.4% to 95.6% in the case of Konya soils, and, most remarkably, from 0.22% to 68.5% in the Koruklu soils. How this affects plant uptake, and, eventually wheat yields, will be discussed in the following tables and explanation of the experimental details. ***Significant at a confidence level of 99.9%

Table 9.2 Effect of Zn fertilization (100 kg $ZnSO_4$/ha) on mean and range of Zn shoot tissue concentration and grain yield among 28 wheat genotypes grown in Konya soils and 25 grown in Gap soils in Turkey

Soil	Tissue concentration (ppm)				Grain yield (kg/ha)				Percentage increase
	Without Zn		With Zn		Without Zn		With Zn		
	Range	Mean	Range	Mean	Range	Mean	Range	Mean	
Konya	7–10	9	10–19	14	1550–4010	2710	248–5900	4570	69
Gap	16–23	19	22–31	25	1530–5260	3500	1550–5300	3510	0

Note: The Zn buffer power is very clearly reflected in grain yield increase. Note that the "b" values (Zn buffer power) for both the adsorption-desorption equilibrium technique and soil incubation technique for Konya soils are 201 and 679 respectively. The corresponding values in Koruklu soils (Gap region) are 17 and 375 respectively. The percentage increase in wheat grain yield in Konya soils is 69 while that in Koruklu soils it is 0. In other words, the Konya soils are highly buffered for Zn bio availability, while Koruklu soils are not

differential in the two soils investigated. The Konya soils, by virtue of their very high Zn buffer power, are able to maintain optimal Zn concentration on the wheat root surface and this positive impact is reflected in grain yield increase. The ***Koruklu*** soils, on the other hand, are unable to maintain the optimal Zn concentration on root surface, hence are unable to sustain high wheat yields (Tables 9.2 and 9.3).

Table 9.3 Inter-relationship between Zn availability parameters and wheat grain yield and shoot tissue concentration of Zn (Turkey)

Parameter investigated	Correlation coefficient ("r" value)
DTPA soil test for Zn versus Zn concentration in shot tissue	0.672***
Zn buffer power versus percentage increase in wheat grain yield	0.860***

Note: Note the very highly significant (at a confidence level of 99.9% ***) "r" value between Zn buffer power and percentage increase in wheat grain yield. It has a Coefficient of Determination of 92.2%. Though DTPA soil test seems to have a positive influence on shoot concentration, it is the Zn buffer power that exerts far greater influence on wheat grain yield

References

Cavdar, A. O., Arcasoy, A., Cin, S., Babacan, E., & Gozdasoglu, S. (1983). Geophagia in Turkey: Iron and zinc deficiency, iron and zinc absorption studies and response to treatment with zinc in geophagia cases. In *Zinc deficiency in human subjects* (pp. 71–79). New York: Alan R. Liss. Inc.

Graham, R. D., & Rengel, Z. (1993). Genotypic variation in zinc uptake and utilization by plants. In A. D. Robson (Ed.), *Zinc in Soils and Plants* (pp. 107–118). Dordrecht: Kluwer Academic Publishers.

Graham, R. D., & Welch, R. M. (1994). Breeding for staple – Food crops with high micronutrient density: Long-term sustainable agricultural solutions to hidden hunger in developing countries. In *IFPRI workshop on food policy and agricultural technology to improve diet quality and nutrition*. Annapolis, Maryland, January 10–12

Nair, K. P. P. (1984a). Towards a better approach to soil testing based on the buffer power concept. In *Proceeding of 6th International colloquium for the optimization of plant nutrition* (Vol. 4, pp. 1221–1228), 2–8 September, Pierre-Martin Prevel, Montpellier, France.

Nair, K. P. P. (1984b). Zinc buffer power as an important criterion for a dependable assessment of plant uptake. *Plant and Soil, 81*, 209–215.

Nair, K. P. P. (1992). Measuring P buffer power to improve routine soil testing for phosphate. *European Journal of Agronomy, 1*(2), 79–84.

Nair, K. P. P. (1996). The buffering power of plant nutrients and effects on availability. *Advances in Agronomy, 57*, 237–287.

Prasad, A. S. (1982). History of zinc in human nutrition. In A. S. Prasad, I. E. Dreosti, & B. S. Hetzel (Eds.), *Clinical applications of recent advances in zinc metabolism* (pp. 1–17). New York: Alan R Liss Inc.

Sillanpaa, M. (1982). *Micronuttrients and the nutrient status of soils. A global study* (FAO Soils Bulletin, No. 48). Rome: FAO.

Sillanpaa, M., & Vlek, P. L. G. (1985). Micronutrients and the agroecology of tropical and Mediterranean regions. *Fertilizer Research, 7*, 151–167.

Welch, R. M. (1993). Zinc concentrations and forms in plants for humans and animals. In A. D. Robson (Ed.), *Zinc in soils and plants* (pp. 183–195). Dordrecht: Kluwer Academic Publishers.

Chapter 10
Case Studies with African Soils with Regard to P and K

Abstract The chapter discusses the importance of the precise determination of both P and K buffer power to formulate precise fertilizer (phosphatic and potassic) recommendations for the cultivation of the important fodder crop white clover (*Trifolium repens*) in The Republic of Cameroon. White clover is an important fodder crop for the emerging livestock industry in the country. Incorporating the P and K buffer power into the computations has resulted in the precise prediction of plant uptake. As against the routine soil tests for P (Olsen's test) and for K (NH4OAc test), which have shown very poor correlation with plant uptake, incorporation of the respective P and K buffer power values into the statistical computations significantly enhanced the correlation between fertilizer application and nutrient uptake, clearly reflecting in crop yield. The chapter further discusses the relevance of the buffer power determination in the case of other heavy metals like Molybdenum (Mo), Iron (Fe), Manganese (Mn) and Boron (B).

Keywords Phosphorus buffer power · Potassium buffer power · Cameroon · White clover

CENTRAL AFRICAN SOILS – THE REPUBLIC OF CAMEROON: The Republic of Cameroon is a small country in Central Africa. Soils are lateritic, highly acidic, and soil pH ranges between 4.5 and 5.0. The country's agricultural scenario is confined mainly to cocoa, coffee and rubber. Among the cereals, both rice and maize are popular. The country has a huge animal population, and good, and nutritious fodder is a great necessity of the country. With this view in background, an attempt was made to introduce white clover (***Trifolium repens*** L.) from the US, and the following investigations refer to the role of "The Nutrient Buffer Power Concept" in the nutrition of white clover.

Experimental Soils:

The soils investigated are widely varying. Their physico-chemical properties are listed in the following Table (Table 10.1).

The main objective of this investigation was to examine how best "Buffer Power Concept" explains soil availability of P and K in very problematic African soils.

K. P. Nair, *Intelligent Soil Management for Sustainable Agriculture*,
https://doi.org/10.1007/978-3-030-15530-8_10

Table 10.1 Physico-chemical properties of the experimental soils

Details	Soil 1	Soil 2
Sand (%)	48	7
Silt (%)	18	41
Clay (%)	34	52
Organic carbon (%)	5.6	8.2
Total nitrogen (%)	0.37	0.48
pH (1 N KCl)	4.4	5.0
CEC (me.equiv./100 g soil)	9.5	20.7
Base saturation (%)	21.8	32.2
Available P (ppm)	2.1[a]	13.0[a]
Available P (ppm)	3.7[b]	4.7[b]
Available P (ppm)	Tr.[c]	3.2[c]
Available K (me.equiv.100 g soil)	0.18[d]	0.27[d]
Available K (me.equiv./100 g soil)	0.17[e]	0.21[e]

Note: All the above mentioned soil tests are "routine" soil tests for available P and K. The only exception is the 0.01 M $CaCl_2$ extraction which gives the P concentration in the equilibrium suspension of soil (Mattingly et al. 1963)

[a]Olsen's test

[b]Bray II test

[c]0.01 M $CaCl_2$ extraction

[d]NH_4OAc test

[e]0.01 M CaCl2 extraction

And, white clover was chosen as the test crop because there is acute shortage of palatable and nutritious fodder for the livestock in the country. What is generally fed to the cattle is the post harvest waste of either rice or maize. This does not ensure good milk yield. It was thought that since the livestock industry had a tremendous potential in the country, white clover was chosen as the test crop. The seeds were obtained through a Memorandum of Understanding between The University Center, Dschang, The Republic of Cameroon, and the University of Florida (Institute of Food and Agricultural and Forest Sciences), in Gainesville, Florida. The investigation was confined to two soil types with very divergent physico-chemical characteristics (Table 10.1). Soil 2 is highly fertile compared to Soil 1.

Data in Table 10.2 clearly indicate that the experimental technique devised by this author has resulted in a remarkable positive inter-relationship between the Q/I measurements. The very highly significant (at 99.9% confidence level) "r" values substantiates this contention. Also, soil 2 has a much higher P and K buffer power compared to soil 1. What follows is the further examination of the question whether integrating the P and K buffer power into the computations enhances predictability of the P and K concentration in plant top, and, eventually, P and K uptake. Answers to this question are presented in Table 10.3.

Data in Table 10.3 unequivocally establish the important fact that integration of the buffer power, both in the case of P and K, enhances predictability of P concentration in plant top and P uptake, so also K concentration in plant top and K uptake, in a very highly significant (at a confidence level of 99.9%) manner. More than this,

Table 10.2 Linear regression functions (Y = bx + a) and correlation coefficients ("r" values) for the inter-relationship between Olsen's P test (Y = P "Quantity") vs. 0.01 M $CaCl_2$ extractable P (x = P "Intensity"), and NH_4OAc extractable K (Y = K "Quantity") vs. 0.01 M $CaCl_2$ extractable K (x = K "Intensity")

Nutrient	Soil	Regression Function	Correlation coefficient ("r")
P	1	Y = 0.41bx + 0.24a	0.994***
	2	Y = 0.47bx + 2.37a	0.890***
K	1	Y = 0.17bx +0.10a	0.960***
	2	Y = 1.28bx +0.17a	0.940***

Note: "b" in the regression function refers to the buffer power of either P or K, a = a constant *** Significant at 99.9% confidence level

Table 10.3 Correlation coefficients for plant top concentration and plant top uptake of P and K without (A) and with (B) the P and K buffer of the experimental soils

Details	A	B
P concentration versus Olsen's soil test	0.61***	0.69***
K concentration versus NH_4OAc soil test	0.05	0.62***
P uptake versus Olsen's soil test	0.17	0.77***
K uptake versus NH_4OAc soil test	0.18	0.78***

Note: In both the cases of P and K uptake, it is clearly shown that it is the buffer power of the respective nutrients that determines the uptake process not the "routine" soil tests. The Coefficient of Determination jumped from a mere 2.9% to 59.3% (jump of 56.4%) by integrating the P buffer power in the computations in the case of P uptake, and, the corresponding values for K uptake are 3.2% to 60.8%, a jump of 57.6%, indicating similar pattern in both the nutrients. Significant at a confidence level of 99.9%

there are other very important additional features of the results, which follow. That is, first, inasmuch as the P and K uptake process is concerned, neither of the widely practiced routine soil test for P, that is the Olsen's Soil test, nor the NH_4OAc test for K, showed any positive statistically significant relationship. This is reflected in the very low "r" values (Table 10.3). But, most remarkably, these "r" values shoot up to very highly significant ones (at a confidence level of 99.9%) when the corresponding P and K buffer power values are incorporated into the computations. In fact, in the case of Olsen's soil test, for P uptake, the coefficient of determination increases from a mere 2.9% in the absence of the P buffer power to 59.3% when the corresponding P buffer power value is also integrated into the computations. In the case of the routine NH_4 OAc extraction for K uptake, the corresponding values are 3.2% and 60.8%. The Coefficient of determination is a precise measure of the accuracy of prediction. In other words, these results unequivocally demonstrate that by integrating the respective P and K buffer power values into the computations, a substantial variance in P and K uptake by plant top can be explained. This is, indeed, a very remarkable observation of this investigation, which has considerable practical significance in white clover production in Cameroon. This is, because, once the P and K buffer power values of the soils in question are determined, to begin with, the application of these important plant nutrients through their respective carriers

Table 10.4 Plant top dry matter production (kg/ha) and P uptake (kg/ha) from the experimental soils as a function of P fertilizer application

Soil	Dry matter production (kg/ha)				P uptake (kg/ha)			
	P application rate (kg/ha)				P application rate (kg/ha)			
	0	12.5	25	37.5	0	12.5	25	37.5
1	29	62	78	76	43	95	152	191
2	67	96	99	93	106	154	176	181

Table 10.5 Plant top dry matter production (kg/ha) and K uptake (kg/ha) from the experimental soils as a function of K fertilizer application

Soil	Dry matter production (kg/ha)				K uptake (kg/ha)			
	K application rate (kg/ha)				K application rate (kg/ha)			
	0	50	100	200	0	50	100	200
1	32	24	27	20	1371	959	1071	737
2	65	73	76	87	2989	3675	3815	4576

(chemical fertilizers) could be appropriately designed. This has very extensive practical consequence in white clover production in Cameroon. Most remarkably, in the case of K, the coefficient of determination for plant top concentration increased from a mere 0.25% to 38.4%. This translates into a 15,360% increase in variance prediction.

Results in Tables 10.4 and 10.5 have important bearing on the practical aspects of white clover production in Cameroon, *vis-à-vis* "The Nutrient Buffer Power Concept." The most important conclusion is that among the two nutrients investigated, namely, P and K, response to P fertilizer application has shown a consistent pattern, with both higher dry matter production and P uptake. With enhanced P fertilizer application rates, there has been consistent increase in dry matter production and P uptake. The magnitude of dry matter production and P uptake in response to enhanced P fertilizer application has been more in the case of soil 2 as compared to soil 1. When we come to K fertilizer application and crop response, it has been quite inconsistent. The consistent and better response in soil 2 as compared to soil 1 can be traced to its much better buffer power (Table 10.2).

What are the implications of this finding in white clover production in Cameroon?

The Republic of Cameroon is an emerging and politically stable country (bilingual –both French and English) on the African continent, with a small population, but, with vast natural resources like oil, minerals (especially bauxite, raw material for Aluminum, which is much sought after by aerospace industry), agriculture (specially cocoa production, second highest producer on the African continent next to Ghana) and a vast potential for tourism. It has a stable livestock industry (cattle production) which has a very bright future. Though on the countryside of the country natural grazing is the common route for cattle feeding, things are giving way to supplemental fodder feeding. It is here the potential for white clover cultivation emerges. This author imported the seeds of white clover from the USA and the results of the experiments reported here are from this imported seeds. The objective behind this investigation was to popularize the cultivation of white clover in Cameroon and it is seen that the "Buffer Power Concept" has a major role to play in its nutrition, inasmuch as P and K nutrition of the plant is concerned, as could be clearly deducted from the experimental results discussed above. Both plant nutrients play a very major role in the nutrition of white clover.

This research formed part of a Master's thesis of one the author's students, while he held the position of Professor and Head of the Department of Soil Sciences at The University Center, and was selected as the best thesis submitted to the University for the Master's degree in 1990 and received a sum of 400,000 CFA (equivalent to more than US$ 1000) awarded in recognition of the academic excellence of the research by the organization for advancement of science on the African continent. The award was given by the Minister of Higher Education and Scientific Research, on behalf of the President Paul Biya, of The Republic of Cameroon, and was celebrated as a very important national event.

10.1 Other Heavy Metals

One other important aspect to be considered, which is of a relatively recent origin, is the heavy metal contamination of agricultural soils and their toxic effects on buffering. An element of particular interest is cadmium (Cd). The main sources of Cd contamination of agricultural soils is through the application of mine wastes, sewage sludge, and other residues. Kuo (1990) has reported on Cd buffering capacity and accumulation in Swiss chard (*Beta vulgaris* L.) in sludge-amended soils. He reported that while DTPA-extractable Cd did not show a high correlation with plant Cd, the Cd buffering capacity and plant Cd content and uptake were very highly correlated.

10.1.1 Molybdenum

Dissolved soil molybdenum can buffer under a wide range (Barber 1995). Values ranging from 9 for a sandy loam to 83 for a silty clay loam have been obtained (Lavy and Barber 1964). If the concentrations of adsorbed molybdenum, which varied from 0.6 to 1350 μg kg^{-1} with solution molybdenum concentrations which varied from 1 to 13 μg L^{-1} are compared, a buffer power range from 1 to 100 is obtained as indicated by the data of Lavy and Barber (1964). By comparison, Australian soils indicated a buffer power range of 10–2000 (Barrow 1970). Such very high buffer power values will greatly affect molybdenum uptake by plants growing in these soils. Since both mass flow and diffusion under differing molybdenum concentrations in the soil solution will affect its uptake, the precise quantification of the buffer power assumes greater relevance in predicting molybdenum bio availability. Since, for a given soil, molybdenum uptake by the plant growing in it will increase as its pH increases, ruminants which consume forage grown on such soils would develop molybdenosis, a disorder related to excessive intake of this element. A prior knowledge of the molybdenum buffer power will help manage better molybdenum fertilization of such soils. Additionally, the excess presence of both phosphate and sulfate will affect molybdenum bio availability in soils, but, their effect on molybdenum buffer power has to be clearly elucidated. There is, but scant, information on these aspects.

10.1.2 Iron

The iron buffer power has to be related to both soil adsorbed iron and soil solution iron. Since redox reactions, precipitation, and presence of iron in organic complexes will affect the equilibrium between adsorbed and soil solution iron, it will be difficult to precisely quantify the iron buffer power of agricultural soils. O'Connor et al. (1971) obtained a buffer power value of 1000 for 59 $Fe2^+$ added to a thoroughfare loamy sand having a pH of 7.9. Since this value increased with time, it might have been affected by precipitation of iron oxides. Inasmuch as iron uptake from soils is concerned, one crucial factor would be root activity. The secretion of Fe-chelating phytosiderophores by Fe-deficient graminaceous species and further the production of siderophores which are low molecular weight iron chelating agents produced by almost all bacteria and fungi under iron limiting conditions (Buyer and Sikora 1990) would make it difficult to quantify Fe flux by mass flow and diffusion, particularly the latter, on which Fe bio availability-based on its buffer power hinges. The effect of both phytosiderophores and siderophores will assume crucial significance in soils of high pH, wherein, these materials will solubilize iron and allow its movement both by mass flow and diffusion to the root surface. Since the ability of plant root to secrete phytosiderophores is both species and cultivar specific, there is an imminent need to understand better how this phenomenon might possibly affect Fe buffering at the soil–root interface.

10.1.3 Manganese

The range in Mn buffer power is greater than those for most other nutrients. Investigations on Mn buffer power are scanty. Though Mn buffer power can be calculated from data on exchangeable Mn and soil solution Mn, the validity of such values is questionable because soil solution Mn has two components, namely, ionic and complexed, and their combined presence will render the Mn buffer power calculations invalid. Large variations have been found between exchangeable Mn and soil solution Mn. Barber et al. (1967) reported values ranging from 527 to 1.1 for these two fractions. Halstead et al. (1968) found narrower (65 to 0.4) ratios. Soils with low soluble Mn have been found to have high values while the reverse is true with regard to acid soils where Mn levels are high (Barber 1995).

10.1.4 Boron

The buffer power for B has been calculated from adsorption curves and values less than 3 have usually been found (Barber 1995). For a loam soil, values of 1.35 and 1.7 have been found (Sulaiman and Kay 1972). The absorption of B by plant roots would differ from that of other ionized species since B is present in solution as undissociated H_3 BO3 (Barber 1995). Since the buffer power values for B are generally small, it implies that the effective diffusion coefficient for B will be large and this in turn results in relatively easy long distance transport of B by mass flow and diffusion to reach plant root (Barber 1995). Obviously, much of the B that is diffusible will reach the plant root during the period of plant growth.

10.1.4.1 Influence of Heavy Metal Contamination on Buffering of Major Elements

With the increased use of sewage sludge and other residues in farming, there is a need to understand the effect of heavy metals carried in these materials on buffering of other major nutrients. Though interaction between heavy metals such as Zn and Fe with P has been studied in great detail (Nair and Babu 1975) there is little information on the effect of heavy metals such as Cd or Cu on the buffering of other major nutrients. Yang and Skogley, (1990) have studied the effect of Cu and Cd on K adsorption and K buffering capacity. When Cu or Cd are introduced into the soil as salts (Yang and Skogley 1989) or when metals accompany sludge applications (Page et al. 1987), soil-solution concentration and plant uptake of these metals is extremely low. Consequently, metal adsorption by soil components is important in regulating soil-solution metal concentration, bio availability, and the toxicity that some of the heavy metals may cause in crop production. Since K is found to be highly vulnerable to adsorptive–desorptive (exchange) reactions, the large application of these metals might affect K dynamics. Yang and Skogley (1990) reported that K buffering was reduced by 20–32% by Cu addition and 7–20% by Cd addition. The rate of application was 400 mg Cu or Cd kg^{-1} soil. Cu and Cd changed the distribution of K, Ca, and Mg between the soil and solution. These results suggest that Cu or Cd addition would influence soil K availability by changing K adsorption kinetics, buffering capacity, and distribution.

10.1.4.2 Possible Buffering Effect on Plant Acquisition of Heavy Metals

There is now growing evidence that root-induced changes are primarily responsible for the differential uptake of heavy metals such as Fe and Zn. In a recent review on Fe nutrition, Marschner and Römheld (1994) reported that two different types of root response strategies to Fe deficiency are identified in all plant species. In strategy I, occurring in all plant species except grasses, a plasma membrane-bound reductase is induced with enhanced net excretion of protons. In strategy II, confined to grasses, there is an increase in the biosynthesis and secretion of phytosiderophores which form chelates with $Fe3^{+}$. Compared to the progress made in identifying the cellular mechanisms of root responses in strategy I and strategy II plants, the understanding of processes taking place in the apoplasm of root rhizodermal cells and the role of low-molecular-weight root exudates and siderophores in Fe acquisition of plants growing in soils of differing Fe bio availability is very limited (Marschner and Römheld 1994); this suggests that there is a need to further investigate the buffering effect of Fe on root-induced differential uptake patterns. Such interactive investigations are conspicuously absent in the published literature. A similar situation exists with regard to the role of HCO_3^- ion in affecting Fe and Zn nutrition of crop plants. Mengel (1994) hypothesized that it is the HCO_3^- ion in association with the NO_3^- ion, found in calcareous soils, that trigger Fe chlorosis. The chlorosis is induced by the impaired Fe uptake from root apoplast into the cytosol of root cells in the presence of excessive HCO_3^- and NO_3^- ions, and a similar situation exists in the case of leaves. Reduction of apoplast pH by spraying chlorotic leaves with an acid, such as $H_2\,SO_4$, led to a regreening of the leaves. The hypothesis of Marschner and Römheld (1994) stresses the physiologic effect, while that of Mengel (1994) stresses the chemical one. But in both cases, the possible Fe buffering effect *vis-a-vis* the root-induced changes merit much closer scrutiny than what has hitherto been attempted.

Parallel to the above studies, the role of excess HCO_3^- ion in the growth medium on Zn nutrition of wheat cultivars has recently been reported (Yang et al. 1994), indicating that excess HCO_3^- ion in the growth medium leads to accumulation of organic acids and their insufficient compartmentation in Zn-inefficient cultivars as compared to Zn-efficient cultivars. These studies point to the lacuna in a proper understanding of the buffering effects *vis-a-vis* organic acid accumulation. It is not known what effects Zn buffering in the medium would have on excess accumulation of the organic acids in the Zn-inefficient cultivars as compared to the Zn-efficient cultivars.

10.1.4.3 Concluding Comments and Future Imperatives

Historically, soil testing has been used to quantify bio availability of essential plant nutrients to field grown crops. However, contemporary soil tests are based on philosophies and procedures developed several decades ago without significant changes in their general approach. For a soil test to be accurate, one needs to clearly understand the physic–chemico–physiologic processes at the soil–root interface, and an understanding of soils and plant root systems as polycationic systems is essential. It is this knowledge that leads to sound prescriptive soil management practices inasmuch as nutrient bio availability vis-a-vis fertilizer application is concerned, because of all the factors which govern sustainability of crop production, the nutrient factor is the most important, yet, it is also the least resilient to management. This book focuses on the buffering power of plant nutrients, especially phosphorus, potassium, zinc, and some other heavy metals such as cadmium, on their availability to growing plants. Experimental evidence is presented from research carried on temperate (Central European and North European) and tropical soils (African and Asian soils) which unequivocally establishes the fact that precise quantification of the nutrient buffer power holds the key to a clearer understanding of the plant bio availability of some of the most important essential plant nutrients, such as, phosphorus, potassium, and zinc. The nutrient buffer power concept attempts to explain the bio availability of these plant nutrients on the basis of the diffusion model, as it is the principal process by which phosphorus, potassium, and zinc are taken up by plant roots. Possibly, other nutrients which are taken up by plant roots by the same process would also conform to the principles of this concept, enunciated in this book. To substantiate this, supportive experimental evidence is needed.

There are similarities between the buffer power approach and quantity–intensity (Q/I) approach in predicting plant bio availability of important plant nutrients such as phosphorus and potassium. While the Q/I approach has met with a fair measure of success in predicting P availability, with regard to K availability the Q/I relationship does not enjoy universal acceptance since a single relationship for all soils between K uptake by a given crop and AR_K does not exist, perhaps due to the nature of the soil components regulating AR_K.

In substance, soil testing in a laboratory is meant to simulate what a plant root would accomplish in a field soil. All soil tests meet this requirement with different degrees of success. But, for a soil test to be universal in scope, it must attempt to quantify, as precisely as possible, factors which most crucially affect plant root uptake. Of all the factors involved in nutrient uptake, the most crucial is the average nutrient concentration on the root surface. In a dynamic state of plant growth, it is virtually impossible to quantify precisely the average nutrient concentration on the root surface. The nutrient's buffer power is an indirect measure of the average nutrient concentration on the root surface, and analytical methods which quantify the buffer power with precision would render the soil test in question dependable; hence, the fertilizer recommendation which follows would be reproducible. It is against this background that the nutrient buffer power concept merits wider application for field use. Fertilizer prices have dramatically escalated following the oil crisis of the mid-1970s, and field experience shows that recommendations emanating from routine soil tests, based on text book knowledge, have met with only limited success. A number of countries on the Asian and African continents spend enormous sums of money on import of chemical fertilizers, which is a great strain on their national exchequers. Most of these countries have no infrastructure for fertilizer manufacture and where there is any, the soil testing support service is far from satisfactory. India is a case in point. The fall of the green revolution is testimony to this truth. Since fertilizer is a crucial input, second only to water, in modern farming, cost-effective fertilizer recommendations are an important prerequisite for successful agriculture. In countries such as India, fertilizer prices, especially of phosphate and potassium, have skyrocketed following governmental policies centered on a market economy. In such situations, farmer confidence can only be instilled and maintained by accurate soil testing and reproducible fertilizer recommendations.

The success of a new approach, to a large measure, rests with the ingenuity of those applying it to suit the demands of a new situation. This principle is no exception in making the "The Nutrient Buffer Power Concept" a universal success. As compared to routine soil testing, this approach calls for the accurate determination of the buffer power of the nutrient in question, to begin with. Once this is done, the buffer power factor can be integrated with the routine soil test data and fertilizer recommendations can be made on the basis of this new information. This implies that, in addition to obtaining routine soil test data, one also needs to know the buffer power of the nutrient in question. This book has shown how the buffer power is determined based on analytical methods varying from a simple adsorption–desorption equilibrium technique, developed by the author, to elaborate electro-ultrafiltration (EUF). Only selected laboratories around the world have EUF facilities. Where they are absent, alternative analytical procedures to precisely quantify the buffer power need to be developed. Developing appropriate procedures to quantify the buffer power and modification of the protocol for routine soil testing to integrate buffer power data into the computations for precise fertilizer recommendations calls for both ingenuity and drive on the part of the researcher, but that will be the future imperative of accountable soil testing.

References

Barber, S. A. (1995). *Soil nutrient bioavailability: A mechanistic approach*. New York: Wiley.

Barber, S. A., Halstead, E. H., & Follett, R. E. (1967). *Significant mechanisms controlling the movement of manganese and molybdenum to plant roots growing in soil* (pp. 299–304). Aberdeen: Trans. Joint Meeting Commissions II and IV. International Soil Science Society.

Barrow, N. J. (1970). Comparison of the adsorption of molybdate, sulfate and phosphate by soils. *Soil Science, 109*, 282–288.

Buyer, J. S., & Sikora, L. J. (1990). Rhizosphere interactions and siderophores. *Plant and Soil, 129*, 101–107.

Halstead, E. H., Barber, S. A., Warncke, D. D., & Bole, J. (1968). Supply of Ca, Sr, Mn and Zn to plant roots. *Soil Science Society of America Proceedings, 32*, 69–72.

Kuo, S. (1990). Cadmium buffering capacity and accumulation in Swiss chard in some sludge-amended soils. *Soil Science Society of America Journal, 54*, 86–91.

Lavy, T. L., & Barber, S. A. (1964). Movement of molybdenum in the soil and its effect on availability to the plant. *Soil Science Society of America Proceedings, 28*, 93–97.

Marschner, H., & Römheld, V. (1994). Strategies of plants for acquisition of iron. *Plant and Soil, 165*, 261–274.

Mattingly, C. E. G., Russell, R. D., & Jephcott, B. M. (1963). Experiments on cumulative dressings of fertilizers on calcareous soils in Southwest England. II. Phosphorus uptake by ryegrass in the green house. *Journal of the Science of Food and Agriculture, 14*, 629–637.

Mengel, K. (1994). Iron availability in plant tissues—Iron chlorosis on calcavcous soils. *Plant and Soil, 165*, 275–283.

Nair, K. P. P., & Babu, G. R. (1975). Zinc–phosphorus–iron interaction studies in maize. *Plant and Soil, 42*, 517–536.

O'Connor, G. A., Lindsay, W. L., & Olsen, S. R. (1971). Diffusion of iron and iron chelates in soil. *Soil Science Society of America Proceedings, 35*, 407–410.

Page, A. L., Logan, T. J., & Ryan, J. A. (1987). *Land application of sludge—Food chain implications*. Chelsea: Lewis Publication.
Sulaiman, W., & Kay, B. D. (1972). Measurement of the diffusion coefficient of boron in soil using a single cell technique. *Soil Science Society of America Proceedings, 36*, 746–752.
Yang, J. E., & Skogley, E. O. (1989). Influence of copper or cadmium on soil potassium availability properties. *Soil Science Society of America Journal, 53*, 1019–1023.
Yang, J. E., & Skogley, E. O. (1990). Copper and cadmium effects on potassium adsorption and buffering. *Soil Science Society of America Journal, 54*, 739–744.
Yang, X., Römheld, V., & Marschner, H. (1994). Effect of bicarbonate on root growth and accumulation of organic acids in Zn-inefficient and Zn-efficient rice cultivars (Oryza sativa L.). *Plant and Soil: An International Journal on Plant-Soil Relationships, 164*, 1–7.

Chapter 11
The Changing Face of Global Agriculture

Abstract This chapter discusses the changing face of global agriculture, where emphasis is laid on the sustainable use of important resources like soil. Discussion in the chapter also relates to the skewed use of global resources, which leads to, as of now, 95% of the global population being deprived of a dignified life. This emphasizes the crucial fact that when agriculture, a crucial source of sustenance of livelihood for the millions of the world's poor and deprived, faces such external threats, the very existence of the millions of poor will be at stake.

Keywords Agricultural production · Sustainability · Global resources

The sustainability of agricultural production, to day, must be viewed against the background of a global context. That intensely driven chemical agriculture, the bed rock of industrial agriculture, which, ironically, was labelled as the "Green Revolution," is no more environmentally sustainable, mainly in the developing world, is now an established fact. The degraded soils, the dried aquifers due to intense water mining to support monoculture, the polluted ground water due to chemical residues from unbridled use of chemical fertilizers, making it non potable, the vanishing bio diversity etc., are but a few of the most disastrous consequences of this so-called green revolution. Punjab State, in India, the "cradle" of green revolution in South Asia, is an outstanding testimony to this. There are hundreds and hundreds of acres in the state where it is difficult to grow even a blade of grass, due to excessive exploitation of soil resources, leading to intense soil degradation, where intense monoculture, the rice-wheat rotation, year after year, led to so much soil degradation. Ground water is no more potable due to excessive accumulation of the nitrate ion, following unbridled use of chemical fertilizers, primarily urea, to boost crop yields.

During the past few decades, the world has seen too rapid and too drastic changes in agricultural production processes and productivity. These changes far exceed the pace and extent of changes that occurred during the past 10,000 years of settled agriculture. Spectacular advances in the science and technology of crop production and the accelerated increase in global population from approximately 1 billion in 1900 to more than 6 billion by the turn of the twentieth century have been mostly

K. P. Nair, *Intelligent Soil Management for Sustainable Agriculture*,
https://doi.org/10.1007/978-3-030-15530-8_11

responsible for such unprecedented pace of change. More importantly, these dramatic changes have not spread uniformly around the globe. Whereas the advanced countries such as the USA, Canada, the European Union and Australia have witnessed very impressive spurts in crop productivity, followed by dramatic changes in economic growth, with controlled growth of population, the developing and least developed countries present a dismal picture, where population growth has outstripped food production rate, leading to food famine. The African continent is a classic example of this very dismal situation, where food famine has led to food riots, as in Sudan. A very dangerous situation is unfolding, and there is much rhetoric, both at the national and global levels, but little concrete and meaningful followup action, at a global level. One must have a clear vision of the inalienable fundamental fact that for any foreseeable length of time, humanity is operating at a fixed mass of soil base that needs to sustain, not just agriculture, but, many other supporting and ancillary activities. When this fixed mass of soil is threatened and deprived of its inherent capacity to sustain crop production, our very existence is threatened. The general tendency globally has been the production of more and more from less and less land, without concern for the ecological irreparable damage, which is a direct outcome of such an approach. In the past 100 years alone, agriculture has transformed from a resource-based subsistence activity to a highly commercial industrially driven chemical agriculture, in particular, in the northern hemisphere. Unfortunately, this model has been attempted to be replicated in the developing and less developed countries consequently with disastrous outcome in many situations. Southern Asia, in particular India and China, and to some extent Pakistan, have had to pay a big price for this thoughtless approach, especially on the environmental front. Global aggregate food production has increased in tandem with global population increase, during the last three decades, and enough food is available today to meet everyone's genuine needs (Pinstrup-Anderson 2002). However, the unsightly ecological and unethical footprints that cover the globe around show that the means adopted to enhance this great leap in food production are, to say the least, questionable. Intensification of agriculture and skewed population growth have created productive, but dysfunctional, agricultural systems worldwide, and some of the major consequences are:

The natural resource base for food production is shrinking as a result of degradation arising from intensive cultivation practices, characterized primarily, by unbridled inputs, the defining feature of the so-called green revolution. In India, Punjab State, the "cradle" of green revolution is a living example of this degradation. It has also had highly negative social consequences, such as, the spread of cancer in districts like Gurdaspur, in the State, which is an outcome of excessive and unbridled use of chemical pesticides, on one hand, and unabated suicides of cotton farmers in rain-fed areas of Vidharba district of Maharashtra State, on the other, where Bt cotton farmers in large numbers went bankrupt due to high input costs (of seeds, primarily and supplemental fertilizer and pesticide inputs, secondarily) and took recourse to suicide to escape from the Shylockian clutches of unscrupulous money lenders, who lent farmers money at high rates of interest.

Cultivable land resources are being taken out of production at the rate of 1% per annum for urbanization, infrastructure development, and needless leisure activities, principal one among them being touristic activities in the most productive parts of some of the individual countries. The State of Kerala in India, with a high density of population, but inadequate land for arable purposes, which does not produce enough food (especially rice being the staple diet of the Keralites) to feed its populace, but, depends on imported rice from neighboring states, is a classic example of this distorted and skewed pattern of development. In all these activities, the role of the "land mafia" or "land sharks" is aided and abetted by vested interests, more often than not, by corrupt politicians who wield the reins of power.

Per capita share of resources is dwindling through exponential population growth in some countries, and in some states of a country, like India, and such population explosion is confined to certain ethnic or religious groups. This is happening especially in certain very vulnerable areas.

Nutrient mining, organic matter depletion (most disastrous consequence of the green revolution), and soil erosion from insufficient inputs and resource-extractive cultural practices have become a major threat to food security in the poorer countries, especially on the African continent. A downward spiral of poverty-environmental degradation-low productivity is sucking millions of poor people around the developing and least developed countries into perpetual deprivation (Saroja Raman 2006). Low and stagnant productivity is widespread in many developing countries. Asymmetric distribution of, and access to, resources is becoming a lot more pronounced than what it used to be in earlier times.

Accelerated population growth predominantly in the low-income developing countries, coupled with these problems, has given rise to unacceptable levels of poverty, hunger and malnutrition, especially in rural areas, primarily on the African and South Asian countries, like India and Pakistan, where more than 50% of the global population is concentrated. Currently, more than 800 million people, globally, have no adequate access to enough food, to ensure food security, though as early as 1948 the United Nations through a covenant has passed the "Right to Food" bill. In India more than 200 million people, that is more than a sixth of the total population, sleep on an empty or half empty stomach each day, and a substantial number of these hungry persons are working women, pregnant mothers, and children. More than 170 million children globally suffer from serious malnutrition, which would adversely affect their mental and physical capabilities, when they grow up. Deficiencies of micronutrients, such as zinc, iron and iodine and vitamin A are rampant among the hungry and under-nourished. Endemic poverty, chronic food insecurity, and poor human development have become major ethical issues globally, yet, they do not receive the immediate and adequate attention they deserve (Saroja Raman 2006).

Perversely, this skewed distribution of global resources has also opened up tremendous opportunities for global trade on a scale not seen earlier for countries with generous endowments of natural resources, high economic development, and technological superiority, such as the United States of America and the nations of Western Europe. It seems, as if, while more than 95% of the global population is deprived of a dignified life, including access to adequate food, less than 10% enjoy a life style, that borders on waste and opulence. Unless the world wakes up to the urgent need of utilizing the global resources, primarily, soil, in an equitable fashion, the future holds only dismay. It is in this context, "The Nutrient Buffer Power Concept," a revolutionary soil management technique, comes center-stage for global attention (Nair 1984a, b).

Agriculture has, thus, emerged as the critical decisive factor in a global context, both to the industrialized western nations, as much as to the far less industrialized southern and eastern nations, but, with conflicting interests. The former, with their unlimited access to recent advances in technological inputs, have adopted intensive, high external input commercial agriculture for production of prolific and cheap food for their own consumption and for economic gains, and they are flush with food surplus for international trade. People in these advanced countries spend at best 15–20% of their personal income on food, and only about 2–5% of their population earn income directly from agriculture. Agriculture is not the sustaining component of their economy. Contrast this with the prevalent situation in the developing and least developed nations, where close to 65% of the population, as in India, necessarily depend on agriculture, in one way or another, as a means for livelihood. This huge population is confined to the rural areas, and, have no access to modernity. This huge population spends as much as, or even more than 50% of their daily income, merely on food. This happens, because, though there has been a quick spurt in food production, following the high input agriculture of the late sixties and early seventies, the key to this bonanza was held by the small and microscopic community of very rich farmers of the states like Punjab or Western Uttar Pradesh, the so-called "green revolution belt," with assured irrigation facilities. When one looks at the situation of the poor or marginal farmers, who are mostly confined to the rainfed areas of the country, the picture is dismal. Debt-ridden and hungry, the story of this huge section of Indian farmer population is far too different and dismal (Nair 2015a, b, c). Even the benefit of the minimum support price for their produce, ensured by the Government of India, primarily for rice and wheat, never reaches these poor and marginal farmers. It is cornered by the rich farmer lobby. And, to cap it all, this "rich farmer lobby," as one witnesses in Punjab, wield a tremendous political clout. Unlike The Republic of China, which harvested in excess of 500 million tons of grain, with a comparable population, India produced only a third of this quantity, last year (2015). In other words, while India's population is already burgeoning, China, on the other hand, through a very strict "one child norm" policy, which is only very recently being relaxed, has kept a tight control on population. In other words, for a burgeoning Indian population, where most are deliberately kept poor due to political chicanery, food has become expensive, and, the country is not producing enough to make food cheap. This leads to widespread hunger, notwithstanding, the roof top claim of the green revolution enthusiasts that the country is "self sufficient" in food production. Accessibility to affordable food becomes a major constraint to their livelihood. Millions in these precarious situations do not know how best to utilize their meager land resources. They are often at the mercy of governmental extension agents, who dish out time worn and "text book" knowledge on how best to farm. This adds to their woes, as they have to, more often than not, to invest all their limited inputs on small parcels of land, and, farming, becomes truly, a "gamble on the monsoon," as has happened more often than not on the Indian sub-continent.

When agriculture, a crucial source for livelihood, faces external threats, the very existence of millions of poor will be at stake. The environmental threats are the gravest. Of these, threats to soil, the core of agriculture, is of paramount significance.

References

Nair, K. P. P. (1984a). Towards a better approach to soil testing based on the buffer power concept. In *Proceeding of 6th International colloquium for the optimization of plant nutrition* (Vol. 4, pp. 1221–1228), 2–8 September, Pierre-Martin Prevel, Montpellier, France.

Nair, K. P. P. (1984b). Zinc buffer power as an important criterion for a dependable assessment of plant uptake. *Plant and Soil, 81*, 209–215.

Nair, K. P. P. (2015a, February 13). The debt story less told. *The Hindu Business Line*.

Nair, K. P. P. (2015b, April 7). When there is no soil left in this world. *The Hindu Business Line*.

Nair, K.P.P. (2015c). The nutrient buffer power concept – A revolutionary soil management technique for sustainable agriculture. In *Keynote address at the National Seminar on "issues, challenges and strategies in sustaining soil health"*. Kerala Forest Research Institute, Peechi, Kerala State, India, December 10–11, 2015.

Pinstrup, A. P. (2002). Towards a sustainable global food system: What will it take? *Keynote presentation for the Annual John Pesek colloquium in sustainable agriculture, March 26–27*, Iowa State University

Raman, S. (2006). *Agricultural sustainability: Principles, processes and prospects* (pp. 13904–11580). Birmingham: The Hawoth Press Inc.

Chapter 12
Sustainable Agricultural Production on a Small Farm

Abstract The sustainability of a small farm is linked to both biophysical and socio-economic factors, and, the various operations on the farm. To be sustainable, the farm should have stable productivity, economic profitability and ecological security. The chapter discusses the above factors, at length.

Keywords Small farm · Socio economic factors · High external input · Developing countries · Sustainable development · Brundtland Commission Report · Economic viability · Hierarchy of agricultural systems · Population density · Production processes

The sustainability of a small farm is linked to both biophysical and socioeconomic factors, and, the various operations on the farm. To be sustainable, the farm should have stable productivity, economic profitability, and ecological security (Saroja Raman 2006).

The farm interacts with the biophysical environment through the use of inputs, such as, fertilizer, water, seeds, and pesticides, normally known as "inputs." If the inputs are not properly managed, they result in serious physical, chemical and biological degradation of production resources. Farm operations such as deep tillage, faulty irrigation, and excessive and unbridled use of chemical fertilizers and pesticides, even water, not only affect crop yields, but, also degrade the soil resources, through erosion, acidification, salinization, water logging etc. They can also impact the surrounding environment, predominantly through pesticide and fertilizer run off and sediment transfer through erosion, resulting in pollution and sedimentation of surface water resources. Unsustainable levels of environmental pollution denote unsustainable farm operation, and, this becomes all the more threatening when the farm size is quite small, as in millions of cases in a country like India. There is an added dimension of uncertainty to the small land holder, which comes in the guise of the newly introduced genetically modified crops, which are now shrouded in secrecy and unpredictable field performance. This is best illustrated in what has happened in the cotton belt of Maharashtra State in India, in the Vidharba district, where thousands of farmers committed suicide, unable to sustain a high input technology with uncertain outcome. Crops (the genetically modified Bt cotton) failed on

K. P. Nair, *Intelligent Soil Management for Sustainable Agriculture*,
https://doi.org/10.1007/978-3-030-15530-8_12

thousands of hectares driving the poor peasants to penury and suicide – a scenario never seen anywhere else in the world, for the sheer magnitude of the tragedy.

The farm also interacts with the socio-economic environment through the purchase and sale of inputs, labor and market interventions. For the farm to be economically viable, the nature and volume of these external inputs have to be properly balanced. The farm is sustainable if its interaction with the biophysical and socio-economic environments is such that it is both economically viable in the short run, and ecologically sustainable in the long run.

Against the background of the sustainability criteria discussed earlier, the farm should fulfill the conditions of high productivity within the resource capacity of the land, adequate profitability, environmental integrity, and on-farm resource conservation to qualify as "sustainable." If one or more of these four boundaries are breached, the system in question turns unsustainable. The conceptual model is given in Fig. 12.2 and 14.1.

12.1 Sustainable Agriculture

12.1.1 Historical Perspective

Discussion on sustainable agriculture has come to occupy center stage ever since the concept of sustainable development was highlighted as a paradigm for intergenerational equity. But, the origin of this concept can be traced to much earlier times. Due credit must be extended to a Roman landowner, Marcus Terentius Varro, of the first century BC, who first propounded the idea of sustainable farming. According to him agriculture is a science which teaches us what crops are to be planted on each kind of soil and what operations are to be carried on, in order that the land may produce the highest yield in perpetuity. This probably explains best the central idea of sustainable agriculture, and forms the core of most subsequent expositions on this topic.

Today, the concept of sustainable agriculture has been transformed beyond recognition from the original definition of Varro. Ever so many parameters have been added. Perhaps, this is a reflection of the manner in which agriculture has evolved globally over time. Evans (1998) has juxtaposed the chronological development of agriculture and population growth to describe the history of the evolution of agriculture in his book Feeding Ten Billion. This treatise provides an excellent basis for understanding the evolution of the concept of agricultural sustainability and its importance to human survival and development (Saroja Raman 2006).

The beginnings of settled agriculture can be traced back to 8000 BC in Central Asia. For more than 99% of its evolutionary history, the human race depended on the bounty of nature for subsistence. The natural mode of sustainability for all other living species worked excellently for humankind also. Agriculture marked the beginning of the manipulation of nature by humans. Subsistence agriculture was practiced for almost 6000 years of the pre-industrial era, from 8000 to 2000 BC, without much disruption of nature. The shifting cultivation practiced in the early days and the traditional agricultural systems of the ancient civilizations of India, China, the Middle East, and so on were durable for thousands of years and provided adequate yield to support the low level of population which existed at the time. Franklin King, a Professor at the University of Wisconsin, in the early twentieth century, wrote a treatise "Farmers of Forty Centuries" (King 1911), on how the integrated use of nutrient recycling, animal manure, and agricultural residues maintained the productivity of the traditional agricultural systems of the Orient for thousands of years. By King's time, the Western world had already come to believe strongly in the power of chemical fertilizers to enhance crop yields and maintain production. More importantly, Justus von Liebig, the father of modern soil chemistry, who started as a chemist in Germany, and went on to formulate artificial chemical fertilizers, greatly contributed to this evolving thought. It should, however, be noted that during the period reviewed by King, the world was almost empty, with just about five million people, to partake of its abundant natural resources. What was sustainable then could no more be logically extrapolated to subsequent times, when demand for food spurted with exponential increase in global population.

A steep increase in the population to 500 million during the period 2000 BC to AD 1500 required the expansion of agricultural activities and intensification of agricultural practices. This forced the extension of cultivation to vulnerable areas, such as, hills, wetlands and drained swamps. The adoption of new innovations in the form of animal power, heavy ploughs, terrace formation, and, an expanded crop portfolio (use of legumes, crop rotations etc.) became widespread during this period. These innovations are known collectively as the "Norfolk Revolution," after changes in East Anglia, England, during the 1700s. The intensification of agriculture led to processes of degradation of natural resources, such as, soil erosion, water logging, and salinization in several places. The earliest recorded instance of flawed management of the soil and water resources leading to a failed agricultural system can be traced to Sumeria in Central Asia during 4000–2000 BC. Salinization, arising from faulty irrigation practices, without drainage, resulted in the gradual destruction of an evolved irrigated agricultural system and brought to an end the advanced Sumerian civilization and its thriving agriculture.

Thus, the battle for sustainability of agriculture started when human numbers and human innovations became agents of disruption of natural resources and processes. Disasters of the Sumerian type were repeated in several other areas of the world, leading to the destruction of more civilizations, such as the Mayan civilization in South America, Harappan civilization in the Indus valley, and the Chinese, Roman, and Greek civilizations, owing to factors including drought, soil erosion, climate change, and water logging. This history of failed agriculture as the agent of destruction of civilization has raised a fear in our minds about the reckless manipulation of natural resources to meet the insatiable human greed for land.

12.1.2 Malthus and His Prediction

During the period 1500–1825, the global population doubled to 1 billion and agriculture turned from a subsistence base to a commercial one. Great strides were made in crop production management such as tillage and crop rotation, including legumes to enrich soil fertility. During this period, the industrial revolution also swept across Europe. Rapid population growth, the industrial revolution, and the agricultural revolution interacted in complex ways to hamper agricultural growth, as there was a shift in favor of industrial activities and urbanization.

The oft-quoted Malthusian specter fell on England during 1794–1800, in the form of poor harvests, inadequate food availability, and high prices for agricultural commodities. This prompted Malthus to write his famous Essay on the Principle of Population, in which he clearly articulated the fear that the mismatch between the productive capacity of the land and increasing population could lead to famines and death on a large scale (Malthus 1798). This is the earliest record of such concern about the limited capacity of the natural resource, principally land, to balance agricultural production and population growth. Though the Malthusian theory has not been proved wholly correct, it is operating in an opaque manner in many countries, such as, India, where the rate of population growth has already begun to outstrip the rate of increase in food production, with an inherent danger of widespread famine taking over the country in some decades to come if vigilance is not observed on the food front. China, by contrast, has coped up well with a very strict population policy, where one child is the norm, though it has now officially accepted the two child norm very recently because of a fast growing ageing population, which if allowed to continue, will severely impact the young work force. The population differential between India and China is about 300 million, but, China harvests more than two-fold grains as compared to India, and, it is the country's strict economic policy to have 60% of its arable land exclusively for grain cultivation. In India, there is no population policy, and ***laissez-faire*** is the order when it comes to certain religious groups, where it is encouraged to have as many children as possible, with one or more than one wife to a man. India, being secular, in terms of the country's constitution, permits this.

The century from 1825 to 1927 ushered in an era of hope in the form of the discovery of a whole series of scientific principles of crop production which helped to forestall the Malthusian prophecy coming true. This marked the beginning of the period of the triumph of human ingenuity over nature that has continued to feed the world since, despite a more than six-fold growth in population. Some of the landmark developments in this period that became the underpinnings of future developments in agriculture are:

1. The discovery of the roles of nitrogen, phosphorus, and potassium in mineral nutrition of plants through the epoch-making contribution of several scientists, principal among them being the great German chemist, Justus von Liebig (the famous Liebig's theory, "Law Of Minimum," 1840) and Humphry Davy
2. Gregor Mendel's law of segregation (1866) and its rediscovery in 1900

3. The discovery of photosynthesis by H. A. Allard and W. W. Garner in 1918, which played a central role in the breeding of high-yielding hybrids adapted to different latitudes
4. The use of chemicals, such as, Paris green and Bordeaux mixture to control pests and plant diseases and
5. The use of DDT (Dichlorodiphenyltrichloroethane) as a wide-spectrum synthetic pesticide against several species of pests and of 2, 4-D (2, 4-dichlorophenoxyacetic acid) as a herbicide in 1940 (Evans 1998)

Rapid technological advances during the period 1927 to 1960, based on scientific principles of the bygone century, helped humankind to make great strides in agricultural production and productivity. The global population attained a level of 3 billion mark by 1960. This was an exciting take off period for agricultural development in the developed countries. The period from the beginning of the nineteenth century until 1950 saw large-scale expansion of the area for agricultural production owing to the availability of cheap fossil fuels, for energy (especially for the manufacture of urea, which, when applied liberally boosted yields of the "miracle" dwarf wheat in South Asian countries like India, Pakistan and Bangladesh), and the adoption of mechanization in the industrialized countries, including the USA and the West European nations. Widespread availability of cheap nitrogenous fertilizers after World War II, coupled with synthetic pesticides, intense mechanization, and high yielding hybrid varieties of crops, brought about a quantum leap in the productivity of cereal farming, in much of South Asia, though, with attendant environmental hazards, principally arising from unbridled use of chemical fertilizers and pesticides. In these countries, agriculture, which had remained almost at subsistence level until the middle of the twentieth century, turned industrial, followed by huge environmental problems, principal among which is the excessive degradation of soil, witnessed in many Asian nations. The exponential jump in cereal yields, principally in wheat and rice, sustained by unbridled use of chemical fertilizers had to pay a huge price in terms of environmental degradation. Soil became the main casuality. For instance, vast stretches of land in Punjab State of India, practicing monoculture of rice-wheat rotation, known in the mid and early sixties of the twentieth century as the "Cradle" of the so-called green revolution, turned to barrenness due to soil degradation. Soil salinity increased. Unbridled use of urea, the principal chemical fertilizer applied to boost cereal yields, led to excessive contamination of ground water due to nitrate loading, making water no more potable.

The "golden era" of agriculture continued from 1960–1975 both in the developed and developing nations (with the exception of African continent on the whole), where science and technology were exploited to the maximum to harness crop yields. The phenomenal increase in the population of the developing countries (especially in South Asia, in countries like India, Pakistan and Bangladesh, with the exception of China, where the "one child" norm was enforced with utmost discipline), and their unpreparedness to meet this contingency, provided ample opportunities of the temperate zone countries, like the USA, Russia, Canada, and European nations (including Australia in Asia) to export large quantities of cereals, principally wheat, and other agricultural commodities of importance like pulses and oilseeds, to these countries in the developing world, and reap a windfall. Agriculture, thus, became a major productivity and profit oriented enterprise for countries such as the USA, Canada, Australia and the countries in Western Europe. In the USA the commodities market and speculation became a way of major global trade, and commodities center in places like Chicago flourished involving millions of dollars in speculative trade each day, especially during the crop season, in crops like wheat.

12.1.3 Commencement of the Faustian Bargain

The unbridled intensification of agriculture with excessive external inputs led to the much written about Faustian bargain between agricultural sustainability and the unparalleled productivity offered by input intensification. By the late 1970s and early 1980s, US agriculture was suffering the consequences of severe land (soil) degradation as a result of soil erosion, depletion of soil organic matter, loss of biological diversity, salinization from excessive use of chemical fertilizers and water, and severe environmental pollution owing to runoff of by products of fertilizer decomposition. In India, Punjab State, known as the "cradle" of the so-called green revolution (an euphemism for "high input industrial type" agriculture), witnessed the worst kind of environmental disaster. Barren lands, dried aquifers, vanished bio diversity due to continuous monoculture of rice-wheat, excessive soil salinization and loading of ground water with high amounts of nitrate ion (a consequence of unbridled use of nitrogenous fertilizers to boost the dwarf Mexican wheat imported into India) making water non potable, led to a disastrous condition (Nair 2014). The safety and health of livestock and humans were threatened by the indiscriminate use of pesticides and antibiotics in livestock, which flooded the food chain. The most thought provoking and revolutionary book "The Silent Spring" by Rachel Carson, published in 1962, alerted humans all over the world to the perils of unbridled use of pesticides, especially DDT, in agriculture, forestry, and public health. High yielding cereals in monoculture (as for example continuous rice-wheat rotation in Punjab State in India), not only led to large scale erosion of genetic diversity, but, also in the proliferation of new pests and pest-resistant strains and the further need for new pesticides against the emerging new pests, coupled with high doses of these pesticides. Agriculture, or more appropriately, the "high input" industrial type of agriculture, unfortunately as the so-called "green revolution" became environmentally unsafe, and provoked strong protests from the environmental bandwagon in the 1970s, who were in favor of the more benign modes of traditional agriculture, which relied far less on chemicals, but, more on age-old practices of farming (Nair 2014). Organic farming or "natural" farming, began getting more and more attention, from both scientists and the society (Saroja Raman 2006).

It must, however, be emphasized that, the failure of the so-called green revolution, is due to a fault line in the approach ***per se,*** and, the way it was practiced. It is common knowledge that crops need external inputs, primarily plant nutrients, since more than 50% of a crop plant's performance is determined by the nutrient factor, which is very crucial, but, also least resilient to faulty management. The "Law of Minimum" of Liebig in 1840 explained this in a rather crude manner. The idea of "how much" and "how best to use" of the nutrient, became a casuality and was replaced by "any" amount, unbridled in almost all of the cases. That is the reason the term became an euphemism. The classic example was what happened in the State of Punjab, in India, the "cradle" of India's green revolution, where the farmers in a frenzy to reap high harvests of the "miracle" Mexican dwarf wheat variety, imported into the country from CIMMYT (the International Research Center for Wheat and Maize, in Mexico) and cross bred with the native varieties, kept applying unbridled amounts of chemical fertilizers. It was not "green," in harmony with nature, but, powered by chemical industry (mainly chemical fertilizers and pesticides) and thus can only be called industrial agriculture. That is the reason it fell flat on its face, in less than a decade since start, when environmental hazards shot up all over the world, be it Europe, north America or South Asia, where it was followed with unbelievable intensity. The African continent was an exception, where, by and large, apathy to change and a total lack of efficient infrastructure kept the continent away from the environmental hazards witnessed elsewhere on the globe. Agriculture still continued to be a primitive way of farming. In Asia, especially in India and China, and to a small extent in Pakistan and Bangladesh, the story was very different. "Green" turned to "Greed." The green revolution, or, more appropriately, the industrial agriculture, had two other dimensions, namely, socio-economic and human health. The first is characterized by the innumerable suicides of poor cotton farmers in Vidharbha district of Maharashtra State in India, who were unable to sustain this "high input" agriculture. And the second was in the Gurdaspur district of Punjab State, again in India, where the local populations fell victim increasingly to cancer due to continuous use of hazardous pesticides, residues of which entered the food chain. Gurdaspur, thus, became the "Cancer Capital" of India.

The theory of agro ecology, which developed into an accepted scientific discipline around the 1970s, sought to highlight the possibility of a scientific, yet, durable agricultural production paradigm based on a synergistic relationship with natural resources and processes. The most noteworthy contribution to sustainable agriculture came from the system of regenerating/alternate agriculture, which evolved parallely with industrial agriculture in the USA. It combined judiciously and selectively the products of newer technologies (such as improved crop varieties, mechanization, and soil testing to allow the measured addition of fertilizers) with natural processes (nutrient cycling, pest control, crop production, etc) to conserve resources and prevent pollution. This was the real birth of a science-based paradigm in sustainable agriculture. The credit for coining the term "sustainable agriculture" in 1970, should go to Eva Balfour, who was a proponent of the alternative agriculture system/model and believed strongly in the need for harmony with nature. Robert Rodale was another pioneer of regenerative agriculture who believed in observing natural soil processes to ensure perpetual productivity (Rodale 1983, Harwood 1990).

12.2 Socio-Economic Fallout of the Industrial Agriculture

High – external – input agriculture not only created environmental hazards, but, also brought about other socio-economic hardships to farming communities. The environmental hazards were compounded by the heavy dependence on purchased inputs, which had made agriculture economically unviable. The hike in oil prices in the 1970s (the "oil crisis") further aggravated this. Farmers in the poor countries became indebted to unscrupulous money lenders, as was the case in India, and there is no parallel in any other country in the world, when lakhs (a lakh = one hundred thousand) of farmers took to suicide, to escape the clutches of Shylockian money lenders. The government sat helpless. Several farmers became bankrupt and either committed suicide or the farms were folded. Opportunities outside the farm sector made farming a far less favored occupation (Nair 2015a, b, c). In a recent survey conducted in India, it came out as a glaring fact that given a dependable alternative more than 40% farmers would quit farming, especially the poor and marginal ones. The skyrocketing prices of inputs was becoming unbearable. To cite an instance, the price of phosphatic and potassic fertilizers have escalated by more than 500% in India during the last couple of years. Heavy government subsidies in the form of minimum support price (MSP) and direct cash incentives to ensure supply of 20 specified agricultural commodities in India under the Commodity Program are actually antithetical to resource conservation and environmental safety (NRC 1998).

12.2.1 The Plight of the Developing Countries

The pre-eminent position of the traditional systems of agricultural production of the developing countries, such as, India, China, Pakistan, Bangladesh and nations of South America, which were once admired by the Western nations for their durability, was lost during the twentieth century. The modest productivity of these systems could no longer match the exponential increase in population in these countries. Agriculture in these regions remained primitive owing to a lack of such enabling technologies, such as fertilizers, efficient plant cultivars, pesticides, and also, appropriate policy directives. Thus, while surpluses in agricultural production and subsidies were making agriculture unsustainable in the Northern hemisphere, shortages were threatening the very survival of the Southern hemisphere in the 1960s.

In the mid 1960s, India faced extensive and protracted drought in addition to its already precarious food situation. China suffered worse situations compared to India until a turn around after the Cultural revolution of the 1960s and 1970s. Severe famines were predicted by the Paddock brothers, William and Paul, in the book Famines- 1975 in various countries of Asia. Thus, total gloom prevailed on the food front in the early 1960s for Asian countries. However, such desperate situations were soon encountered by setting up effective national agricultural systems to strengthen agricultural education, research and extension at an accelerated pace in

India. The Indian Council of Agricultural research (ICAR), established in July 1929, was reorganized in 1966, to unify agricultural research, education and extension in the country, conforming to the Land Grant Pattern of USA. Earlier, agricultural education in India was modeled along British pattern, (also similar to the European pattern) without an interlink between all the facets. It has evolved into a model for many developing countries to emulate. Generous and long-term financial assistance/support from donor agencies, such as the Ford and Rockefeller Foundations, and technical cooperation from international research centers, including the International Maize and Wheat Improvement Center in Mexico (CIMMYT), the International Rice Research Institute (IRRI) in Philippines, the International Institute for Tropical Agriculture in Nigeria (IATA) and the International Potato Center (CIP) in Cali, Colombia, provided tremendous boost to agricultural productivity in Asia, Africa and Latin America during the 1960s and 1970s.

The introduction of the "miracle" dwarf wheat varieties from CIMMYT and the dwarf rice varieties from IIRI, with high harvest index, and high response to applied fertilizer, in particular, nitrogen, irrigation, and pesticides, without doubt, led to an exponential growth in cereal productivity in South and South East Asia. Enlightened and enabling support in the form of the supply of inputs and policies by the governments of the day played a key role in the agricultural "revolution" that took place in these countries. In some ways, it was akin to the "cultural revolution" launched by Mao Zedong in China, but, each had its own color. The former was "green" while the latter was "red," and that, perhaps, is the reason why Dr. William Gaud of the United States Department of Agriculture (USDA), who was then the Director of the United States Agency For International Development (USAID), appropriately named it the "green revolution," and, interestingly the term stuck, and, was bandied around, in countries such as India, as though coined by an Indian scientist! Agricultural production tripled from the level attained in 1950 to that of 1975 in India. The addition of one more billion people to global population, between 1975 and 1987, mostly in the developing countries, passed off without raising much concern about food security, owing to the enhanced food production resulting from the so-called green revolution, though in countries like India, the extreme adverse fallout on the environmental front, such as soil degradation, salinization, dried aquifers, ground water contamination with excess nitrate making it non potable, and vanished bio diversity due to continuous monoculture of rice-wheat rotation, especially in Punjab, the "cradle" of the green revolution in India, had become quite visible even in the early 1980s. The adverse effects have begun to show up in more and more places in India, and, farmers suicides, on a scale unheard of anywhere else in the world, is an offshoot of this "high input technology." Indeed, the term green revolution has really become an euphemism! (Nair 2014).

True to its inherent technical flaws, the so-called green revolution, which was an extension of the high-external-input agriculture devised in the Western conditions, and blindly aped in the developing countries, started leaving its dirty ecological mark in the developing countries beginning the 1980s, and is continuing to do the same at an accelerated pace in many regions. In countries such as India, Pakistan, Bangladesh and Indonesia, problems also started to develop with regard to resource degradation and pollution owing to unbridled use of chemical fertilizers and water to boost crop yields (mainly the imported dwarf wheat and rice, subsequently crossed with the native cultivars, but retaining most of their inherent traits). Along with this resource degradation, secondary problems also emerged, of which the incidence of new pests and diseases became very common. The emergence of rust in wheat and blast in rice are classic examples of this severe pestilence. The most recent example is that of Bt cotton in Maharashtra State of India, in particular, Vidharbha district, the "cotton belt" of India. Initially the Bt cotton seemed to be a solution to control the American bollworm, for which it was evolved, but soon, the crop gave way to sucking pests and aphids, which became uncontrollable. Farmers' suicides was on the upsurge on a scale unheard of anywhere else in the world, all due to the failure of the high input technology of the green revolution. At the time of writing this book, Bt cotton was totally devastated in Punjab, neighbouring Haryana and Rajasthan states due to severe attack of white fly. The farmers are in total disarray and many have committed suicide due to bankruptcy. It has become a national crisis.

When crops, boosted by high inputs, such as fertilizers, started to fail, farmers became bankrupt and the only route to escape the Shylockian grip of the unscrupulous money lenders, who provided the cash to buy the high cost inputs, including expensive Bt seeds, was suicide. This became a national shame in India. And those concerned elsewhere in the world were aghast. It became quite apparent that the so-called Western-inspired, and blindly aped green revolution, had fallen completely on its face in India, by the early 1990s. Poor farmers were shattered and a recent survey in India, National Sample Survey, showed that as much as 40% of the farmers chose to quit agriculture. On one hand agriculture was no more attractive as in the early seventies, due to the high yields obtained, and on the other, prices of agricultural commodities plummeted making farming an unviable economic proposition.

The economic viability of production was compromised by the high prices of inputs. Price of phosphatic and potassic fertilizers in India escalated by more than 500% in just 2 years (2010–2012). This was in sharp contrast to the price of urea, a main supplier of nitrogen, where the price was kept from jumping, through fertilizer subsidies. This led to farmers making a scramble for urea, to compensate for the high prices of the phosphatic and potassic fertilizers, in the hope that what was lost in one could be compensated by another. This is leading to an escalating nutrient imbalance in Indian soils, which will have an extremely deleterious effect in the long run due to the skewed use of the three principal fertilizers. To stem all these, the government (Indian) unfolded a policy of large-scale subsidies for inputs, an assured minimum support price (MSP) for the produce (specially for grains like rice and wheat), and subsidized distribution of grain to the people. The Food Security Act of 2014 of the Government of India is an attempt in this direction.

The green revolution had completely bypassed the resource starved farmers in the low potential complex, diverse, and risk prone (CDR) areas and farmers with small holdings in India. This was very characteristic of the rainfed areas of the country, and, it is also in these regions where the maximum farmers'suicides which took place. The lack of livelihood opportunities in the non agricultural sector resulted in massive rural poverty and food insecurity. The Supreme Court of India, the apex judicial body, intervened to pass strictures on the government. "Right to Food" like "Right to Work," enshrined in the basic human rights, endorsed by the United Nations, as early as 1948 through Article 25 of the Covenant, to which more than a 100 countries were signatories, including India, was only on paper. Despite public outcry, the government of the day turned a deaf ear to the misery of the poor.

Both scientific and public criticism mounted against the so-called green revolution. The "Silent Spring" of Rachel Carson seemed to come full circle in India. But the earlier proponents of the green revolution, who wielded much political clout, ignored the danger signals, and called for a "second green revolution," and some even touted for an "ever green revolution" to bring into the country the transgenics as a panacea of all ills faced by Indian agriculture. At the time this book is being written, the specially constituted Parliamentary Committee of the Government of India, appointed with the task of looking critically at all the aspects of transgenics, clearly mandated a halt to all field trials with GM crops, which has now been endorsed by the Supreme Court's Technical Expert Committee (TEC) which has mandated a 10 years moratorium on all field tests of GM crops. The entire question of the so-called green revolution and the impact of high input industrial type agriculture has come full circle in India. The need for agriculture to develop holistically, with due care for the environment, economic viability, societal responsibility, justice and equality, became most urgent in the developing countries.

12.3 Sustainable Development as a Harbinger of Sustainable Agriculture

The concept of sustainable development became an appropriate basis for defining future development goals for agriculture. Although the Brundtland Commission Report mainly targeted the environmental damage caused by intense industrialization, it made explicit reference to deficiencies in world agricultural systems also and the need for a new holistic approach. It observed that to take care of current and future needs for food, several other factors were critical in addition to adequate global food production:

1. Reducing the scale of environmental degradation brought about by modern technologies and intensive agriculture
2. Reducing the distortions in the structure of the world food market (especially with regard to commodity markets, involved in large-scale speculation)
3. Tackling problems of population, poverty, and food security at the national level for developing countries
4. Maintaining, enhancing, and restoring the resource base for sustaining agricultural production

In short, the report advocated agricultural systems which focus as much attention on people as they do on technology, as much on resources as on production, as much on the long terms on the short term. This was the harbinger of a new paradigm of sustainable agriculture.

Sustainable agriculture was quickly acknowledged as the most important component of sustainable development. International organizations, such as, the Food and Agriculture Organization (FAO), the World Bank, and the Consultative Group on International Agricultural Research (CGIAR), made sustainability the basic objective to be pursued in all their future programs in agriculture. An explicit definition of sustainable agriculture and the agenda for its operationalization were formalized by CGIAR in 1988. It made sustainability one of the target criteria for all of its research projects and adopted the following definition:

> Sustainable Agriculture is the successful management of resources for agriculture to satisfy changing human needs, while maintaining or enhancing the quality of the environment and conserving natural resources (CGIAR/TAC 1988).

The CGIAR definition did not include the economic and social aspects of sustainable agriculture overtly, because, the mandate of CGIAR was focused primarily on improving production and productivity. The American Society of Agronomy in 1989 adopted a more holistic view of sustainable agriculture, and came up with the following definition:

> Sustainable agriculture is one that over the long term, enhances environmental quality and the resource base on which agriculture depends; provides for basic human food and fiber needs, is economically viable; and enhances the quality of life for farmers and society as a whole.

Sustainable agriculture, thus, operates within the boundaries of physical and biological resources on the one hand and socioeconomic viability and equity on the other.

The definition that G.K. Douglas (1984) formulated for sustainability a little earlier than when the first definition in 1988 was formulated by CGIAR, captures the essence of sustainable agriculture in its totality. It goes as follows:

> Sustainability must be regarded as long term food sufficiency which requires that agricultural systems be more ecologically based and do not destroy their natural base... Sustainability as stewardship means that agricultural systems are based on a conscious ethics regarding humankind's relationship to future generations and to other species in nature... Sustainability as a community means that agricultural systems are equitable. Agricultural systems cannot be sustained if there is gross misdistribution of land, wealth and power. Social tensions and upheavals in many parts of the world are often manipulations of such inequalities.

12.4 Elements of Sustainable Agriculture

Human, economic and social developments are considered the most critical areas for the attainment of sustainable development in the course of the next two generations (NRC 1999). Food security is at the core of human development, which includes several other parameters such as, health, education, shelter, human longevity, and basic amenities for a descent living. At the core of all this is the question of right to food. What does it imply? The following discussion is pertinent in that context:

In 1948, the Universal Declaration of Human Rights was passed by the United Nations and the right to food was acknowledged. The international declaration, asserting the right to food does not imply that states shall be responsible directly for fulfilling the individual's need and right to food. Article 25 of the declaration maintains that "Everyone has the right to a standard of living adequate for the health and well-being of himself and his family, including food and the right to security in the event of unemployment, sickness, disability and widowhood, old age, or other lack of livelihood in circumstances beyond his control." In 1966, the UN General Assembly adopted the International Covenant on Economic, Social and Cultural Rights, which further formalized the right to food, as a basic human right. Article 11 of the Covenant further endorsed Article 25 of the 1948 Declaration cited above.

By 1989, as many as 85 States, including India, had signed the Covenant. Yet, nearly after close to two-thirds of a century, after adopting the original declaration in 1948, and after three world food summits in Rome, discussing hunger and how to overcome global hunger, and close to the end of the 2015 Millennium Goal when universal food security had to be achieved, in India, about 6% of the total global population, most of whom are working women, children and pregnant mothers sleep on an empty or half empty stomach each day. And this, despite the country harvesting in excess of 200 million tons of grains on average for the last close to half a decade! There is no comparable shame on a scale like this anywhere else in the world! The UN in adopting the Universal Declaration of Human Rights in 1948, had expected human rights to go through the legal rigmarole before becoming assimilated in the national and international legal systems. Though this happened to some extent in the case of political and social rights, economic and cultural rights have not come within its ambit. Right to food has been treated as an "aspirational right," which has to be achieved through economic development. It thus gets pushed to the periphery, while it should have been at the center of all rights. It is an irony that while all States, big or small, continue to endorse the right to food, they have not translated this endorsement into legally binding methods. There is a "Food Security Bill" gathering dust in the Indian Parliament since 2010 with no objective discussion taking place, while millions of poor starve, and grain "mountains" stored in improper conditions in the Food Corporation of India go downs (procured grain from the farmers giving them the "Minimum Support Price" mentioned earlier in the above discussion), are beginning to rot and subjected to rodent attack. The apex court of India, the Supreme Court, had decreed in 2011 that the stored grain be immediately distributed among the poor and needy, but, the government is inactive and dragging its feet, while millions starve!

There is no national or international mechanism to supervise the implementation of this right (Right to Food). In general, international economic laws have separately evolved and are not based on human rights. The General Agreement on Trade and Tariffs (GATT), the World Trade Organization (WTO) and the North American Free Trade Agreement (NAFTA), generally uphold the "domino theory" and there is no place for human rights in these economic laws. Take for instance the Trade Related Intellectual Property Rights (TRIPS). How does it view the contributions of poor peasants of the bio resource- rich developing world towards the development of new crop seeds, the sale of which is a multi-million dollar trade filling the coffers of the multinational and transnational companies?

The Human Development Report 1999 puts on record the question of TRIPS as follows – "New patent laws pay scant attention to the knowledge of indigenous people. These laws ignore cultural diversity in the way innovations are created and shared – and diversity in views on what can and should be owned, from plant varieties to human life. The result: A clandestine theft of centuries old knowledge from some of the poorest communities in the developing world.

The record of some of the men from developing countries, who in the past were willing accomplices to this clandestine theft, is far from flattering. Those who played the "hand-in-glove" game stood to benefit enormously from this blatant piracy. The question of "proprietorship" of third world genetic material and of products originating in the third world as raw material, but, ending up in the developed world as "finished goods," to be recycled back to the developing countries for marketing, has defied a satisfactory resolution., despite engaging the attention of international agencies and concerned individuals.

For instance, if a wheat variety grown on the African continent, in a particular country, were to show a high degree of resistance to a particular disease, and if a multinational company (MNC) were to identify the gene which imparts this trait and then transfers the same to another wheat variety that is already in its kitty, and then goes further to market the new wheat variety with the resistant gene, in the African market, under a new brand name, who would be the rightful "proprietor" of the new variety? – the particular African country or the MNC? Or, more appropriately, how would the MNC compensate the African country/its farmers, in question? During a decade of professional life on the African continent, this author has helplessly watched such clandestine theft, with the local counterpart playing second fiddle to the piracy – of course, for a price, a very lucrative job in comfortable settings.

There is, yet, another route through which this theft of rare genetic material from developing countries takes place. The most recent example is that of the evolution of Bt brinjal in India by an alien giant multinational agribusiness company in seed business. The company forayed into the native brinjal germplasms to scout for transferring the Bt gene, without seeking the permission of any Indian authority for their use. The finished good was about to be okayed for public commercial cultivation, when owing to a concerted public outcry, the minister of environment and forests had to put a moratorium on the release of the Bt brinjal from commercial cultivation. Now the Technical Expert Committee (TEC)empowered by the Supreme Court of India has put a 10 years moratorium of field testing of all GM crops, until all bio safety norms are respected. This is a move in the right direction.

The Geneva Conventions of 1949 and their protocols of 1977 provide a legal framework for safeguarding the right to food of prisoners and war victims, but, the same right is not safeguarded during times of peace. Despite the International

Covenant on Economic, Social and Cultural Rights adopting a procedure for countries to report progress on human rights to a working group of Government experts, replaced by a Committee on Economic, Social and Cultural Rights, the working group was unable to monitor the progress of various States, because of the vagueness of rights and of the obligations of the States, including insufficient guidance to the signatories on the mode of reporting.

Some of the arguments that States put forth in justifying their inability to enforce a legal right to food are:

1. High cost of enforcement
2. Inability to define economic and social rights in legally binding terms
3. Protecting such rights would involve redistribution of private resources and such rights can be abused by repressive regimes
4. The failure of the State to target development towards the poorer sections, despite the UN declaration, has convinced agencies and individuals of the need to create a legally binding obligation for the State to act against hunger.

It is this author's firm conviction that hunger must be felt and not only debated or written about. Only a hungry human will know the pangs of hunger. It is the firm belief of the author that it is a lack of empathy on the part of those who wield power, sometimes enormous power, mostly political, which stands in the way of banishing global hunger. Sustainable agriculture must be hunger centric, if it should really carry the true meaning of the word.

12.5 Key Aspects of Elements of Sustainable Agriculture

Human development, economic development, and social development are considered the most critical areas for the attainment of sustainable development in the course of the next two generations (NRC 1999). Food security is at the core of human development and its ramifications have already been explained above. In view of the continuing escalation of world population, and because of the limitations for expansion of the productive cropland, food security is implicit with a concomitant increase in agricultural productivity until population increase stabilizes, which, seems most unlikely. The sustainable development paradigm helped to develop a holistic perception of sustainable agriculture, with due care for all the factors involved, which are impacted by agriculture, such as productivity, ecological safety and integrity, economic viability, and social responsibility toward the farming community. The basic tenets of sustainable agriculture can therefore be summed up as follows:

1. Plant and animal productivity adequate to meet the needs of a growing population
2. Ecological safety and integrity and environmental quality
3. Economic viability
4. Social responsibility and acceptability

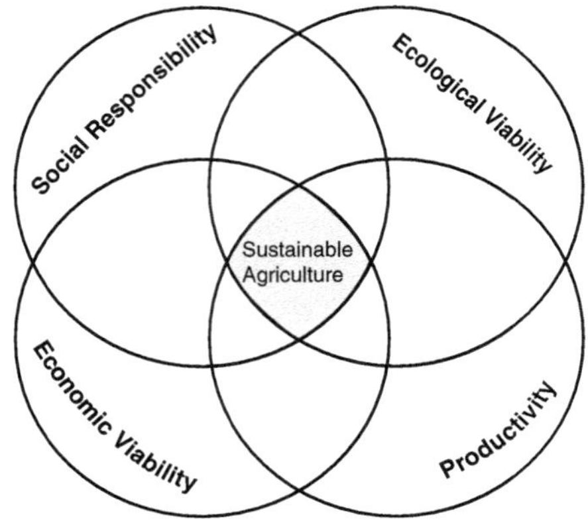

Fig. 12.1 Sustainable agriculture

All of the above parameters must be fulfilled for optimum development of agriculture to become sustainable. It is graphically illustrated in the following figure (Fig. 12.1).

If a system is not ecologically sustainable, it cannot persist in the long run and cannot be productive and profitable. Similarly, if a system is not productive and profitable, it cannot be sustained economically, no matter how ecologically sound it is. This simply demonstrates how closely knit these important parameters are to make agriculture sustainable. Social harmony, acceptability, and equity are critical to the sustainability of agriculture or any other human development system.

12.5.1 Productivity

There can be no dispute that per capita food adequacy at the global level is most essential, non compromisable, and the minimum component of sustainable agriculture. This is calculated as the ratio of total food production (the aggregate of the component production systems of global agriculture: national, regional, farming, cropping systems, and individual plots/farms) to global population. However, this measure is not a sufficiency condition because it does not guarantee food security at the individual level. A whole lot of other dimension exists to this aspect of individual food security that is not covered by food sufficiency at the global level.

Owing to the combined efforts at the individual farmer, regional and national levels, global food production has so far been quite sufficient to feed everyone if the food were distributed according to needs (Pinstrup Anderson 2002). This has been achieved mainly through productivity increases since the 1950s, through the intensification of agriculture by the land-saving technologies of fertilizers, pest control chemicals, irrigation and time- and labor-saving machinery. Unbridled use of these external inputs in the post- World War II years has given rise to an unprecedented scale of environmental pollution: degradation of agricultural lands through soil erosion, salinization, water logging etc.; and economic challenges as a result of the increasing cost of purchased inputs. It is now clear that there are ecological, environmental, economic, and even social limits to intensification of production in many of these overexploited areas, where lower rates of growth, lowering of factor productivity, and even yield stasis are surfacing. In addition, the high cost of production and low profitability are threatening the sustainability levels of productivity of these intensive cultivation systems (Cassman et al. 1997, Ladha *et al.* 2003). This led to the realization that maintainability of productivity per unit area of land at sustainable yield level, that is within the productive capacity of the land, is a basic condition for the sustainability of an agricultural production system. The twin challenges of decreasing per capita arable land and limits to intensification of cultivation, through current practice of incremental addition of external inputs, have necessitated a search for newer, alternative paradigms for increasing both production and productivity, with acceptable levels of adverse ecological (as it will be well nigh impossible to eliminate total adverse environmental fallout) and sociological impacts. This is the current concern in terms of the productivity component of sustainable agriculture. "The Nutrient Buffer Power Concept" fits well within this vision.

12.5.2 Ecological Viability

Conservation and enhancement of the quality of the natural resources of land, water, air, and bio diversity to sustain biological productivity and ecosystem services is basic to sustainable agriculture. The unsustainable use of natural resources has become rampant because natural resources are limited and only conditionally renewable, but, the pressure on them is increasing with time, as a result of the increase in population and aspirations of people for better livelihood conditions. Maintaining the ecological viability of agricultural production systems is more complex than the sustainability of natural ecosystems, because an agro ecosystem is a system of human intervention with mandatory goals. Sustaining the environment simply by preserving it in its natural state cannot support humans in their current numbers. Agricultural operations impose additional anthropogenic stresses and shocks induced by cultivation practices. Hence, agro ecosystems have to develop resilience to contend with these perturbations, in addition to natural stresses and shocks.

Conserving the production resource base and maintaining environmental safety and quality are the basic criteria for the ecological sustainability of agricultural production. The natural environment is a source of resources and a sink for waste materials. To be sustainable, the consumption of renewable resources cannot exceed their regeneration and the dumping of degradable waste materials should not exceed the capacity of the environment to absorb them. Short-term biological productivity alone is not an indicator of ecological health or the integrity of the ecosystem. Ecological processes, to use a metaphor of the agriculture factory (Vandermeer 1992), need to be overhauled and serviced to remain fit for use through proper management techniques involving efficient use of inputs. Apart from guarding the resource base, this will also reduce environmental pollution and degradation and keep them within the absorptive capacity of the environment. Attributes such as soil quality, fertility balance, and other indicators of sustainability are manifestations of healthy ecological processes and of natural resources that have not had their physical, chemical, or biological integrity compromised.

If biological productivity becomes the only goal of natural resources and ecosystem management, sustainability may become a casuality as a result of retrofitted management strategies that exacerbate the deterioration of ecological processes and the environment. High-input agriculture, with synthetic nutrients and pesticides, powerful machinery, narrow crop biological diversity, and so on, though more profitable in the short run, is not sustainable over the long term. It requires the continuous racheting up of inputs at the expense of compromising the ecosystem processes that actually undergird the capacity for sustainable production (Miller 1998).

Agricultural systems operating on agroecological principles with reduced use of external inputs and non renewable resources are more ecologically sound than those that depend predominantly on external inputs, which have the potential to damage resources, the environment, livestock, wildlife, soil microorganisms, useful insects, pollinators, and pest predators. Increasing biologically favorable inputs such as crop residues, manures, green manures, legumes, crop rotations, biological pest control, and minimum tillage facilitates better use of ecosystem services and enhances ecological soundness. A wide variety of farming practices developed at different times and in different regions, such as organic farming, alternative farming, natural farming, zerobudget farming, and biodynamics, represent a greater degree of ecological prudence and soundness in terms of resource conservation and environmental safety than modern farming systems.

The human element in agriculture makes it somewhat subjective and value oriented. Agricultural systems have the imperative to sustain biological productivity as their primary ***raison d'etre.*** In addition, they have to fulfill other aspects of human welfare such as sustaining life-support services, adequate profitability, and social responsibility toward rural farming communities. Hence, in addition to the need to maintain ecological integrity, which is essential to sustain biological productivity, human welfare has to be factored into productivity and cannot be replaced by higher yield and profitability. This is the essential conundrum in terms of the ecological viability component of sustainable agriculture.

12.5.3 Economic Viability

The economic dimension of agricultural sustainability is quite complex, indeed. Economic viability in the case of a farm, the lowest level in the economic hierarchy of agricultural systems, can best be quantified as profit for the farmer. Simply stated, the monetary value of the output from the farm should be substantially more than the value of all the inputs, put together for production, such as, chemical inputs (both fertilizers and pesticides), machinery use, labor, and the cost of land conservation, so that the farmer can afford a fairly comfortable lifestyle typical of local socio economic levels.

Although, ***prima facie,*** economic viability is directly linked to productivity, it is neither static nor stable, as demand for farm products varies with the dynamics and specific needs of a human population and with the agricultural and non agricultural policies of local, State, and national governments. Even at the farm level, how economic viability/profitability works for the farmer is a complex and highly contextual process. It is guided directly by governmental policies in terms of the input subsidy towards the cost of seed, fertilizer, pesticides, irrigation water, and electricity used to pump water etc., and in terms of the procurement prices for the farmer's produce by the government, crop insurance, credit policies, etc. It is also dependent on factors such as technology, market facilities, infrastructure, including roads and storage facilities, market demand and consumer preferences, and cooperative group farming or corporate farming facilities. The availability and prices of commodities in international markets and government policy on the import and export of agricultural products, including import tariffs, export incentives, international trade policies, and the balance of payments, influence economic viability at the national level. All of the above factors are crucial to a country like India.

Often governments are forced to export even below their national prices or production prices, to maintain favorable trade balance and foreign exchange reserves. Economic viability is also influenced by certain political and social factors such as the supply of agricultural commodities to economically and socially challenged groups of people at preferential rates. Politically motivated, skewed policies can often distort the economic viability of agriculture at the national level. However, the compulsion to maintain socioeconomic balance and stability under certain circumstances may make this unavoidable. Additional factors are the funds available from international agencies and the interests of multinational corporations in specific food-processing industries in the local market or in export activities. Economic viability will also depend on the value system, of the local people in terms of the stewardship of natural resources and the ecological goods and services that derive from them. The bottom line is that economic profitability for the farmer is essential if agriculture is to be sustainable.

At the global level, countries may be guided by different sets of circumstances in setting the price of their agricultural products. For example, countries such as the United State and nations of Western Europe, including France and the Netherlands, produce very much in excess of their internal needs, and turn agriculture into an export advantage. Hence, their policies are different from those of the countries which need to import food. These countries offer huge subsidies to farmers to facilitate exports and to keep local food prices low. For countries such as India and China, where a large percentage of the population is dependent on agriculture, and, where the share of agriculture in the national economy is disproportionately large, typically like India, pricing policies have to be different to balance the affordability for local people, export competitiveness, and import rates for deficiently produced commodities.

12.5.4 Social Responsibility

Social justice and equity, the fourth component of sustainability, is even more complex. In short, it means that sustainable agriculture should take into account the welfare of farmers and be responsive to the basic needs of the farming communities that produce the food. It should be in tune with social values and the traditional social cultural background. It includes a plethora of definable as well as vague parameters such as poverty, cultural factors, education, social capital, justice and equity, value systems, food security at the household, regional, and national levels, livelihood opportunities, and government policies.

Similar to economic viability, it is hierarchical. Social equity at the rural small farmer level will mean food security, livelihood opportunities, and access to good health and education. The same does not apply to large farmers with adequate income and opportunities for a comfortable lifestyle. The social acceptability of a particular mode of farming relates to the economic conditions of the farmer. Broad differences exist depending on whether the farmer is predominantly a subsistent farmer or a commercial farmer.

The acceptability or otherwise of specific modes of cultivation also depends on the physical resources available. For example, poor farmers and farmers who own complex and risk-prone lands such as the arid and semiarid regions of India and Africa could not benefit from the green revolution technology, which requires level, healthy soil with assured irrigation and other inputs such as fertilizers, pesticides, and tractors for timely crop operations. At times, the notion of social equality tends to include utopian ideals, without taking into account their ecological and economic implications.

12.6 Priorities and Trade-offs among Elements of Sustainability

All the four elements discussed in the preceding sections together form the basis for agricultural sustainability which must be considered *in toto*. But, trade-offs and compromises are often necessary in the pragmatic pursuit of sustainability. If one were to set priorities among these elements from a global perspective, productivity and ecological soundness would take precedence over economic and social viability, because, food security is contingent on adequate per capita productivity. Permanence of productivity is dependent first and foremost on the ecological base for production. Unless this is kept in sound health, not only productivity, but also all other life-support functions will be affected. Sustainability is basically a resource issue. Ecological limits exist to biological productivity and hence to sustainable intensification of agricultural production systems. This is corroborated by the fact that a number of intensively cultivated systems are showing signs of yield fatigue or saturation (yield plateauing). When humans try to tilt the balance too much in their own favor, a backlash takes place, in the form of yield plateauing, a decrease in factor productivity, or, even decrease in yield. In extreme cases, even the life support systems are in danger of collapse (Cassman and Harwood 1995).

Economic viability will follow productivity and ecology because it is a strong mobilizer and enabler of agricultural productivity. Globally, except in small and remote pockets of the world, notably on the African continent, agriculture has metamorphosed from subsistence to a strong commercial and economic activity with a powerful profit motive. Adequate economic returns are required for agricultural sustainability because all human interventions such as ploughing, pest control, and harvesting of the crop produce and further processing before marketing, cost money. Sustainable systems have to be economically viable either naturally or by human intervention. Since a healthy ecosystem is productive and profitable, farmers generally have a stake in adopting ecologically sound farming practices. However, conflicts arise between the short-run interests of individuals and the long-term interests of communities, and, finally, the whole of humanity. Economics has come to be such a dominant factor in agriculture that it has become the agent of destruction of ecological health, not only at the level of individuals, but, even at national levels. The widespread resource degradation and environmental pollution of intensive cultivation are a consequence of the "profit-first" mindset of man. This can be countered through such policy measures as providing economic incentives to individuals to follow good practices in ways consistent with longterm societal interests.

Economic benefits are not the only rewards of agriculture; many other intangible benefits can be derived. However, human civilization is such that people cannot persist in actions inconsistent with their economic survival, regardless of their values, preferences, or principles. Enterprises that lack economic viability will lose control over the use of ecological resources to their economically viable competitors. Agriculture cannot survive, however, if the system degrades the resource itself for merely economic gains.

The conflicts between the short-term interests of individuals and longterm societal interests have to be solved in a pragmatic manner. Essentially, social norms, values and rules, which help to shape human behavior and are primary products of any human society, solve conflicts between individual and collective social benefits and costs. Public policy can enable ecologically and socially to help faming to be profitable by providing non market economic incentives for individual actions to conform to societal interests. Public policy generally relies on a variety of subsidies, penalties, rules, and regulations that reward those actions which contribute to societal good and punish those that cause societal harm. But, these provisions, are, more on paper in many instances in the developing world, especially in countries like India, Pakistan and Bangladesh. For instance, the land mafia in India is out to grab the country's fertile fields for constructing expensive dwellings, which, at the time of writing this book, has become a national shame threatening the very continuance of the government in power, which has been found to be in a hand-in-glove understanding with the unscrupulous land mafia. Perhaps, examples could be multiplied in other parts of the world. Making ecologically sound faming profitable, through appropriate policy support, will benefit agricultural sustainability, but then society will have to bear the cost of such policies, through their consequences in terms of the availability and prices of food or governmental taxing and funding. By one means or another, farming systems will have to be made ecologically sound and economically viable, if they are to be sustainable. Both are necessary and neither is sufficient, singly.

Social acceptability, equity, and justice are the bedrock of sustainable agriculture, which, in no case can be compromised. Agriculture systems need to benefit all humans, at least to the extent of providing their basic need of food, and other components of human development, as explained in the section on "Right to Food" earlier. Unfortunately, current food systems are quite harsh, devoid of empathy for the under privileged or voiceless of the society, especially in countries like India, Pakistan and Bangladesh, where despite the availability of enough food, accessibility is a big casuality, depriving the poor of a healthy meal. Societal indifference to a large extent makes hunger an endemic situation in these countries. This is a totally unhealthy and undesirable situation. It is foolhardy and unethical to expect food systems to last out in the absence of social justice, even if they are ecologically sound and economically viable for a few. Agriculture is the creation of human society and it can be also destroyed by human society, when thoughtless actions, without a long range perspective, take the upper hand.

The absence of economic security and social justice is not sustainable over long periods of time and will lead to social conflicts, which may lead to irreparable economic and ecological damage. Failed social systems aggravate events such as famines and social conflicts and can even cause such natural calamities as droughts and floods by their assault on natural resources, including forests, mangroves, and rangeland. Agriculture that fails to sustain a society will not be sustained by that society. Hence, socially responsible agriculture that meets basic human needs equitably and provides social equality is critical to the long-term sustainability of agriculture. It is up to society at the global, national, regional, community, family, and individual levels to find answers to such problems. In this, those who are producers and those who are consumers have equal stakes. Science will not provide definitive answers, though ecological and economic soundness will help steer us toward social justice.

Agricultural systems need to be constantly monitored if they are to overcome the various types of stresses they develop, including social impacts of farmers and rural communities, economic strains on consumers, and pressures on the environment. Destroying any one of the critical elements of the agrosystem will damage the other elements as well as essentially destroy the whole system. The risks arising from the lack of sustainability at one place can be counteracted elsewhere in the global agriculture system temporarily and to a limited extent. If the stress caused by one subsystem exceeds a certain threshold, however, the danger arises of subverting the whole system.

12.6.1 *Hierarchy of Agricultural Systems and Trade–Offs*

Sustainability is a system issue. Systems are defined in terms of boundaries, thereby establishing the distinction between internal components and external influences and defining, in particular, inputs and outputs which cross the boundaries. Agriculture systems are generally defined and assessed in terms of a hierarchy ranging from soil systems to cropping systems, to farming systems, to regional, national and global (supranational) systems (Izac 1994). It is graphically represented below (Fig. 12.2).

Agricultural systems can also be aggregated ecologically as production systems (fields), watersheds, and finally river basins. Dynamic interactions take place between the components of a system as well as between different scales or levels of the hierarchy. Many of the determinants of sustainability have an impact on all

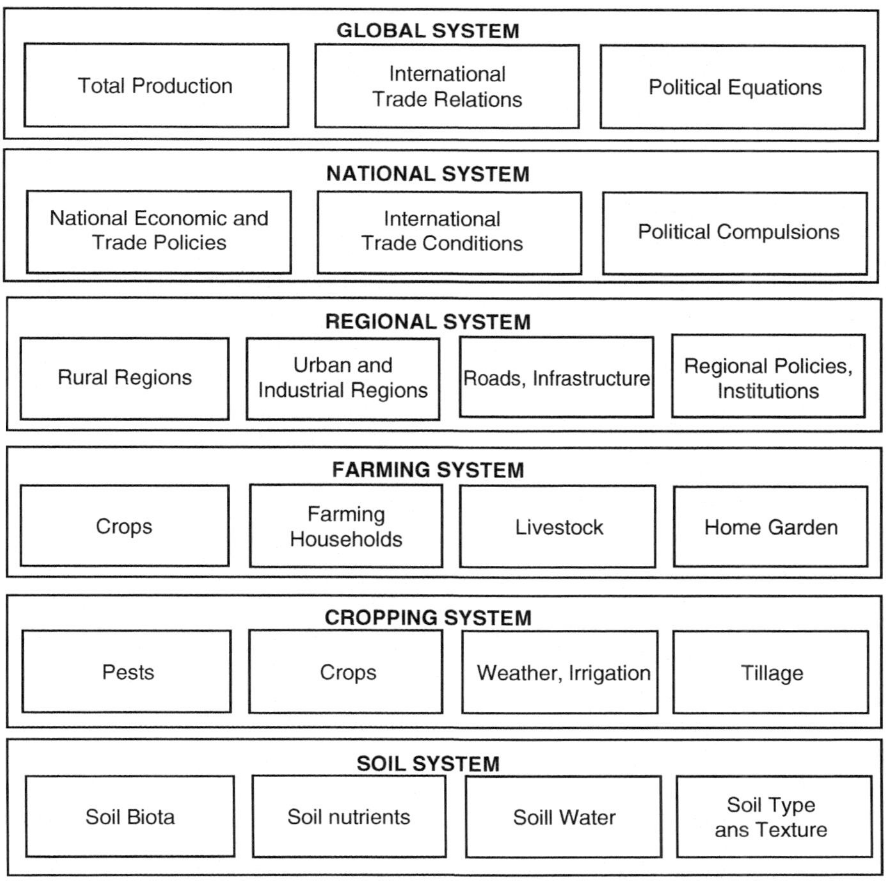

Fig. 12.2 Hierarchy of agricultural systems. (Source: Adapted from Izac 1994)

scales, but, their relative importance and the form in which they are expressed differ (Scholes et al. 1994).

The ecological aspects of productivity and sustainability are best addressed at the cropping system or farming system level. Above these levels, fewer new biological, physical, and chemical influences intervene. The hierarchical structure shown is increasingly dominated by economic aggregation from level to level. Social, political, and economic influences become increasingly strong at higher levels. In addition, because so many external factors influence the sustainability of agricultural systems at the regional level, determining the sources of these impacts becomes very complex (Norman et al. 1997).

12.6.2 Hierarchical Linkages and Sustainability

The problems confronting the sustainability of agricultural systems cannot be addressed at one hierarchical level in isolation – local, regional, national, or international. The different levels are linked and impact each other. Local production systems are linked through the market to the regional and national levels and to the outside world through various factors, including international trade, the balance of payments, political power equations, ideology and strategy. Thus, shifts in world prices for agricultural commodities or national agricultural policies can affect profoundly the livelihoods of farming households. Similarly, changes in global climate, major technological changes, wars, droughts and floods, and pest and disease epidemics can have severe impacts on local production. For example, genetically modified organisms (plants), which enjoy intellectual property protection (Intellectual Property Right, IPR or what is called "Trait Value"), may destabilize local production systems in developing countries, as has happened in the cotton belt of India in Vidharbha district of Maharashtra State, where due to the introduction of Bt cotton, not only did native cotton varieties disappear, but, thousands of farmers committed suicide owing to the fact that they were unable to economically sustain a "high input" technology, to sustain the Bt cotton, and chose suicide rather than becoming bankrupt and victims to the Shylockian grip of unscrupulous money lenders, when the crop failed. Similarly, the opening up of agricultural markets in developing countries has put the livelihoods of small and marginal farmers to great risks. Although the question is not pondered seriously, what happens to food security at local or national levels, if most farmers decide to quit farming, in search of more secure opportunities elsewhere? The rural-urban migration in India is a classic example of this malady. A recent national survey in India (National Sample Survey, NSS) showed that, given a better alternative, as many as 40% of farmers would quit agriculture.

The linkages between hierarchical levels are not simple. The behavior of higher levels of the hierarchy cannot be reduced to the sum of behaviors at the lower levels, nor are the latter simple disaggregates of the former. Agricultural sustainability will depend on the sustainability of each level in the agricultural hierarchy, both in its own right and in relation to other levels above and below. Agricultural policy and planning have to take into account the impact at each level separately, along with its ramifications across hierarchies. The uniqueness of each production system in the agricultural hierarchy and hierarchical linkages between the different levels make it necessary to tackle the goal of sustainable agriculture in a holistic manner at all levels – farm, regional, national, and international (Conway and Barbier 1990).

12.6.3 Temporal Dimension of Agricultural Sustainability

Sustainability implies a number of component steady states with a dynamic equilibrium among them. When changing conditions or demands require them, corresponding readjustments occur in the system to reorient it toward a new steady state. Sustainability is relevant for different systems over different time scales. Although agriculture is to be sustained indefinitely, mainly for the benefit of humankind, the time frame for sustainable agriculture should be more realistic and in conformity with the changing conditions and readjustments. No system is infinitely sustainable. Natural ecosystems change and evolve through a highly variable set of quasi-stable conditions over periods of time that are generally longer than the human life span. Agricultural production systems are not static; they undergo changes over time, in tandem with changing demand scenarios such as population growth, demographic transitions, evolving tastes, aspirations, economic strength, technological development, awareness, and attitudes. It is generally the case that important trends affecting the sustainability of production systems become apparent over a span of two to four decades. Hence, it is relevant to review the sustainability of agricultural systems every two to three decades. For agricultural production systems, a life span of about one generation, which is approximately 20–25 years, is realistic (Herdt and Steiner 1995).

12.6.4 Contextual Nature of Agricultural Sustainability

Sustainable agriculture has evolved as a complex and diffuse concept because of its multidimensional character. There have been hundreds of definitions of sustainable agriculture, and more are still being articulated to accommodate individual perceptions. This has resulted in a decline in clarity and rendered it a biased and value-laden concept. Inasmuch as sustainable agriculture has been recognized as the entry point for sustainable development, for the goal of future development, it becomes essential to set out the path approaching it. This is possible only with an unambiguous expression of its intent and the terms and conditions for attaining it. In addition to the complexities of providing a definition, the fact that the types of agricultural systems prevailing in different parts of the world are themselves diverse and depend upon local ecological and socioeconomic factors have contributed to the prevalent confusion in thinking over the topic. The contextual nature of agricultural systems needs to be clearly understood before proceeding to analyze the conditions and means to make these systems sustainable (Saroja Raman 2006).

12.6.5 Diversity of Agricultural Systems

Crop production systems have undergone dramatic changes since the beginnings of settled agriculture, some 10,000 years ago in the plains of Mesopotamia in Central Asia. What started as foraging was transformed into a variety of production systems at different places, and at different times. Today, the world knows a wide spectrum of agricultural production systems, such as, hydroponics, green house cultivation (especially for vegetables and flowers), high external input industrial type of agriculture, euphemistically labeled as "green revolution," traditional low input agriculture, passing as "sustainable agriculture," "zero input farming," "shifting cultivation" (called "***Jhum***" cultivation in Assam State of India), and natural ecosystems. The most important and fundamental difference among all these is the degree of human control/intervention that goes with them. For instance, whereas hydroponics is completely controlled by humans, natural ecosystems are completely under the control of nature. Human intervention decreases progressively in the above sequence.

12.7 Determinants of Agriculture

Both supply and demand factors shape the type of agricultural systems and the practices which are followed. Supply factors include climate, availability and quality of resources, market opportunities and infrastructure, enabling policies and political environment, availability of credit, technology, and other services, and opportunities for specialization. Demand factors comprise such determinants as population density, ecological constraints, economic considerations, social and cultural expectations, and most important of all, ethics. Together, these factors determine the kind of production systems that can be used and the types of crops which can be grown in an area at a given time.

Four groups of determinants – physical, socioeconomic, and cultural – are considered important in the structuring and management of agricultural systems. Physical determinants include climate (temperature, rainfall and radiation), land availability, and soil quality parameters, such as physical, chemical, biological, erosion, water holding capacity, depth and slope. Biological determinants include energy and biotic factors such as insect pests and predators, crop diseases, weed species, functional diversity of the soil's biotic community, photosynthetic efficiency of crop plant, crop rotations, and cropping patterns. The importance of socioeconomic and cultural determinants, such as population density, economic and market opportunities (capital, credit, prices of inputs and outputs), policy issues, political relationships, availability, accessibility and affordability of technology, inputs and labor, work culture, traditional knowledge systems, socio cultural beliefs and ethics have been discussed in the earlier sections.

Agroecosystems have coevolved with the changing technological, biophysical, socioeconomic, and cultural determinants, all of which, independently and interactively, influence their evolution. Conditions outside the area of production, including industrial development, economic conditions, opportunities for employment in non agricultural sectors, and capital availability, also affect the way production systems are designed and operated upon. Just as different parts of the world have different combinations of ecological and socioeconomic factors, so do their agricultural profiles.

12.7.1 Population Density and Production Processes

Population density and demographic changes influence the regional and global nature of production systems profoundly. A small and scattered population with scarce capital has to be self-sufficient because markets to cater to its needs or to sell its products are not easily accessible. Hence, specialized products will have no value and production has to be geared toward products for local use and toward locally available resources. Labor-intensive practices will not be popular where population is scarce and land is plenty. Such conditions prevailed in the early agricultural period and prevail even today in some parts of the African continent, where shifting cultivation continues to be practiced. Land and labor intensification are not meaningful under these conditions, and, shifting cultivation with long fallow periods and free-ranging domestic livestock are more practical.

As per capita land decreases, because of population increase, labor intensification becomes more meaningful if capital is scarce, because, older methods fail to produce enough good results. If markets are attractive, capital can be generated from other sources and production improved through capital-intensive measures such as the use of external inputs. Beyond this, the future of agriculture is intimately linked to the path of the non agricultural sector. Where there is significant economic growth, specialization occurs and opportunities in industry and services become abundant along with an increase in urbanization. Under these conditions, rural labor becomes scarce, capital more plentiful, and land relatively cheap, because it is in less demand as a source of livelihood. Farms grow larger due to consolidation, and labor saving through mechanization of farm operations and liberal use of synthetic inputs becomes inevitable. This is the current scenario in industrially developed countries such as the USA and the nations of Europe and in some parts of Asia and Latin America.

If economic opportunities do not develop outside the farm sector, and the rural population remains large and dense, land increases in value because per capita land decreases. This is the unfortunate scenario that prevails in most of the developing countries of Asia and Latin America, where a duality has emerged in agriculture; a small number of large-scale, highly capitalized operations owned by the rich exist side by side with a multitude of small, fragmented holdings owned by the poor. Returns to labor become very non competitive and inadequate under these circumstances and poverty becomes endemic.

References

Cassman, K. G., & Harwood, R. R. (1995). The nature of agricultural systems: Food security and environmental balance. *Food Policy, 20*(5), 439–454.

Cassman, K. G., Olk, D. C., & Dobermann, A. (1997). Scientific evidence of yield and productivity decline in irrigated rice systems of tropical Asia. *International Rice Commission Newsletter, 46*, 7–16.

CGIAR/TAC. (1988). *Sustainable agricultural production: Implications for international agricultural research*. Washington, DC: Consultative Group for International Agricultural Research.

Conway, G. R., & Barbier, E. B. (1990). *After the Green Revolution: Sustainable agriculture for development*. London: Earthscan.

Douglas, G. K. (1984). *Agricultural sustainability in a changing world order*. Boulder: Westview Press.

Evans, L. T. (1998). *Feeding the Ten Billion: Plants and population growth*. Cambridge: Cambridge University Press.

Harwood, R. R. (1990). A history of sustainable agriculture. In C. A. Edwards, R. Lal, P. Madden, R. H. Miller, & G. House (Eds.), *Sustainable Agricultural Systems* (pp. 3–19). Ankey: Soil and Water Conservation Society.

Herdt, R. W., & Steiner, R. A. (1995). Agricultural sustainability: Concepts and conundrums. In V. Barnett, R. Payne, & R. Steiner (Eds.), *Agricultural sustainability: Economic, environmental and statistical considerations* (pp. 1–17). New York: Wiley.

Izac, A. M. N. (1994). Ecological economic assessment of soil management practices in tropical countries. In D. J. Greenkand & I. Szabolcs (Eds.), *Soil resilience and sustainable land use* (pp. 77–96). Wallingford: CAB International.

King, F. H. (1911). *Farmers of forty centuries*. Emmaus: P.A. Rodale Press.

Ladha, J. K., Pathak, H., Padre, A. T., Dave, D., & Gupta, R. K. (2003). Productivity trends in intensive rice-wheat cropping systems in Asia. In J. K. Ladha, J. Hill, R. K. Gupta, J. Duxbury, & R. J. Buresh (Eds.), *Improving the productivity and sustainability of the Rice-Wheat systems: Issues and impact* (ASA Special Publication 65) (pp. 45–76). Madison: American Society of Agronomy.

Malthus, T. R. (1798). *An essay on the principle of population, as it effects the future improvement of society*. London: J.J. Johnson.

Miller, F. (1998). Soil-land degradation: Interventions for sustaining civilization. In R. Lal (Ed.), *Soil quality and agricultural sustainability* (pp. 13–32). Chelsea: Ann. Arbor Press.

Nair, K. P. P. (2014, August 18). Green revolution or silent spring? *The Hindu Business Line*.

Nair, K. P. P. (2015a, February 13). The debt story less told. *The Hindu Business Line*.

Nair, K. P. P. (2015b, April 7). When there is no soil left in this world. *The Hindu Business Line*.

Nair, K.P.P. (2015c). The nutrient buffer power concept – A revolutionary soil management technique for sustainable agriculture. In *Keynote address at the National Seminar on "issues, challenges and strategies in sustaining soil health"*. Kerala Forest Research Institute, Peechi, Kerala State, India, December 10–11, 2015.

Norman, D., Janka, S., Freyenberger, B., Schurle, B., & Kok, H. (1997). *Defining and implementing sustainable agriculture* (Kansas Sustainable Agriculture Series Paper. I). Manhattan: Kansas State University.

NRC. (1998). *Alternative Agriculture*. Washington, DC: National Academy Press.

NRC. (1999). *Our common journey: A transition towards sustainability*. Washington, DC: National Academy Press.

Pinstrup, A. P. (2002). Towards a sustainable global food system: What will it take? *Keynote presentation for the Annual John Pesek colloquium in sustainable agriculture, March 26–27*, Iowa State University

Raman, S. (2006). *Agricultural sustainability: Principles, processes and prospects* (pp. 13904–11580). Birmingham: The Hawoth Press Inc.

Rodale, R. (1983). Breaking new ground: The search for a sustainable agriculture. *Futurist, 1*, 15–20.

Scholes, M. C., Swift, M. J., Heal, O. W., Sanchez, P. A., Ingram, J. S. I., & Dalal, R. (1994). Soil fertility research in response to the demand for sustainability. In P. L. Woomer & M. J. Swift (Eds.), *The biological management of tropical soil fertility* (pp. 1–14). Chichester: John Wiley.

Vandermeer, J. M. (1992). *Thoughts on Agriculture and the Environment in a Post-Modern World.* Symposium on Enhancing the Future of the Land Grant System. Irving C.A: National Academy of Sciences Board on Agriculture.

Chapter 13
General Profile of Current Agricultural Systems

Abstract The profound differences in the way in which agricultural systems and practices are shaped by biophysical and socioeconomic factors can be appreciated by looking at three typical scenarios: (1): the highly mechanized and external-input-intensive industrial model of the USA (also countries like Canada) and Western Europe (2): the green revolution model of the fairly developed countries of South Asia, for instance India and China, also Argentina and Brazil of South America (Latin America) and (3) the low-input resource extractive agriculture of the Sub Saharan Africa , like Ethiopia, Tanzania in Eastern Africa and even countries of Central Africa like The Republic of Cameroon, Senegal, Gabon etc. The chapter discusses these aspects I detail.

Keywords Industrialized countries · Fairly developed countries · Least developed countries · Sustainable agriculture · "Blanket recommendations" · Extension specialists

The profound differences in the way in which agricultural systems and practices are shaped by biophysical and socioeconomic factors can be appreciated by looking at three typical scenarios: (1) the highly mechanized and external-input-intensive industrial model of the USA and Western Europe (2) the Green revolution model of the fairly developed countries of Asia and Latin America like India, China, Argentina, Brazil and (3) the low-input resource-extractive agriculture of the countries of sub-Saharan Africa.

13.1 Industrialized Countries

The croplands of the USA and Northern, Central and Western Europe, including Scandinavia, fall mostly in the temperate region and are generally endowed with favorable ecological factors: cooler climate, evenly distributed rainfall, fertile soils, with high organic matter, and high potential productivity. The climate is generally suited to annual crops, but, the native pool of plant biodiversity is limited and most species have been imported from the tropics. These countries possess advanced

K. P. Nair, *Intelligent Soil Management for Sustainable Agriculture*,
https://doi.org/10.1007/978-3-030-15530-8_13

technologies and economic resources and also stable socioeconomic policies. Their agricultural production, productivity, and policies are resilient to environmental and economic changes. Overall, these countries are free from demographic pressures, with a fairly stable population, favorable per capita natural resources, and excellent opportunities for employment in the non agricultural sector. Technologies and policies are geared toward export-oriented commercial agriculture and low-cost food for domestic consumption. Both Europe and USA produce far in excess of what they consume, in terms of their internal requirements, through intensive high-external-input agriculture. There are, however, subtle differences between them. The European countries are short on land and use almost double the fertilizer per hectare compared with the USA, to boost land productivity. The USA concentrates on labor productivity; farms are generally large, at an average of 64 ha per person, and agriculture employs only 2% of the total of the country's population compared to 6–10% in Europe. Both Europe and the USA offer large financial incentives to improve farm productivity and economic viability of farming and to make living standards of farmers comparable to those of the workers in the secondary (manufacturing) and tertiary (services) sectors. Agriculture contributes an insignificant 3–10% to the Gross Domestic Product (GDP) in these countries, and hence, it is possible to maintain high level of support to farmers through direct or indirect subsidies.

13.2 Fairly Developed Countries

The agricultural profile of the Asian and Latin American countries which have adopted the so-called green revolution technology in the past few decades is quite different because the ecological and socioeconomic features of these countries are less favorable. They are situated in lower latitudes with generally hot and sub humid or arid climate. Soils are low in organic matter, which have now been further reduced due to unbridled use of chemical fertilizers, and, rainfall is unevenly distributed. However, these countries enjoy plenty of sunshine and possess a vast pool of native plant bio diversity, capable of adapting to different ecologies to provide stable crop yields and a varied basket for food security. On the other hand, pest and disease problems are more serious, and pressure on land is high because of population density. This is particularly so in a tiny State like Kerala, in southern India, where the hunger for land is almost insatiable.

The example of India, which is the second most populous country next to China, in the world, is a picture in contrasts. Although the natural endowments of India are not overly generous, and, the pressure on them from the country's huge population (at more than 1.2 billion at the time of writing this book) is immense, India is predominantly an agrarian economy, with more than 65% of its population connected to agriculture in one way or another, with agriculture contributing 25–30% of the country's GDP. India more than tripled its grain production, through a combination of the "imported" dwarf wheat and rice varieties (the first one from CIMMYT, Mexico and the second one from IRRI, Philippines), unbridled use of chemical fertilizers and pesticides and irrigation water, during a relatively short span from the late sixties to mid seventies. By the second half of the seventies, this industrial "high input technology" of farming took a terrible toll on the environment, in the form of degraded soils, dried aquifers and vanished bio diversity (principally due to the rice-wheat monoculture in States like Punjab, known as the "cradle" of India's green revolution). The so-called green revolution bypassed the vast majority of India's farming community (almost 60% of the total engaged in agriculture in the country), whose lands depend on rainfall, the infrastructure being next to nothing, with poor roads, inaccessibility to markets, and very low capital base (both land and monetary resources) of the farmers. In short, the farmers in these complex, diverse, and risk-prone (CDR) areas were left in the lurch. The moneyed rich farmers of Punjab, and the neighboring State of Haryana, and also Western Uttar Pradesh, became the ***nouveau riche,*** while those in the CDR regions went bankrupt and took to mass scale suicide, unheard of in any other parts of the world.

13.3 Least Developed Countries

The most unedifying scenario in terms of agricultural production is found in Sub Saharan Africa, which has multiple constraints on ecological, economic and socio-economic fronts. The countries of the region have harsh, fragile, tropical, humid/arid ecologies with very low soil fertility and uneven terrain, coupled with very adverse economic indicators such as high population growth, weak infrastructure and marketing facilities, and inefficient and unstable political regimes. Additionally, risky, unsuitable, and unaffordable technologies and reforms have often been imposed on poor farmers by a proselytizing West that does not appreciate local constraints, forgetting that the mores and traditions of these poor people have a hoary past. These abrupt externally imposed changes conforming to Western norms have had a greatly destabilizing effect on the local populace. This author observed such explosive changes in The Republic of Cameroon, where he spent a decade teaching and researching at The University Center in the central part of the tiny nation in a place called Dschang, dominated by an earlier French influence, which was being attempted to be replaced by American influence. These technologies have a very unimpressive track record in agricultural production and technological realization, despite considerable efforts and investments by international donor agencies, such as, the United States Aid For International Development (USAID). Widespread degradation of land has occurred as a result of negligible or very low fertilizer use and resource-extracting agriculture. This is the only region in the world where agricultural productivity has not kept pace with the population growth since the 1980s. Poverty and hunger have remained chronic and have become the root cause of the very low human development status of these unfortunate countries, leading to untold misery.

13.4 Impacts of the Changes in the Determinants of Agriculture

The determinants of agriculture are not spatially or temporally static. Changes can occur over comparatively short or long periods. For example, climate, topography, and population density change slowly, whereas soil quality, erosion, chemical runoff, and pest incidence can alter dramatically over a cropping season. Socioeconomic determinants can be even more volatile than biophysical ones. The cost and availability of technology and inputs are not only dependent on the regional or national context, but also, subject to international trends. Similarly, government policies on commodity prices depend on political perceptions, regional and national demands, and international trade relations. Changes in these determinants change the type of agricultural system. The technological, socioeconomic, and political group of determinants is generally the primary engine of change, although biophysical changes such as resource degradation and environmental pollution can also trigger changes in production systems.

Sustainable agriculture seeks permanence to agricultural production systems to enable food security for all in an ecologically sound, economically viable, and socially responsible manner. In the current globalized and liberalized trade environment, changes that are favorable or adverse can occur over short periods to affect local and international prices and demand for agricultural commodities. Access to such new technologies, as genetically modified seeds, can affect the profits and acceptability of agricultural systems. A market can emerge locally or internationally for a new exotic or fashionable commodity or for a preferred process, as for instance organic farming. An adverse change for production in one country can create a boom market for another. Changes in policy regarding subsidies of various types can overnight create a favorable or unfavorable effect within and outside the country of origin. The liberalized conditions of trade for agricultural products under General Agreement on Tariffs and Trade (GATT) have created several such problems for small farmers in the developing countries.

Agricultural systems have to adjust to these changes to be sustainable. The kinds of adjustments required depend on the context, and a generic description for sustainability along the lines of "one size fits all" cannot be recommended. It is in this context that the crucial importance of **"The Nutrient Buffer Power Concept"** emerges. The text book knowledge, for instance, that the DTPA extraction for "bio available" soil Zn is universally pertinent/applicable has been proved to be more than false in the context of the Kerala soils, as substantiated by the experimental results discussed in the earlier sections. If one were to proceed further with the approach of "one size fits all," as in the case of Zn fertilizer application discussed in earlier sections, it can lead to great pecuniary loss to the farmer. Thus a farmer might be "advised" to apply Zn fertilizer in a quantity more than what is truly required. Alternatively, a farmer who would stand to gain by applying more of Zn fertilizer might be ill advised by a wrong recommendation, not to apply enough. An analogy of this "one size fits all" approach is the "blanket" recommendations extension specialists make, based on routine soil test analysis, in classical soil-test crop response relationships. Diverse pathways are needed to make agriculture sustainable, depending on the context, and this is what makes sustainable agriculture a continually learning process, requiring constant monitoring and online changes to match changing circumstances. "The Nutrient Buffer Power Concept" aims to fulfill such an objective.

Chapter 14
Sustainability Conundrums

Abstract Sustainable agriculture requires the simultaneous maintenance of several complex dimensions: productivity, ecological integrity, economic viability, and social relevance. Each of these dimensions by itself, is complex and highly contextual. The chapter addresses these aspects.

Keywords Sustainable agriculture · "**The Nutrient Buffer Power Concept**" · Adaptive process · Long term productivity · Resource conservation · Resources · Natural capital · Physical capital · Social capital

Sustainable agriculture requires the simultaneous maintenance of several complex dimensions; productivity, ecological integrity, economic viability, and social relevance (Saroja Raman 2006). Each of these dimensions, by itself, is complex and highly contextual, both spatially and temporally. In addition, interconnections and interdependencies of various types and degrees exist among them that are also dynamic. More importantly, both synergistic and antagonistic relationships may be expressed simultaneously among core tenets. On account of this, sustainability has become a highly complex and contested concept. A number of definitions have emphasized different values, priorities, and goals and definitions of sustainable agriculture. According to Andrew Campbell, sustainability, like beauty, lies in the eyes of the beholder, and hence, it is inevitable that assessments of relative sustainability are socially constructed, which is why there are so many definitions.

The contextual relevance of the above statement becomes obvious when we critically look at "**The Nutrient Buffer Power Concept.**" For instance, given the external constraints, a relevant fertilizer recommendation for a farmer in India for a specific crop, using "The Nutrient Buffer Power Concept," for a specific nutrient like Zn, might turn out to be totally irrelevant to another in a European context. This is because of several variable factors, like the price of the Zn fertilizer, the soil conditions, climate, so on and so forth. However, the core hypothesis does not change. The specific fertilizer recommendation for the specific plant nutrient (in this case Zn) will still be linked to the nutrient's (in this case Zn) "buffer power." This "buffer power" is the core parameter in the new concept, and, indirectly reflects the mean concentration of the nutrient (in this case Zn) on the root surface of the plant in

K. P. Nair, *Intelligent Soil Management for Sustainable Agriculture*,
https://doi.org/10.1007/978-3-030-15530-8_14

question. This is not static for a growing root. It is so very dynamic, and is related to the "quantity" of the nutrient (in this case Zn) in the soil matrix and the rate of release of the nutrient (in this case Zn) from the soil matrix into the soil solution, which is related to the "intensity" of the nutrient (in this case Zn).

Much confusion abounds in the scientific literature about approaches to sustainable agriculture. The nebulous and value-laden approach to agricultural sustainability can be attributed mainly to (1) not relating sustainability to specific types of agricultural systems and practices in the context of the ecological and socioeconomic determinants and (2) overlooking the fact that the determinants of agricultural systems are not static and will undergo changes over time and space.

The confusion also arises because sustainability is a multidisciplinary concept and carries the baggage of the individual biases of different disciplines. Although sustainability is a holistic expression of ecological, economic and social components, each group of specialists often highlights its own bias. Ecologists, soil scientists, agronomists, economists, social scientists and Green activists have, more often than not, peddled their own narrow and truncated versions as "holistic and sustainable," leading to much confusion when viewed from different viewpoints. The concept of sustainable agriculture has also been distorted and is often made more confusing and less exact by practitioners of various forms of agriculture vying with one another to stake their claims of sustainability. Thus, organic agriculture, regenerative agriculture, alternate agriculture, low-input traditional agriculture, bio dynamic agriculture, and ecological agriculture, have all been promoted under the banner of sustainable agriculture. It should be made clear that sustainability is contingent upon many factors; productivity, ecological, economic, and social viability and acceptability; and the enabling factors of technology, policy, and service appropriate to local needs. Sustainability is not inherent in any particular set of practices or farming system, except that some forms of agriculture have more enabling factors for ecological sustainability than others. Biased interpretation of sustainability and failure to consider all its elements ***in toto*** and in their correct perspective have done great disservice to the understanding and operationalization of the principles and processes of sustainable agriculture.

14.1 Sustainable Agriculture: An Evolving and Adaptive Process

It is important to note that sustainable agriculture is not about finding the right technology or protocol for the production process, but, about successfully adopting it to changing conditions. The essential benchmark or crucial test of agricultural sustainability is that when any of the determinant changes positively or negatively, agricultural practices, farmers, and communities should be able to adapt to the new circumstances and conditions. Typical examples of positive changes are advances in science and technology, whereas global warming, water shortages, fossil fuel depletion, and erosion of genetic base, both of plant and animal origin, are negative developments.

Harwood (1990) suggested a framework definition for sustainable agriculture that could be filled in with appropriate details according to the prevalent conditions in any country and the desired time frame. According to the author "Sustainable Agriculture is agriculture that can evolve indefinitely toward greater human utility with greater efficiency of resource use and in balance with the environment that is favorable to both humans and most other species." The general framework is heavily value ridden. Translating this concept into functional reality depends on the context, the enabling factors, and the policies and political will of the country in question.

The following definition provided by FAO (1991) captures the multidimensional elements of sustainable development, especially with regard to the maintenance and management of the resource base. It's definition is as follows:

"Sustainable development is the management and conservation of the natural resource base and the orientation of technological and institutional changes, in such a manner so as to ensure attainment and continued satisfaction of human needs for present and future generations. That such sustainable development (in the agricultural, forestry and fishery sectors) conserves land, water, plant and many other genetic resources, is environmentally non degrading, technically appropriate, economically viable and socially acceptable."

The above definition clearly rules out a rigid protocol for approaching sustainable agriculture. Explicit in this definition is the need for continuously evolving adaptive technologies and institutional support for conservation and sustainable use of natural resources within the framework required for adequate productivity, ecological security and environmental safety, economic viability, and social responsibility. Hence, sustainable agriculture does not pertain to a permanently defined set of technologies, practices, or policies, as is often claimed. It is a dynamic and process-oriented approach governed by the spatio-temporal and socio-economic context.

14.2 United Principles, but Different Approaches

The differences in approach to sustainability of different systems can be seen in the way ecological sustainability is enforced in different forms of cultivation. Conservation of the natural resources of land, water, air, and bio diversity is the ***mantra*** for ecological stability in all systems. Notwithstanding the fact that problems of agricultural sustainability are contextual and deserve differential consideration, agroecology has defined a set of basic criteria that are essential in all kinds of situations in the pursuit of ecological soundness. The following points aptly summarize this:

Strengthening the immune systems of agricultural operations: stabilizing pests and nurturing natural pest control

Decreasing toxicity in the system and environment through optimum use of external synthetic chemicals

Optimizing metabolic functioning such as nutrient cycling and organic matter-recycling
Balancing regulatory systems –nutrient cycles, water balance, energy flows and population regulation
Enhancing conservation and regeneration of soil and water resources and bio diversity and
Increasing and sustaining long-term productivity (Altieri 2002).

14.3 Conceptual Framework for Sustainability of Agricultural Production Systems

Biophysical endowments and socioeconomic factors are the determinants of sustainable production on a farm. In this chapter, a general framework for agricultural sustainability for farms operating within four combinations of favorable and unfavorable biophysical and socioeconomic endowments, based on a model developed for the sustainability of business enterprises operating under similar circumstances (Ketola 1997) is discussed. This is followed by an illustration of specific factors for validating the sustainability of a small farm.

14.3.1 General Framework for Sustainable Agriculture

A framework for conceptualizing the characteristics and the role of the farm is given in Fig. 14.1 (below).

This framework can be used to analyze the sustainability of the farm in relation to its socioeconomic and ecological environments as illustrated in the following figure (Fig. 14.2).

As shown in the below figure, a farm has its own natural resources and other human, economic, technological and knowledge resources. The biophysical attributes of the farm are controlled by the prevailing geographic, physical, chemical

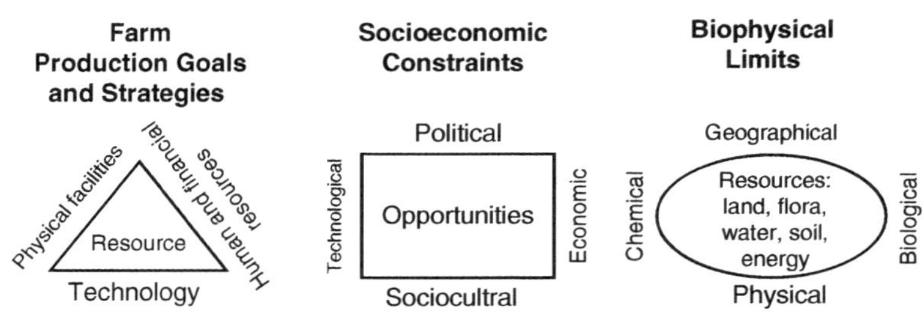

Fig. 14.1 Components for analyzing the sustainability of a production system. (Source: Conceptualized from Ketola (1997). After Saroja Raman (2006))

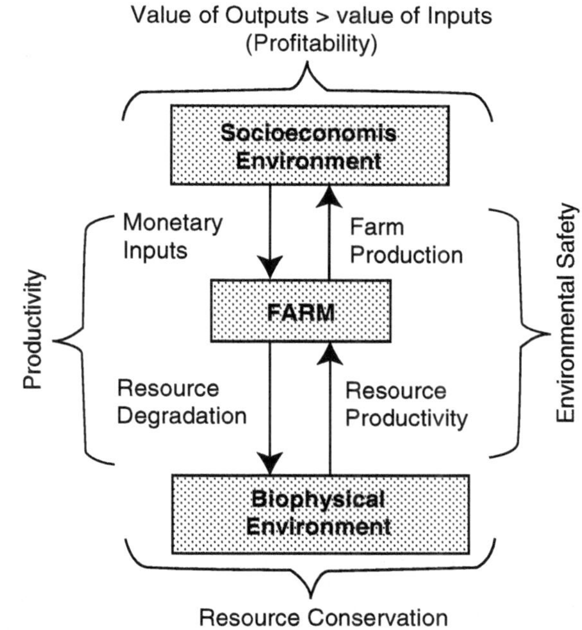

Fig. 14.2 Framework for a sustainable production system. (Source: Adapted from Hart and Sands 1990)

and biological attributes which determine the climate, topography, native biota, soil type, water availability, and so on. These determine such factors as soil quality and productivity characteristics, the plant and insect gene pool, surface and ground water, soil moisture storage, and solar energy. These ecological factors also set certain limits to farm operation and productivity. Exceeding these limits can result in environmental stresses and shocks and affect the sustainability of the farm.

The human imposed socioeconomic environment within which the farm operates has also a number of control factors, broadly categorized as "PEST" factors:

Political (agricultural and non agricultural policies and priorities) of the government: imports, exports, free market etc.
Economic (price of inputs and outputs, ex free market etc).
Sociological (regional preferences, population, labor constraints, skill and awareness of the cultivator, value system etc).
Technological (level of available technology and support services such as timely availability of inputs, infrastructure and markets).

The farm interacts with this system through the purchase of inputs and sale of outputs and is facilitated or constrained according to the PESTenvironment. The socioeconomic environment also sets limits to the sustainability of the farm's production.

Describing the prevailing biophysical and socioeconomic environments as favorable and unfavorable, four combinations are possible, according to the four scenarios described below

Scenario 1: The socioeconomic and biophysical environments are both favorable for production

This is the ideal situation for carrying out sustainable agricultural production. Such win-win situations with respect to ecological and socioeconomic determinants exist for farmers in the fertile, temperate regions of countries, such as the USA and the nations of Western Europe, where soils are fertile and resilient, contain high levels of organic matter, have good structure, and rainfall is generally adequate and well distributed. Easy availability of inputs, enabling services, infrastructure, markets, and favorable economic policies, such as, subsidies, minimum support prices for farm commodities and export incentives (all of which exist in India), provide a socioeconomic environment conducive to sustainable farming operations as in the figure depicted above.

Scenario 2: The socioeconomic and biophysical environments are both adverse

This is the worst nightmare for any form of productive agriculture, let alone, sustainable. Such desperate conditions exist in many areas of Sub-Saharan Africa, with hostile ecological conditions, including low-fertility soils and a harsh political climate with either very low or very high humidity.

Added to this are adverse socioeconomic conditions, such as, high population growth, poverty, adverse governmental policies, and political volatility. The result has been chronic low productivity, uneconomic returns, and resource degradation (as for instance soil degradation), resulting from mining of soil nutrients, loss of organic matter, severe soil and water erosion, leading to loss of top soil. Consequently, the rate of productivity growth in these areas has been slowing continuously and food production has not been able to keep pace with population growth.

Scenario 3: Biophysical factors are supportive but the socioeconomic environment is not conducive to sustainable production

Typical examples of this scenario can be found in many parts of South Asia, where traditional farming in fertile areas have good soil and water endowment and well-adapted crop varieties, but are not economically profitable. Small, non viable farm size, coupled with the lack of a positive socioeconomic environment to allow remunerative prices, leads to unprofitable farming. Nevertheless, these farms have remained ecologically viable for long periods as a result of farmers' stewardship of the land and commitment to future generations, despite generations of farmers having lived under extremely austere and difficult conditions.

Scenario 4: Ecological factors are adverse but the socioeconomic environment is supportive

Examples of such production systems can be found in fragile, ecologically stressed areas such as mountain slopes, forest-cleared lands, and shallow soils with low fertility and poor structure. These systems may be temporarily economically remunerative through the production of high-value cash crops for export or profitable products for a niche market. Farmers often exploit such opportunities for short-term gains without concern for ecological sustainability, especially if the lands are not under permanent ownership, but, are leased out or are common property resources.

14.4 Operationalization of Sustainable Agriculture

14.4.1 Resources and their Importance for Sustainable Agriculture

The perpetual availability of resources and their effective management are essential to the sustainability of any system or activity or process. This is the basic rule of thermodynamics. The natural ecosystems of our planet owe their sustainability, permanence, perpetuity, and durability to the sun, an infinite source of energy that supports the functioning of the myriad ecosystems and ecological communities. Behind this well-ordered machinery of nature are the many hierarchies, controls, and communication and feedback mechanisms built into the system.

Humans are the only species that can and has breached ecological rules and regulations and dictated its own agenda for multiplication, survival and development through unmatched ingenuity and innovation. Whereas all other organisms live within the natural resource endowment, humans have their own monetary, human-made, social, and human capital to augment or complement natural resources to enable them to pursue their self-determined development and welfare goals. A thorough audit of available resources and their efficient management lies at the core of any form of development, including agricultural development and sustainability. Sustainable agriculture has the basic responsibility of fulfilling the food requirements of the growing human population according to its emerging needs and of improving standards with due concern for ecological, economic, and social viability. The interdependence of these dimensions necessitates the explicit and implicit recognition and synergistic management of the different types of resources/capital available to sustain agriculture.

The stock of materials or information that exists at a given time is referred to as "capital stock." Capital stock can take distinct physical forms or intangible forms such as stored information in computers and human brains or genetic information stored in species and eco systems. Each form of capital stock can be used to generate goods and services, for human welfare, either autonomously or in conjunction with other capital stocks. Human use of this flow of services may or may not leave the original capital stock intact. The types of capital for sustainable agricultural development can be identified as:

A. Natural capital
B. Physical capital (monetary and human made capital)
C. Human capital
D. Social capital

14.4.1.1 Natural Capital

Humankind is totally dependent for its survival and well-being on the biosphere, which is a thin layer of soil, water, and air approximately 20 km thick surrounding the globe in which all life on the earth is concentrated. This segment of the earth provides all the physiological necessities of life, including oxygen, water, food, energy, and raw materials. The earth's resource base, consisting of land, water, the atmosphere, minerals, fuels, and the complete array of flora, fauna, and microorganisms, is referred to as "natural capital." It generates an indispensible array of goods and services.

14.4.2 Ecosystem Goods and Services

Ecosystem goods are items that can be given monetary values in the market place. These include food, timber, construction materials, medicinal plants, wild genes for domestic plants and animals, tourism and recreation. The magnitude and importance of natural products can be gauged from the fact that in 25% of countries in the world, crops, timber, and fish contribute more to the economy than industrial goods. Global agriculture alone produces US $ 1.3 trillion worth of food and fiber each year. In traditional and primitive societies, the products of natural ecosystems provide food and livelihood. Agriculture, forestry, and fisheries provide 50% of all the jobs in the world and 70% jobs in sub Saharan Africa, East Asia, and the Pacific (WRI 2000).

Several authors have described various forms of ecosystem functions and services essential for human welfare in detail. Daily (1997b) provided a detailed account of ecosystem services and their indicative values and methods for evaluation according to their benefits to humanity. She classified ecosystem services as follows:

- Purification of air and water
- Mitigation of floods and droughts
- Detoxification and decomposition of wastes and renewal of soil and soil fertility
- Pollination of crops and natural vegetation
- Control of the vast majority of agricultural pests
- Dispersal of seeds and translocation of nutrients
- Maintenance of the bio diversity that has provided humanity with agricultural, medicinal, and industrial inputs
- Protection from ultraviolet rays of the sun
- Partial stabilization of the climate
- Moderation of temperature extremes and the force of winds and waves
- Support of diverse human cultures
- Provision of beauty and intellectual stimulation that lift the human spirit (Daily 1997a)

14.4.3 Importance of Ecosystem Functions

Natural capital (other than minerals and fossil fuels) and ecosystem services are renewable, at least conditionally, which makes them perpetually available to all living beings, including humans. Maintaining the ecosystem cycles and functions is critical for sustaining the integrity of the planet. The repercussions of tampering with them can be enormous because humans lack the knowledge and ability to generate substitutes for ecosystem functions. Engineering to replace the services that ecosystems now provide, if not completely beyond current technology, will be prohibitively expensive even on a trivial scale.

Humans are already faced with the colossal problem of having altered the composition of the atmosphere through uncontrolled discharge of the greenhouse gases, carbon dioxide, methane, and nitrous oxide. This has triggered global warming and the consequent adverse climate change. Anthropogenic climate change is feared to be the biggest challenge to sustainable agriculture, the safety of human and animal health, and the quality of natural resources. Similarly, the discharge of chlorofluorocarbons from aerosols and refrigeration industries has depleted the ozone layer in the atmosphere, which acts as a shield against ultraviolet rays. The hole in the ozone layer over Antarctica presents grave danger to all living beings from exposure to unsafe doses of ultraviolet rays. Reversing this damage may be a difficult job, perhaps, even impossible.

14.4.4 Value of the World's Ecosystem Services and Natural Capital

The natural capital stock and the ecosystem goods and services that it generates are critical to the functioning of the earth's life support system because it is not feasible to substitute purely non natural capital for natural capital (Costanza and Daly 1992). To this extent, their value is infinite. They should be considered part of the total economic value of our planet, as they contribute directly and indirectly to our welfare. Until the science of ecology was fully evolved, ecological services remained unacknowledged and undervalued, even in terms of their functional importance, let alone their monetary value, because, unlike physical assets, such as, land, water and minerals, they could not be traded in the market. They were perceived to be unlimited, free, and without a price tag, because of their regenerative and non monetary nature.

Costanza, d'Arge et al. (1997) estimated the value of the world's natural capital (excluding non renewable fuel and minerals) and the goods and services it provides to be in the range of at least US $ 16–54 trillion per year, with an average of US $ 33 trillion per year. This is almost double the value of the global gross national product (GNP), which is approximately US $ 18 trillion per year. This shows the relative magnitude and importance of the natural capital stock and ecosystem goods and services. To replace the services of ecosystems at the current margin, global GNP would need to increase by at least US $ 33 trillion annually to cover both services already captured in GNP and those not captured in GNP. This impossible task would not increase welfare, as we would simply be replacing existing services, and it ignores the fact that many ecosystem services are literally irreplaceable. Policy decisions continue to ignore or undervalue ecosystem services because they are not fully reflected in commercial markets and cannot be precisely quantified in the same way as of manufactured goods and economic services. Recent methodologies for the valuation of ecosystem services can fill this lacuna. Environmental and ecological economics is concerned with this problem. Cobb and Cobb (1994) initiated the concept of green national product to include ecosystem health as an indicator of growth or progress. Recent studies have confirmed that considerable deterioration has taken place in ecosystem health owing to the mismanagement of natural resources by humans (WRI 2000a).

14.4.5 Sustainable Agriculture and Ecosystem Services

Sustainable development depends largely on the renewable natural resources of air, water, soil, and sunlight; communities of plants and animals; and the services they render. These, however, are not infinite. Tampering with their quality by over stretching or over stressing them can seriously impair these lifesupporting systems. The limited carrying capacity of the earth and the concept that emerged in the 1970s of earth as a "Spaceship" are, in particular, relevant now and for the future because the natural ecosystem is under great stress from human interventions.

The potential of these natural systems to produce goods and services has to be durable to sustain humankind and other life on earth. Further, the demand for food, energy and power will increase several-fold with rising human and cattle populations and increasing aspirations for better lives, especially in the developing countries. Production of these goods and services needs to be increased to maintain per capita flows. This is among the prime challenges of sustainable agriculture and sustainable development, irrespective of how economic development, social aspects such as intra- and intergenerational equity, and value judgments are made, defined, or practiced. Conserving the ecological integrity and environmental safety of our planet to the extent required to take care of our life-support systems in this and future generations is at the core of sustainable agriculture (Carpenter 1998).

The sustainability of agricultural production systems and agricultural development is contingent on such ecosystem services as soil fertility control, genetic diversity, pollination, nutrient cycling, hydrological cycles, quality of soil, water, and air and pest stabilization. Subsidies and artificial controls, including fertilizers, irrigation, pest control chemicals, and tillage, have so far succeeded in increasing and sustaining productivity of the natural resource base to the desired extent, but, not without substantial costs in terms of the quality of these resources and the services provided by ecological communities. Very clear indicators point to the fact that such external inputs may not be able to sustain and increase productivity indefinitely.

Although the intrinsic value of natural capital for life on earth has remained undiminished since life began on this planet, the issues at stake now are the quantity and quality of the current levels of natural capital stock and ecosystem services. The evidence is unassailable that human activities of different types, including agriculture, forestry, and urbanization, have transformed more than half of the earth's surface depleting natural resources and diminishing their potential (WRI 2000a).

Agricultural production is the most important and uncompromisable economic activity that is explicitly dependent on all the constituents of the biosphere. Human ingenuity and innovation have succeeded in manipulating almost 30% of the natural biomass to meet the food demands of an ever increasing global population since the time of the Neolithic Revolution around 8000 BC. The pressure on the natural resources for agriculture began to intensify in the middle of the twentieth century, when the limits of extensification of the land for cultivation were reached, but, the population growth and the demand for food continued to accelerate. In the past half a century, the global population has increased almost three-fold. This has led to a severe drop in the per capita share of natural resources and genuine fears about human numbers exceeding the carrying capacity of the biosphere and cutting into the needs of future generations. In a way, this is validating the Malthusian theory of population growth overtaking the global food productive capacity.

It is based on these conditions that the Brundtland Commission (WCED 1987) advocated a development paradigm of strict conservation of natural capital and natural ecological processes and services. The concept of sustainable development was rooted in the perception that natural capital had already reached a critical low and hence future activities should not erode it further. Maintaining a constant natural stock through sustainable use and conservation is basic to sustainable development because below a certain minimum level natural stock increases exponentially in value/price toward infinity. As degradation accelerates, the economic value of the next unit of natural stock at risk becomes higher than that of the unit already degraded.

The basic strength of natural capital lies in its diversity, which creates many routes for the flow of energy from the sun to maintain the planet and sustain it against stresses and shocks. Resilience, which is an indicator of undiminished natural capital, is the key to sustainability. It is important that agriculture and other human-dominated activities preserve this resilience. Advances in ecology have reinforced the scientific basis of sustainable development and sustainable agriculture.

14.4.5.1 Physical Capital

Physical capital consists of monetary, capital and tangible forms of wealth associated with industrial, natural biological, and agricultural production, infrastructure, minerals, and so on that can be financially valued and marketed. It is the best understood form of capital and the one most obviously relevant to economic development. Its valuation in the form of GNP/GDP is used as indicator of economic progress in neo classical economic theories, which deal with the demand and supply of goods and services through the exchange of money. This economic system operates as a closed system in isolation from such external processes as how it affects other environmental factors or human welfare and equity issues such as health, education, livelihood security, and food security. Physical capital is the most easily understood, tangible, and coveted form of capital for agriculture. The spectacular production and productivity gains made in the past five decades through capital-intensive, high –external - input agricultural technology are testimony to its importance. Without the intensification of agriculture through technological innovations, food security for almost half of the global population in the populous developing countries of Asia, Africa, and South America would have remained an unfulfilled dream.

Huge investments in irrigation, fertilizers, machinery, seeds, infrastructure, subsidies, and so on in countries such as China, India and Bangladesh have yielded more tangible and intangible benefits from agriculture than from any other sector. Machines, fertilizers, pest control chemicals, and genetically superior, high yielding, and stress-resistant cultivars together practically tripled overall food production from 420 million tons to 1176 million tons between 1950 and 1985 (FAO 2000), with even higher gains in productivity. This has also helped to keep global food prices relatively affordable even in the poorer and less developed countries.

14.4.5.2 Human Capital

Human capital broadly refers to such attributes as education, knowledge, skills, physical and mental well being, value systems, and the institutions and infrastructure to enhance them. The value and power of human capital have been widely recognized by society in the spectacular progress achieved in almost all spheres of life. Investing in human capital has paid rich dividends in terms of all-round social well-being through its contribution to economic, technological, and ecological progress. Although human capital is vested largely in individuals, their individual and collective efforts have contributed to meeting societal needs for comfort and happiness, to enhancing physical and mental health, and to preparing us to face any adverse contingencies, physical or biological.

A massive expansion of human, technological, and institutional knowledge has taken place in agricultural systems. Traditional agriculture depended largely on experimental knowledge and wisdom gathered over a long period from various combinations of climate, biophysical, socioeconomic, and cultural contexts. The knowledge embodied in people, societies, and cultures was responsible for creating and nurturing the traditional systems of cultivation, which were highly enduring and sustained production over thousands of years. The knowledge of peasant farmers with regard to ecological factors such as climate, soil, pest population, land use capability, and plant diversity has been recognized to be of a very high order, to be natural resource sustaining and upgrading, and more importantly, to be contextually appropriate.

In the mid-twentieth century, traditional knowledge was supplemented by scientific breakthroughs in fertilizer use, the use of high yielding crop varieties, efficient chemical pest control etc. However, progress towards sustainable agriculture is largely guided by the wisdom of the past. Such knowledge will continue to be important for resource-poor farmers for whom the green revolution technologies have been irrelevant. The vast African continent is its best example.

Humans are already facing several challenges to sustainable agriculture in the future, including climate change, loss of bio diversity, ozone depletion in the atmosphere, environmental pollution, and unabated population growth. Natural capital yields a flow of natural resources and services that enter the process of production, and human capital serves as an agent in the process of translating the resource inflow into a product outflow. Thus, the two are complementary and not totally substitutable. The most pressing need to day is to stem further depreciation of natural capital so that threshold limits are not exceeded, which would render rehabilitation or enhancement difficult or even impossible, both ecologically and economically. Knowledge will be the most important resource of the future. Expansion of knowledge on the requisite scale, at the requisite level, and of the requisite kind will be essential to sustain future growth in agricultural productivity. Investing in knowledge is prudent because the elasticity of the supply of knowledge relevant to agricultural production is high relative to the elasticity of the supply of natural resources. The distribution of investment to expand the supply of knowledge among technology, human capital, institutional innovations, natural resource enhancement, and so on will depend on contextual needs (Crosson 1994). Surviving future constraints depends solely on the adaptability of agricultural production systems, which can largely be ensured by extending the frontiers of human capital, especially since per capita natural capital is rapidly dwindling.

14.4.5.3 Social Capital

Sustainable development and sustainable agriculture are human development goals for sustained overall well being. The enabling role of society at large is the most essential catalyst for achieving these goals. Sustainability is a collective and multifunctional process with a number of actors and consequences. It has to benefit the producer, the consumer, the environment, the economy, and the society by creating livelihoods and a range of public goods, such as, clean water, carbon sequestration, and conservation of biodiversity. This cannot happen through individual efforts alone, but, only through social cohesion, which can enhance private and public benefits (Pretty 1999).

Natural resource management, which at the core of sustainable development, depends primarily upon social capital because a significant tranche of natural resources are public goods and common property. Without basic social ethics and discipline, their management will be chaotic, diffuse, unproductive, and unsustainable. Science and technology are futile in an atmosphere of social instability and turmoil.

Social capital, or the structure of a community, is a broad term for such cohesive forces as community and political stability, safety, family values, interpersonal trust, and cooperation, which contribute to economic progress and human well-being. Coleman (1990) was the first to use the term and to write about its value for productive human endeavors. Social capital has four core components: trust, norms and sanctions, reciprocity, and connectedness. All are part of the social structure necessary to realize personal and social interests. Trust enables cooperation and confidence among individuals and groups and reduces transaction costs. It builds confidence in the prevailing social structure. Reciprocity and connectedness also cement trust and induce cooperation and joint action. Norms are mutually agreed upon and binding rules of behavior and transactions that place group/community interests above individual interests. They generate collective interest and investments in group activities geared towards the long term interests and stability of the community.

Connectedness through various types of networks, organizations, cooperative and mutual aid societies, non governmental organizations, credit groups, and professional societies is vital for social capital formation and maintenance, as such connectedness are conducive to cohesive and effective institutional performance.

Institutions, both formal and informal, are necessary to overcome several obstacles that farmers face in terms of unavailability of credit support, inherently weak enforcement of contracts, and a lack of adequate and timely information, markets and infrastructure. Reducing transaction costs through institutions is necessary for small agricultural markets to operate and is key to reducing poverty and overcoming food insecurity. Unfortunately, the building of institutions for agricultural markets is challenged by factors of geographic dispersion, urban bias in public policy, and uncertainties resulting from the vagaries of climate. The World Development Report 2000 specifically emphasized the need for informal institutions for resource-constrained rural farmers.

The facilitating role of social capital in agricultural development is clearly visible in the different responses of various developing countries to the development assistance rendered by international agencies in cash, kind, technology, training, and education in the past five decades. Whereas India and several countries of South Asia have reaped rich benefits, the sub Saharan Africa has lagged behind and has totally failed to achieve the continent's potential, to a large extent because of a deficit of social capital.

Similar to natural capital, social capital tends to be a public good with no proper market value and largely unsubstitutable by human made capital. Technology and science can become totally incapable of improving agricultural production and productivity and conserve natural capital if the social environment is not supportive. Just as nature owes its resilience to its diversity, society also gains its durability and strength from plurality, which enables it to withstand stresses and shocks of various kinds. Social capital is more an abstract or non material form of capital, but, its material consequences are quite significant –all the more in the modern interconnected world. Although economic indices, such as gross national product GNP and gross domestic product GDP, which exclude changes in the value of social, human and ecological capital, continue to engage attention of many governments, the world over, more holistic and realistic indices, such as the genuine progress indicator (GPI), and gross welfare index (GWI), which include all forms of capital, are being developed to measure sustainable economic progress. Studies have revealed that although global GNP and GDP have shown a continuous increase since the 1950s, GPI and GWI have shown a leveling off or even a dip, because, beyond a certain threshold, economic activities tend to adversely affect ecological, social and human capital (Costanza, Norgaard et al. 1997; Max-Neef 1995). Perhaps, this is also the prime reason why a tiny state of the Indian Republic, Bhutan, opted for the Gross Happiness Index (GRI). Bhutan, by any western standards, would be classified as an economically "backward state," but, the people, by and large, are happy and contended. This is an antithesis of what one sees in the "developed" western world today, which, most Indians, especially the elite, and upwardly rich, mistakenly attempt to emulate.

References

Altieri, M. A. (2002). Agroecological principles for sustainable agriculture. In N. Uphoff (Ed.), *Agroecological innovations: Increasing food production with Participatory Development* (pp. 40–46). London: Earthscan.

Carpenter, R. C. (1998). Can sustainability be measured? In H. S. Sharma & S. K. Chattopadhyay (Eds.), *Sustainable development: Issues and case studies* (pp. 46–55). New Delhi: Concept.

Cobb, C., & Cobb, J. (1994). *The Green National product: A proposed index of sustainable economic welfare*. New York: University Press of America.

Coleman, J. S. (1990). *Foundation of social theory*. Cambridge, MA: Harvard University Press.

Costanza, R., & Daly, H. E. (1992). Natural capital and sustainable development. *Conservation Biology, 6*, 37–46.

Costanza, R., Norgaard, R. B., Goodland, R., Daly, H. E., & Cumberland, J. H. (1997). *An introduction to ecological economics*. Boca Raton: CRC Press.

Costanza, R., d'Arge, R., de Groot, S., Farber, M., Grasso, B., Hannon, K., Limburg, S., Naeem, R. V., O'Neill, J., Paruelo, J., Raskin, R. G., Sutton, P., & van den Belt, M. (1997). The value of world ecosystem services and natural capital. *Science, 387*, 253–260.

Crosson, P. (1994). Conditions for achieving a sustainable global agricultural system. In *Environmental agriculture: Rethinking development issues for the 21st century. Proceeding Symposium in honor of R. D. Havener, May 5–6, 1993*. Morriltown: Winrock International Institute for Agricultural Development.

Daily, G. C. (1997a). *Nature's services: Societal dependence on natural ecosystem*. Washington, DC: Island Press.

Daily, G. C. (1997b). What are ecosystem services? In G. C. Daily (Ed.), *Nature's services: Societal dependence on natural ecosystem* (pp. 1–7). Washington, DC: Island Press.

FAO. (1991). *The den Bosch declaration and agenda for action on sustainable agriculture and rural development*. Rome: Food and Agriculture Organization.

FAO. (2000). *The State of Food and Agriculture: Lessons from the past 50 years*. Rome: Food and Agriculture Organization.

Hart, R. D., & Sands, M. W. (1990). *Sustainable land use systems research and development.* In: Proceedings of the International Workshop on Sustainable Land use Systems research. February 12–16, New Delhi. Oxford IBH, pp. 1–11.

Harwood, R. R. (1990). A history of sustainable agriculture. In C. A. Edwards, R. Lal, P. Madden, R. H. Miller, & G. House (Eds.), *Sustainable Agricultural Systems* (pp. 3–19). Ankey: Soil and Water Conservation Society.

Ketola, T. (1997). Ecological Eldorado: Eliminating excess over ecology. In R. Welfod (Ed.), *Hijacking environmentalism: Corporate response to sustainable development* (pp. 99–136). London: Earthscan.

Max-Neef, M. (1995). Economic growth and quality of life: Threshold hypothesis. *Ecological Economics, 15*, 115–118.

Pretty, J. N. (1999). Value of natural and social capital for sustainable agriculture. In G. S. Dhaliwal, R. Arora, & A. K. Dhavan (Eds.), *Emerging trends in sustainable agriculture*. New Delhi: Commonwealth Publishers.

Raman, S. (2006). *Agricultural sustainability: Principles, processes and prospects* (pp. 13904–11580). Birmingham: The Hawoth Press Inc.

WCED. (1987). *Our common future*. Oxford: Oxford University Press.

WRI. (2000). *World resources 2000–2001. People and ecosystems, the Fraying web of life.* Washington, DC: World Resources Institute.

Chapter 15
Land Management for Sustainable Agriculture

Abstract Of all the natural resources for agriculture, land/soil is accorded the first status among equals. Early civilization, the world over, venerated land as the "mother", and, many continue to do so. India is the most illustrious example. The coinage by the author of this book for soil stands as follows: SOIL: S-Soul, O-Of, I-Infinite-, L-Life.

Keywords Soil · Land · Per capita land availability · Pre historic farming · Global soil degradation · Water and wind erosion · Chemical degradation · Physical degradation · Biological degradation · Human induced soil degradation · Crop production · Sustainable agriculture · Sustainable land management · Conservation · Sustainable use of natural resources · Appropriate land use patterns · Agroecological zones · Soil resilience · Soil degradation · Soil erosion control · Maintaining soil fertility

Of all the natural resources for agriculture, land/soil is accorded first among equals status. Early civilizations, the world over, venerated land as the mother, and many continue to do so. India is the most illustrious example. The Vedas (ancient Indian scripture, written in Sanskrit, the age old language of the country), even prescribes that before one gets up from the bed and stands on one's feet, one should touch the ground and place the palm on the head, beseeching mother earth for forgiveness for all the wrongs one might do, during the course of the day. The belief of ancient Indians is that the mother earth (symbolized by the name Bhoodevi – Bhoo/Bhoomi (earth) devi (goddess) carries the weight of all humans day in and day out, and one should seek forgiveness for this. This reverence is vindicated when the role of land is analyzed scientifically without any emotional or abstract bias. This is also the reason, why this author termed soil/land as the Soul of Infinite Life as described below:

SOIL: S- Soul, O- Of, I- Infinite, L- Life

Land/soil is a critically important component of the biosphere from which all living organisms, including humans, derive shelter, food, growth, and all other activities. It is also a mediator for many life-support systems. It provides food, fod-

K. P. Nair, *Intelligent Soil Management for Sustainable Agriculture*,
https://doi.org/10.1007/978-3-030-15530-8_15

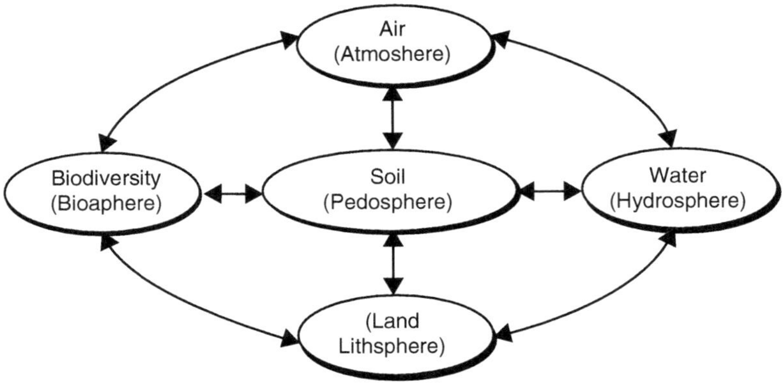

Fig. 15.1 Soil as an environmental integrator

der, fiber, fuel, timber, medicinal and ornamental plants and other biotic materials for human use and consumption in their natural forms or through agriculture, animal husbandry, aquaculture, or inland and coastal fisheries. More than 96% of food for human consumption comes from land, and only 4% from marine sources.

Soil has another, much wider and overreaching role to play in the sustainability of the biosphere. It acts as an integrator of the environment, namely, the lithosphere, (land), hydrosphere (water), atmosphere (air), and biosphere (living organisms), as depicted in Fig. 15.1.

Soil provides several major ecosystem services essential for the planet's life-support systems and the maintenance of environmental quality and integrity. Karlen et al. (1997) classified the essential functions of soil more explicitly and completely as follows:

Sustaining biological activity, diversity, and productivity
Regulating and partitioning water and solute flow
Filtering, buffering, degrading, immobilizing, and detoxifying organic and inorganic materials, including industrial and municipal by-products and atmospheric deposition
Storing and cycling nutrients and other elements within the earth's biosphere
Providing support for socioeconomic structures and protection for archaeological treasures associated with human habitation

Soil is also a major source and sink for global gases and plays an important role in the regulation of radiatively active gases (green house gases) in the atmosphere. Soils constitute the largest terrestrial pool of carbon, estimated at approximately 1550 Pg, which is central to the global carbon cycle. The world's soils also contain approximately 95 Tg of nitrogen (Lal et al. 1995). Both of these pools contribute actively to atmospheric greenhouse gases through decomposition and oxidation as a consequence of deforestation and faulty land management practices. In the past decade, 1.6 billion tons of carbon is estimated to have been emitted annually as a result of land clearance, compared with 6.4 billion tons released into the atmosphere by the combustion of fossil fuels (Saroja Raman 2006).

15.1 Land and Crop Production

The terms "land" and "soil" are often interchangeably used in agricultural science, although, "land" has a broader connotation as it includes diverse environmental factors, such as, climate, population, soil, components below and above the soil layer, topography, natural and planned vegetation, and hydrological and mineral characteristics. These factors are interlinked with productivity and other attributes of the soil in the landscape (Latham and Syers 1994). The formation of soil, its nutrient cycling capacity, and its resilience, that is, its capacity to "regenerate" after set backs and shocks, the "Buffer Power," which is the central theme of this book, are all largely a function of the land's original characteristics combined with important factors, such as, the hydrological cycle, climate, bioenvironmental function, and, land management elsewhere. In addition to anchoring plants, and facilitating root growth of the plant and its proliferation, land/soil performs a wide range of functions related to plant growth described earlier. This chapter is concerned primarily with the role of soil and its management for sustainable agriculture.

15.2 Global Distribution of Arable Land

The earth's total "icefree" land area is estimated to be approximately 13.1 billion ha, of which less than one-third is potentially arable. The actual area in agricultural use is approximately 1.475 billion ha (FAO 1990). This includes land with permanent crops and arable land with temporary crops, temporary meadows, market and kitchen gardens, and temporary fallow. The remaining potentially cultivable land is marginal for agricultural use, because, most of it is either inaccessible or located in ecologically fragile regions. Steep slopes, shallow soils, extremes of climate, such as, very high temperatures or precipitation, very low soil fertility, and lack of available inputs are additional constraints to crop production.

Recent estimates of the continent-wise distribution of potentially arable land under rainfed conditions are given in Table 15.1. It is clear from the data in the table that potentially arable lands are unevenly distributed not only geographically, but also, demographically. The distribution is extremely unfavorable to Asia, which accounts for almost 50% (and growing) of the global population. Asian continent has only 26% of the world's potentially arable land, of which 64% is currently in use (in percentage terms, this incidentally is also the highest global use of arable land). Large areas of potentially arable land are present in the African and Latin American continents, but, 70% of this land is affected by unfavorable conditions, such as, steep slopes, poor drainage, and shallow soils.

Forty two per cent of the land in Sub Saharan Africa and 46% of the land in Latin America faces management problems and suffers from low inherent fertility.

Table 15.1 Distribution of actual and potential available arable land for rainfed agriculture

Region	Gross potential arable land (rainfed)/(10^3 ha)	Net potential arable land (rainfed cultivation)/(10^3 ha)	Actual arable land in use (10^3 ha)[a]	Percentage of potential arable land (rainfed)
Sub-Saharan Africa	1,119,492	1,050,083	157,608	15
North Africa (East)	50,017	44,815	71,580	160[b]
North Asia and East of Urals	286,800	275,902	175,540	64
Asia and Pacific	812,551	742,672	477, 706	64
South and Central America	1,048,071	979,946	143,352	15
North America	463,966	431, 488	233,276	54
Europe	363,120	323, 903	204,322	63
World	1,411,017	3,848,809	1,463,384	38

Source: Modified from Koohafkan (2000)
[a]Figures from 1994
[b]Includes irrigated land

Table 15.2 Estimated and projected world population (million)

Region	Estimated population		Projected population (2050)		
	1950	2003	Low	Medium	High
World	2519	6301	7409	8919	10,633
More developed regions	813	1203	1084	1220	1370
Less developed regions	1706	5098	6325	7699	9263
Africa	221	851	1516	1803	2122
Asia	1398	3823	4274	5222	6318
Latin America and the Caribbean	167	543	623	768	924
Europe	547	726	565	632	705
North America	172	326	391	448	512
Oceania	13	32	40	46	52

Source: Adapted from World Population Prospects

15.3 Per Capita Land Availability

Arable land is a limited resource, whereas global population growth has been accelerating since the beginning of the twentieth century. The global human population is currently above 6 billion and is expected to stabilize between 10 and 11 billion towards the end of the twenty-first century (Table 15.2). Future increases in population will take place in the already over populated developing countries, leading to a drastic reduction in their per capita land availability. This will have serious

Table 15.3 Per capita actual and potential arable land for agricultural and total population (ha)

	Agricultural population			Total population		
	Actual			Actual		
Region	1965	1995	Potential[a]	1965	1995	Potential[a]
Africa	0.86	0.47	2.8	0.62	0.26	1.64
Europe	2.25	4.23	4.4	0.34	0.26	0.59
South America	1.17	1.88	13.8	0.49	0.37	2.87
North and Central America	4.95	5.41	11.7	0.85	0.59	1.35
Asia	0.36	0.24	0.5	0.25	0.15	0.27
World	0.78	0.59	1.6	0.42	0.23	0.74

Source: Adapted from Koohafkan (2000)
[a]Figures from 1994

repercussions for food security, which may be further exacerbated by their low economic development and inadequate access to modern technology and other resources leading to degradation of currently productive land.

Recent data on per capita potential and actual arable land in use are given in Table 15.3, which shows a decrease in per capita actual arable land globally of approximately 46% from 1965 to 1995 across all continents. In the US, Europe and Latin America, per capita arable land for the agricultural population increased as a result of consolidation of land holdings. In contrast, it decreased quite perceptibly by 40–50% in Asia and Africa, indicating fragmentation of land. Political expediency also has contributed to this phenomenon. The classic example is in the State of Kerala, in India, where the "Lad Reforms Act" of 1953, under the first democratically elected communist government (at the time there was only the Communist Party of India –not the present Communist Party of India (CPI) and the Communist Party of India (Marxists), (CPI-M) as of now. The Act abolished "Land Lordism," the "Absentee Owner" situation in the State where some rich families could hold hundreds of acres of land. The government decreed that a family cannot hold more than 15 acres. This, led to fragmentation of land, and in some parts of the State, quite severally, to make farming economically viable and "Community or Group Farming" came into practice, through land consolidation, in the southern parts of Kerala State. But, when a family further fragmented, the land also did, giving in to such acute land fragmentation in the State, where a farmer tilled on a few cents one – hundredth of an acre) of land. The inevitable outcome is the spread of poverty and landlessness, because of a lack of livelihood means in the non agricultural sector, which has no land to depend upon. Encroachment onto agricultural land by industry, urbanization, infrastructure development, recreational facilities, and so on, are also making huge dents in per capita land availability. This is especially a very serious problem in China and India (Saroja Raman 2006).

15.4 Land Quality and Distribution

Eswaran et al. (1999) divided land capability into nine land quality classes and estimated the distribution among countries. Class I land, which is the most productive and free from any constraints, is extremely limited, forming only 3.2% of the "ice-free" land surface (Table 15.4). Large contiguous areas of such prime lands are found only in the US, Argentina, Uruguay, southern Brazil, and to a lesser extent in Europe and northern China. South Africa is the only country with significant areas of Class I land in Africa. Class I, II and III constitute 12.7% of the total land area globally and generally are suitable for most of the agricultural purposes. Such land is preset mainly in the temperate areas and accommodates approximately 24.2% of the global population. Class IV, V, and VI lands are located primarily in the tropics and carry 52% of total population. Class IV lands can be managed with medium or high input levels for agriculture. The presence of mostly low productivity lands in the tropics, coupled with high population density and other resource constraints, leaves poor farmers in these areas with no other option other than to cultivate the fragile lands, most often in a low input –low output mode. This has led to a large-scale soil degradation over the past few decades, especially in the ecologically harsh areas of sub Saharan Africa. In contrast, Class I, II and III lands are highly productive and respond very well to soil conservation measures to preserve their premium attributes for good farming, of which the buffer power, is the most critical.

Table 15.4 Estimates of land area and population in designated land quality classes

	Land area		Population	
Land quality classes	Million km^2	Percentage	Million	Percentage
I	4.09	3.2	337	5.9
II	6.53	5.0	789	13.7
III	5.89	4.5	266	4.6
IV	5.11	3.9	654	11.4
V	21.35	16.3	1651	28.8
VI	17.22	13.2	675	11.8
VII	11.65	8.9	639	11.1
VIII	36.96	28.3	103	1.8
IX	21.78	16.7	625	10.9
World	130.60	100.0	5,739	100.0

Note: The world population density map is limited to latitudes 72° N to 57° S
Source: Adapted from Adams and Eswaran (2000)

15.5 Soil Degradation and Its Impact on Agriculture

Human-induced soil degradation resulting from agricultural operations has become a major concern for sustainable agriculture since 1990. Soil degradation refers broadly to a quantitative or qualitative diminution of the soil's current or potential capacity to produce goods and services. In the context of agricultural production, it implies a diminution of productive capacity through adverse changes in one or more of several attributes, such as, soil organic matter, structure, nutrient status, drainage, and accumulation of salts and toxic chemicals.

15.5.1 Pre-historic Farming and Soil Degradation

Scientific literature is replete with examples of pre-historic agriculture and its impact on soils and the environment of the Middle East, and in the valleys of the Nile river, in Egypt, Yangtze river in China, and Indus (Ganges) in India. However, only little is known about pre-historic farming and soil quality in other sites of ancient cultures. Yet, maintaining soil fertility and agronomic productivity have been serious issues facing humanity ever since the dawn of settled agriculture. For instance, it has been long recognized that catenal processes result in large fertility contrasts between summit/shoulder slopes and footslope landscape positions (Scholes 1990), and, these contrasts were accentuated by the activities of pre-historic cultures, such as, addition of charcoal to create the "*terre preta de Indio*" (Mann 2002). Two important soil related constraints faced by pre-historic farmers, drought stress and soil nutrient depletion, were alleviated by use of the innovative technologies of both supplemental irrigation and nutrient recycling, since long, or, better still, precise soil nutrient management, such as through the "Buffer Power Concept," as has been proved very recently (Nair 2013).

Irrigation and water management played an important part in the so-called "hydric civilization." The Indus valley civilization is the most classic example. In fact, the term "*Hindu*" derives its origin from the original Greek term "*Indoo*," which the Greeks lent to the highly intelligent civilization on the banks of the Ganges in India, where the hydric civilization played a major part, both in human activity and agriculture. In addition to the well known use of irrigation in Asia (Knorzer 2000; Premathilake 2006), there are other illustrious examples of using irrigation in the pre-historic Americas (Park 1983; Lange 1992a, b). Central America has been the cradle of agriculture in the new world (Clement and Horn 2001). In the Lake Titicaca Basin, covering the Andean Highlands of Peru and Bolivia, archaeological investigations have shown a massive landscape modification throughout pre-historic times to intensively cultivate marginal lands. These modifications, over 82,000 ha of land area, included raised beds and large earthen platforms to improve drainage, increase soil fertility by recycling nutrients and improve micro climate (Erickson 1992). In the Pampa de Chapari hyper-arid region of Peru, pre-historic irrigation canals and furrowed fields have been preserved for 500 years (Nordt et al. 2004). On Easter Island, Chile, cultivation of yams (*Dioscrorea spp.*) and taro (*Colocasia esculenta*) began in the twelfth century and continued through the fifteenth century (Stevenson et al. 2006).

Archaeological evidence of early irrigation systems has also been discovered in southeastern Arizona. Palacios-Fest et al. (2001) have identified several stages of water management in the Santa Cruz River Valley of Southeastern Arizona in the pre-historic era. These researchers reported that pre-historic people first operated their irrigation system in a simple mode involving diversion of ephemeral flows following storms, and later in a complex mode involving diversion of perennial flows. Berger (2004) reported the use of canal irrigation from 1600 years to 800 years ago (400 AD to 1200 AD) in the flood plains of Salt River in Phoenix, Arizona. Masse (1981) used aerial photographs to reveal complex and extensive remains of ancient Hohokam irrigation systems in the Salt River Valley near Tempe, Maricopa County, Arizona. These systems were probably constructed and used between 850 AD and 1450 AD, and consist of over 2100 km of canals in the north and south of the Salt River.

Nutrient depletion was also a major problem in settled agriculture, except in the flood plains where soil fertility was renewed by alluvial deposition. Pre-historic depletion of soil nutrients was documented after centuries of indigenous agriculture in Hawaii. Hartshorn et al. (2006) observed that farmers began growing dryland taro and sweet potato (*Ipomoea batatas*) using a digging stick on the leeward slopes of East Mauri beginning approximately 500 years ago. Centuries of this extractive farming method lowered concentrations of calcium, (49%), magnesium (28%), sodium (75%), potassium (37%) and phosphorus (32%) in cultivated as compared to uncultivated soils. Similar to the problem of accelerated water erosion faced by the Mayan culture in Central America, wind erosion was a serious constraint faced by the pre-historic farmers of the Kalaupapa field system, Molakai Island, Hawaii (McCoy and Hartshorn 2007). In New Zealand, there is strong evidence of pre-historic cultivation of four introduced Polynesian plants: bottle gourd (*Lagenaria siceraria*), paper mulberry (*Broussonetia papyrifera*), sweet potato and taro (Horrocks et al. 2004). These data identify combinations of early Polynesian crops, including both field – and tree – cropping systems, and provide evidence of pre-historic taro cultivation in the South Island. Lepofsky (1995) observed the pre-historic agricultural production and human-induced environmental changes in the Society Islands dating back to the seventh–tenth century AD. Valleys with the greatest arable potential were cultivated earlier than less preferred sites. Similar to the pre-historic civilizations, the problem of nutrient depletion, due to soil degradation, has aggravated the food insecurity in Sub Saharan Africa continuing even during the twenty-first century (Smaling et al. 1993).

Population increase, because of the transformation of the "hunter-gatherer" system into settled agriculture, necessitated development of an "*ard*," or prehistoric plow (Highman et al. 1981; Lal et al. 2007). Introduction of the plow exacerbated the problem of soil erosion due to soil displacement in the field and depletion of soil organic matter (SOM). The plow-induced soil degradation plagued mankind even during the pre-historic times. Accelerated soil erosion had a devastating effect throughout the Middle East and Central American Mayan culture. With world population projected to rise to 9.5 billion by 2050, the challenge of producing not only more food to meet the exponentially growing world population, but, also making it equally nutritious, the basic human right, as clearly enunciated in the UN Covenant of 1948, has a far greater dimension than that faced by the pre-historic farmers, when population was sparse, and land was large.

15.5.2 Extent of Global Soil Degradation

Concern about widespread degradation of arable land led the United Nations Environment Program (UNEP), in collaboration with the Food and Agriculture Organization (FAO), to initiate the Global Assessment of Soil Degradation (GLASOD) project to establish realistic estimates of the state of soil degradation worldwide. It published a 1:10 million scale world map depicting the status of human-induced soil degradation in 1990 (Oldeman et al. 1990). Exact quantification of the extent and seriousness of various soil degradation processes was later carried out through digitization and geographical information systems (GIS) analysis (Oldeman 1994).

The GLASOD project found that agricultural activity, deforestation, and overgrazing were the major causes of soil degradation. The global levels of degradation of croplands, forest and woodlands, and pasture during the four decades from 1950 to 1990 were estimated to be 38, 18 and 21% respectively. The continent-wise figures show that degradation has been quite extensive in Asia and Africa (Table 15.5).

Table 15.5 Extent of human-induced soil degradation of agricultural land, permanent pasture, and forest and woodland

	Agricultural land		Permanent pasture		Forest and woodland	
Region	Total (million ha)	Degraded (%)	Total (million ha)	Degraded (%)	Total (million ha)	Degraded (%)
Africa	187	63	793	31	683	19
Asia	536	38	978	20	1273	27
South America	142	45	478	14	896	13
Central America	38	74	94	11	66	38
North America	236	26	274	11	621	1
Europe	287	25	156	35	353	26
Oceania	49	16	439	19	156	8
World	1475	38	3212	21	4048	18

Note: mha = million hectare
Source: Adapted from Oldeman (1994)

The project also graded the severity of degradation based on the change in prevailing conditions after World War II and produced global figures of 38.1, 46.3, and 15.6%, respectively, for its light, moderate, and strong and severe categories.

15.5.3 Causes of Soil Degradation

Soil degradation is caused by natural and human-induced factors. Natural degradation factors consist of climate characteristics, such as, temperature and rainfall (intensity and distribution) and soil terrain characteristics, such as, topography, geology, type of soil, vegetation, and hydrology. Thus, lands in tropical regions with high rainfall are very vulnerable to erosion as a result of excessive runoff and water erosion, especially on slopes. Sub humid and arid regions are susceptible to desertification and wind erosion. Thus, vast areas of the African continent are prone to soil degradation, which is a serious consequence of wind erosion.

Human-induced factors are related mainly to socioeconomic conditions. Land-use patterns, and management options, land degradation, population, economic growth and poverty, social structure, farm policies, and common resource policies are inextricably linked and related. Poverty is often the root cause of soil degradation in the developing countries. The poor often resort to exploitative agriculture, resulting in over mining of nutrient reserves and the land being rendered unproductive in the long run. The vicious cycle of poverty-exploitative land use and agricultural practices –land degradation – increased poverty (Pinstrup Anderson and Pandya-Lorsch 1994; Raman 2003) is exacerbated by inappropriate policies concerning agricultural production and management of common property resources.

15.6 Types of Soil Degradation in Agriculture

The GLASOD project identifies four major groups of soil degradation; water erosion and wind erosion, which involve displacement of soil constituents, and chemical degradation and physical degradation, which pertain to *in situ* deterioration of soil.

15.6.1 Water and Wind Erosion

Erosion is the displacement of soil physical constituents, resulting in soil loss, and in severe cases, reduction of rooting depth. Water erosion often causes loss of soil depth and an adverse rooting environment – a process known as sheet erosion. Terrain deformation also occurs, particularly in sloping lands through the formation of rills and gullies. Water erosion is the most widespread form of erosion, affecting

56% of global land area. Deforestation, removal of vegetative cover through over grazing, and mismanagement of agricultural land use are the main causes of water erosion.

Wind erosion occurs widely in arid and semiarid regions and in coarse-textured soils without vegetative cover. It affects 26% of the global area. Top soil depletion is the most widespread effect of wind erosion. Terrain deformation occurs to a limited extent through over blowing. Wind erosion is generally light or moderate and is caused mainly by over grazing, over exploitation of dry agricultural lands, and removal of vegetative cover. The extent and severity of degradation through water erosion and wind erosion are summarized in Table 15.6.

Accelerated erosion is the most destructive form of soil degradation and impacts all the major properties and attributes of soil related to cultivation, with the exception of certain types of *in situ* degradation including salinization, acidification, and chemical pollution. Accelerated erosion results in depletion of organic matter and the finer clay and silt fractions, which are the major sources of plant nutrients in all soils. Severe erosion leads to reduction in effective plant rooting depth, which is the major impediment to optimum crop growth. Erosion also sets off several other degrading effects, such as, leaching (loss of inherent plant nutrients), acidification (rendering important soil nutrients non bio available or toxic to plant growth), soil compaction, hard setting, and biological degradation (rendering many favorable microorganisms non active), all of which lead to threats in sustaining crop production. The Global Agro-Ecological Zones (GAEZ) program of the FAO has estimated that some 16% of global land area is at the risk of severe soil erosion.

Table 15.6 Effect of water erosion and wind erosion (mha)

	Light		Moderate		Strong and Extreme			
Region	Water	Wind	Water	Wind	Water	Wind	Total water	Wind
Africa	58	88	67	89	102	9	227	186
Asia	124	132	242	75	73	15	439	222
South America	48	26	65	16	12	–	123	42
Central America	1	+	22	4	23	1	46	5
North America	14	3	46	31	–	1	60	35
Europe	21	3	81	38	12	1	114	42
Oceania	79	16	3	–	+	+	83	83
World	343	269	526	254	223	26	1.092	549

Note: mha = million hectare
Source: Adapted from Oldeman (1994)

15.6.2 Chemical Degradation

Chemical degradation is defined as the cumulative negative impact of applied chemicals (mainly through application of inorganic fertilizers) and chemical processes on those soil properties and attributes of soil that regulate crop growth, and are responsible for its life-supportive processes (Logan 1990). Both natural and anthropogenic processes lead to chemical degradation. The former includes depletion of organic matter and plant nutrients as a result of water erosion, water logging, and anaerobic conditions in low lands and deposition of volcanic ash. Human-induced chemical degradation is mainly the result of agricultural management, primarily, soil management. Examples are unbridled use of chemical fertilizers, as has happened during the chemically powered industrial agriculture, euphemistically called the "green revolution" in India, where excessive use of chemical fertilizers led to severe land degradation. Punjab State, in India, known as the "cradle" of green revolution is the classic example of this. Overuse of chemical pesticides, water and pollution from industrial activities, especially when these industrial wastes are dumped into the soil, also lead to soil degradation. Extremely exploitative agriculture, powered by high input technology, especially of chemical fertilizers, leads to mining of soil nutrients and organic matter, leading to loss of soil fertility.

In this connection a relevant discussion is on what happens when excessive chemical fertilizers are used, especially nitrogenous fertilizers, to global warming. This aspect merits discussion, because, the impact of high input (especially chemical fertilizers) agriculture on global warming has been neglected for a very long time. One needs to focus on this, as well.

The conventional view on global warming is that it is, primarily, due to CO_2. But, there are two other gases, nitrous oxide, commonly known as "laughing gas" N_2O and methane CH_4 whose role is never taken into account, but, which contribute much to global warming. Why it is important to include these in this discussion is, because, both are by products of the most important human activity, which is agriculture and soil management comes into sharp focus in this light. Both are "long lived" gases, along with CO_2. If we take into consideration the total anthropogenic radiative forcing since 1750, as W per sq. m, CO_2 is ranked highest at 1.68 (warming intensity), while CH_4 ranks second at 0.97, and N_2O comes third, in order, at 0.17. N_2O is a byproduct of urea hydrolysis. In 2011 N_2O accounted for about 5% of total gas emissions in the US (Green House Gases, GHGs). N_2O is naturally present in the atmosphere as part of Earth's N cycle, and has a variety of natural sources. However, human activities, primarily agriculture, fossil fuel combustion, waste water management, and industrial processes are increasing the amount of N_2O in the atmosphere. N_2O molecules stay in the atmosphere for an average of 120 years before being removed by a sink or destroyed through chemical reactions. The impact of 1 pound of N_2O on global warming is over 300 times more compared to 1 pound of CO_2.

N_2O is emitted when people add N to the soil through the use of synthetic chemical fertilizers. Urea application is the best example. Agricultural soil management is the largest source of N_2O emissions in the US, accounting for about 69% of total emissions in the US, in 2011. Emissions can be reduced by reducing N-based synthetic fertilizers and using N fertilizers more efficiently, which the "Buffer Power Concept" can help. In the US, N_2O emissions from agricultural soils have increased to 80% higher in 2011 compared to 1990. N_2O is also emitted during the breakdown of N in livestock manure and urine, which contributed to 5% of N_2O emissions in the US in 2011.

Another important "long lived" gas is methane (CH_4). This is produced from paddy (rice) fields in wetlands in the standing water. This especially is the case in tropical countries like India and China where irrigated paddy (rice) is common. In the fields with standing water, one can smell the produced methane. This also substantially contributes to global warming, which is second only to CO_2 in terms of anthropogenic radiative forcing, measured by warming due to W per square meter.

The above clearly shows that soil management, especially when it is wrong, can substantially contribute to global warming. "The Nutrient Buffer Power Concept," which aims at optimizing chemical fertilizer use, can mitigate these ill effects not only on soil environment, but, the ambient environment, as well.

Four types of chemical degradation were identified in the GLASOD project: loss of soil nutrients and organic matter, salinization, acidification, and pollution. Nutrient depletion is often a result of exploitative agriculture or depletion of organic matter and primarily clay particles, which are the repository of plant nutrients, as a result of erosion and surface runoff of water from rainfall. Salinization is essentially a human-induced degradation resulting from improper irrigation management in arid and semi arid regions, over irrigation and lack of drainage in wet areas, and saline water intrusion of ground water reservoirs in coastal areas. Acidity and aluminum toxicity arise from over use of chemical fertilizers, and, more importantly, from excessive leaching leading to low soil pH, low levels of exchangeable calcium, potassium, and magnesium, and toxic levels of aluminum in extreme cases, soil conditions which are limiting to plant growth and inhospitable to beneficial soil microorganisms. Excessive acidity can result in acid sulfate soils through the oxidation of ferrous ions from iron pyrites under anaerobic soil conditions.

Chemical pollution of soil arises from over use of pesticides and also industrial wastes, when dumped into the soil. And this becomes a problem when land disposal of chemical wastes occurs. Whereas chemical degradation such as acidity and nutrient and organic matter depletion are reversible to some degree, the accumulation of ions, such as, lead and cadmium at toxic levels becomes irreversible.

15.6.3 Physical Degradation

The physical properties and attributes of soil (soil structure, water retention and infiltration, porosity, aeration etc) are central to crop growth and to the normal functions of soil, such as, control of hydrological cycles and microclimate, and nutrient recycling and detoxification by microorganisms. Physical degradation of soils impairs one or more of these properties or functions and reduces the potential for crop production and ecosystem services.

The GLASOD project identified compaction, sealing and crusting, water logging, and subsidence of organic soils as the major forms of physical degradation. Sealing and crusting are mostly due to over grazing and insufficient natural vegetation or crop cover. Water logging is generally the result of excessive irrigation, without proper drainage, or excessive rainfall. Subsidence of organic soils is a problem mostly of coastal swamps in Southeast Asia and some parts of Europe. Compaction occurs through over use of agricultural machinery. Severe depletion of organic matter can destroy soil structure, which in turn, can trigger events, such as, crusting, surface sealing, and densification and result in a low infiltration capacity and increased surface run off and erosion. The global extent of chemical and physical degradation is summarized in Table 15.7.

15.6.4 Biological Degradation

The biological characteristics of a soil are strong indicators of its productivity and sustainability. Soil biota are responsible for the biological health of agricultural soils. Soil microorganisms facilitate most of the biochemical processes important for crop production, among them the bio geo chemical processes involving the mineralization of carbon, nitrogen, sulfur, and phosphorus, which are crucial for the maintenance of soil fertility, and hence, good crop growth. Detoxification of soil pollutants and toxic organic chemicals, suppression of some pathogenic organisms, and nutrient enhancement are other important services rendered by microorganisms.

Biological degradation is the impairment or elimination of one or more significant populations of microorganisms and their role (Sims 1990). It is not visually manifested or easily measurable and is often transient. The GLASOD project did not mention this type of soil degradation. Very high external input commercial agriculture, in fact, industrial type of agriculture, euphemistically called the "green revolution, where unbridled use of chemicals to soil took place, as in northern India in the wheat belt of Punjab State, which became known as the "cradle" of green revolution, in the long term runs the risk of reducing the soil to a biologically impaired or "dead" physicochemical medium. This is so very evident in the State of Punjab in India, where hundreds of acres can no more be cultivated due to degraded soils, dried aquifers and vanished bio diversity. In fact, many knowledgeable scientists have unequivocally concluded that in Punjab State, "*the green evolution has fallen on its face*." It is in places like this where "The Nutrient Buffer Power Concept" offers a ray of hope to the native farmers.

Soil physical and chemical degradation indirectly bring about soil biological degradation, through depletion of organic matter, and ruin of soil structure consequently. Proper crop management practices such as tillage, application of fertilizers and pesticides, and irrigation induce transient changes to soil microorganisms and

Table 15.7 Extent of chemical and physical degradation of soil (million ha)

Region	Chemical soil degradation					Physical soil degradation			
	Loss of nutrients	Salinization	Pollution	Acidification	Total	Compaction, sealing, crusting	Water logging	Subsidence of organic soils	Total
Africa	45	15	+	1	62	18	1	–	19
Asia	15	53	2	4	74	10	+	2	12
South America	68	2	–	–	70	4	4	–	8
Central America	4	2	+	–	7	+	5	–	5
North America	–	+	+	+	+	1	=	–	1
Europe	3	4	19	+	26	33	1	2	36
Oceania	+	1	–	–	1	2	–	–	2
World	136	77	21	6	240	68	11	4	83

Source: Adapted from Oldeman (1994)

soil quality. High doses of pesticides can be detrimental to soil microbial communities, but, the effect is often transient, except with unusually high doses. Nitrification, nitrogen fixation, and ammonification are the processes most sensitive to pesticides (Sims 1990). Soil microorganisms are extremely resilient and reestablish themselves once the adverse conditions are reversed.

15.7 Soil Degradation and Crop Production

The adverse impact of soil degradation on crop production is complex. Human-induced soil degradation of arable land is among the most pressing problems at the present time, because, it lowers the productive capacity of the land and is a major threat to sustainable agriculture and food security. Continued soil degradation may impair the productive capacity of soil beyond technological, economical, and managerial redemption. Depending on the type and severity of impairment, yield reduction can vary from complete loss of productivity to less severe production constraints such as economic, technological, managerial and environmental problems. The latter are manifested as reduced response to inputs, lower flexibility and land management, irrigation problems, diversion of labor, and additional cost of reclamation. All can translate into socioeconomic problems, such as, landlessness, poverty, food insecurity, and loss of livelihood. Soil degradation also leads to water and air pollution, which can, in turn, lead to serious environmental and human health related problems. Loss of soil bio diversity is another serious threat to the productivity and sustainability of crop land. Local soil degradation can escalate into global impacts, including climate change (as earlier discussed in this chapter). Alterations in water and energy balances, and disruption in the bio geo chemical cycles of carbon, nitrogen, sulfur, and other important plant nutrients (Lal and Stewart 1992). The impact of land degradation on agricultural production is difficult to clearly assess, because of the overlapping effects of various other favorable and unfavorable factors, but, its cost in terms of loss in productivity and sustainability is high (Crosson 1995; Pimental et al. 1993). An overview of the potential damage can be gleaned from the reviews of Scherer and Yadav (1996) and Scherr (1999). Past statistics and future prognoses are quite depressing.

In Africa, yield reductions owing to past erosion ranged from 2 to 40%, with means of 8.2% for the continent and 6.2% for Sub-Saharan Africa, which is likely to rise to 16.5% by 2020. Similarly, if erosion is not well controlled, South Asian and Middle Eastern countries may suffer a decline of more than 20% in food production, and large-scale nutrient depletion will result in Nepal and northeastern India, Thailand, Africa, and Latin America. Salinization of the irrigation systems in India, China, Thailand and Mexico and agrochemical pollution of cotton producing areas in Turkey and Southeast Asia and of the banana plantations of Central America are causes for concern.

A 13% cumulative productivity loss for cropland, a 4% loss for pasture, a 17–30% increase in world prices, and a 1–7% lowering of agricultural income are anticipated in South and Southeast Asia by 2020. Food insecurity, poverty, malnutrition, and loss of livelihoods, will result in the case of people living on marginal lands in Africa and Asia. The replenishment costs for the nutrients that are lost annually from soils are estimated at US $ 1billion. In South Asia, the production loss/replacement cost is estimated to be US $ 10–11 billion a year. This is equivalent to 2% of GDP and 7% of agricultural GDP of the region (FAO 1994).

A 17% cumulative loss in productivity is estimated to have occurred during the period between 1945 and 1990, as a result of soil degradation, but, has, not halted the more than two-fold growth in global food production during this period because of other favorable factors.

15.8 Sustaining Land Productivity

Sustainable agriculture has to ensure production and productivity growth in line with the future scenario for population growth, demographic change, and economic growth, while simultaneously maintaining environmental safety, economic viability, social acceptability, and social responsibility towards growers. Land management for agricultural production has garnered maximum attention in the past half a century and has dominated all other factors of production and productivity. Agenda 21 of the Rio Earth Summit (UNCED 1992) made its largest number of recommendations for land management, with specifics for agriculture in Sect. 12.2 on "Sustainable Agriculture and Rural Development." It should be acknowledged that similar, but, less explicit efforts have also been under way since the 1970s in the form of such initiatives as the Framework for Land Evaluation (FAO 1976), the Agro-Ecological Zones (AEZ) project (FAO 1978–1991), and the assessment of Potential Population Supporting Capacities in the developing countries. All addressed issues of matching land potential in specific environments to appropriate and compatible long-term use.

Dumanski (1993) provided the comprehensive definition of sustainable land management for agriculture as follows:

Sustainable land management combines technologies, policies, and activities aimed at integrating socioeconomic principles with environmental concerns to:

1. Maintain or enhance production/services
2. Reduce the level of production risk
3. Protect the potential of natural resources and prevent degradation of soil and water quality
4. Be economically viable
5. Be socially acceptable

The concept of sustainable land management is broader than that of sustainable agriculture because it encompasses agriculture within a framework of environmental resource optimization and land-use planning. Sustainable agriculture is generally viewed as dealing with on-site constraints at the field level, whereas sustainable land management deals with off-site constraints as well and therefore with wider areas at the watershed level (Syers and Latham 1994). Sustainable land management also has to address the wide diversity of on-site and off-site environmental and socioeconomic conditions. Offsite effects, such as, adjacent land use, hydrology, water quality, atmosphere, and loss of bio diversity cannot be wholly addressed by the individual farmer and require community involvement using such instruments as incentives, regulations, laws, and institutions to implement them.

15.9 Challenges to Sustainable Land Management

The conservation and sustainable use of all natural resources and the equitable distribution of benefits, are the basic tenets of sustainable development. The management of land resources for agriculture has become particularly challenging because of the following reasons:

1. The availability of land for agriculture is finite, or is even dwindling owing to encroachment from non agricultural uses (providing housing, infrastructure provision, providing recreation facilities etc. Building sprawling mansions, and state–of–the–art recreation facilities, even in a tiny State like Kerala, where land availability is acutely limited and population density pressure is intense, is a classical example of this. In particular, is this adverse development seen taking place on the paddy wetlands, threatening the already fragile food security of the State, and posing environmental hazards.
2. The distribution of productive land is severely asymmetric, to the disadvantage of the most populated and poorest areas of the globe
3. Per capita land availability is rapidly declining, especially in the developing world, as a result of exponential population growth, and the pace of degradation of arable land is accelerating through poverty-driven resource-extractive cultivation practices.
4. Accelerated degradation of arable land during the past five decades owing to intensive cultivation like the chemically driven agriculture practices, euphemistically known as the "green revolution," is a very grave problem. In India, the northern States like Punjab, Haryana and Western Uttar Pradesh are classic examples of this very unhealthy development.
5. The physical renewal and restoration of the performance potential of soil over desired time frames is strictly conditional. Although productivity is the primary function of arable land, its potential to maintain life-support services and eco system functions is vital for the continued health and quality of life for all living creatures, including humans.
6. In contemporary agriculture, land is one resource which cannot be moved around, unlike other resources, such as capital, technology, information, genetic resources, agrochemicals, and fossil fuels, which can be freely distributed around the globe to satisfy the needs of different locations and communities

and to maximize capital formation and accumulation. Even water can be moved around, to a limited extent.

7. Advances in science and technology have worked wonders to enhance productivity several-fold and to lay to rest the Malthusian prophecy of large-scale starvation and death, of course, with a heavy toll on the environment, as the so-called green revolution, especially in South Asia, in particular, in India, proved. Suffice to say, it produced huge amount of foodgrain for some time, at an enormous cost to the environment. The environmental gloom and deceleration of the economy predicted by the Club of Rome (Meadows et al. 1972) might only have postponed, not really averted the impending catastrophe. Environmental degradation through agriculture may become the forerunner of a slow down in economy, as predicted by Brown (1997), which is unfolding now in the twenty-first century. In South Asia, in particular India, has already begun to experience this catastrophe.
8. While other natural resources such as freshwater, energy, and even biodiversity (albeit to a lesser extent) are reasonably uniform in terms of quality and performance criteria, both spatially and temporally, land and soil are highly heterogenous resources. They have widely varying biophysical characteristics and performance potentials and require site-specific, customized management protocols for conservation and productivity
9. The vagaries of climate and topography and the variations in availability of other natural resources such as water, energy, and plant species and cultivars also impact land management. More importantly, various "soft" and highly contextual issues such as specific value systems, cultural ethos, social and economic priorities and opportunities, also guide agricultural patterns. Factors related to the availability, affordability, and appropriateness of technology, services, and policies specific to each context also have an impact.
10. Given the current status of land care and land degradation, it is a tall order to maintain agricultural productivity growth at the current rate for the next three decades, to meet the projected increases in food demand (FAO 2000).

15.10 Framework for Land Management for Sustainable Agriculture

Advances in science and technology in recent years have contributed enormously to coaxing more and more out of limited land, in synergy with other measures. The unifying principles of science and technology and their general universality make it possible to formulate a broad set of guidelines for sustainable land management, leaving the specifics to contextual requirements. Sustainable land management requires long-range planning in terms of enhancing its potential to produce and to be sustainable. Adopting a proper land-use pattern and implementing synergistic soil and crop management practices are necessary to conserve the potential of land in perpetuity. Sustainable land management consists of the two key issues of what to use the land for and how to use it to obtain maximum production in perpetuity. Generic guidelines related to sustainable land management are discussed here under the broad headings of appropriate land use patterns and core sustainable soil management strategies.

15.10.1 Appropriate Land Use Patterns

The FAO (1976) provided a framework of two orders for land evaluation based on the concept of land suitability for a specific use: suitable (S) and non suitable (N). Classes 1–3 under S and 1 and 2 under N reflect the degree of suitability. The appraisal of classes within an order is carried out in accordance with such limitations as climate [c], topography [t], wetness [w], salinity [n], fertility [f], and soil physical limitations [s]. Sys (1976) proposed a scheme for evaluating the degree of limitation with the following ratings:

0 = No limitation: Optimum plant growth
1 = Slight limitation: Nearly optimum plant growth
2 = Moderate limitation: Moderate decline in plant growth
3 = Severe limitation: Uneconomical for suggested use
4 = Very severe limitation: Not fit for suggested use

The United States Department of Agriculture (USDA) introduced the concept of land use according to land capability in the 1990s to curtail the highly prevalent land degradation in several parts of the world, owing primarily to inappropriate crop choices. The land capability classification is an interpretative grouping of different soil units to facilitate land use planning according to the suitability of the soils for cultivation of crops, pastures, forestry, and so on. It focuses primarily on the constraints and thus provides directions for the management of and improvement of different soils to increase production (Dent and Young 1981).

The basic USDA classification consists of eight classes (I–VIII), Classes I–IV which are suitable for cultivation, with further sub division according to the number and severity of constraints such as erosion risk, wetness, rooting depth, and climate limitations. Different crops require different types of soils and climatic conditions. The emphasis of this system is on environmental factors that cannot be easily modified in practice, for example, soil physical characteristics, such as, soil structure as opposed to such properties as nutrient status, soil pH, and drainage, which can be ameliorated. Non adherence to the recommendations for land use has often resulted in early deterioration and unsustainable use, especially during the green revolution phase, in, South Asia, primarily India, and to some extent in Africa (where the green revolution is only making a beginning).

Eswaran et al. (1999) identified nine quality classes to evaluate land's potential to sustainable agricultural production and maintain the integrity of the environment.

The objective is to seek a balance between agricultural use and environmental management.

15.10.2 Agroecological Zones and Land Use

The grave implications of the impact of agriculture on natural resources globally has triggered a focus and emphasis on eco regional and land-use systems. The concept of land use according to land capability has been further broadened to include other environmental parameters impacting crop growth and crop suitability, such as, climate, soil, and physiography (land form), in the form of agro ecological zones (FAO 1978–1991).

Climate, soil, and physiography are major determinants of agricultural ecosystem productivity; a detailed knowledge of these factors leads to the proper understanding and evaluation of the soil-plant –climate relationship at a particular site. An agro ecological region or zone is a homogenous agricultural environment with homogenous temperature and rainfall characteristics, bio physical environment, and natural resource problems and land use capabilities. This approach helps to plan for specific crops suited to particular eco regions and to ensure, appropriate, effective, and sustainable land use with better prospects to improve productivity. This leads to the more efficient and appropriate development of resources and their sustainable and efficient use within the framework of resource constraints and potential for each region. Land-use planning according to agro ecological characteristics, which is a holistic appraisal of the various parameters of climate, inherent soil quality, physiography, and hydrological characteristics in relation to specific crops, is the best method for sustainable land use.

The agroecological zones methodology developed by FAO and International Institute for Applied Systems Analysis (IIASA) provides a standardized framework for characterizing climate, soil, and conditions of the terrain related to agricultural production. It has been used to evaluate crop production potential in several countries. The availability of an extensive digital database has made it possible to produce global assessments of agricultural potential.

15.11 Soil Management and Sustainable Agriculture

The performance potential of a soil is termed "soil quality." Soil quality has emerged as a quantifiable attribute and has been acknowledged as a parameter which is valid to evaluate the sustainability of agriculture. Basically, soil quality is a holistic expression of a set of soil properties and attributes that collectively quantify the ability of the soil to perform its functions effectively. Its components are discussed in the following chapters.

Sustainable soil management is maintaining or upgrading the potential performance of a soil over several crop cycles. Failure to do so often results in some form and degree of degradation of the soil. The term "Degradation" is used here in a broad sense to imply any kind of temporary or permanent impairment to performance potential. Degradation is a manifestation of a disturbance, stress, or shock as an impairment of one or more soil properties or attributes necessary for optimum crop growth and productivity, for example, as detailed below:

1. Depletion of nutrients or organic matter
2. Impairment of soil structure, rooting depth, and tilth
3. Water infiltration as a consequence of erosion
4. Depreciation of soil ecological functions such as nutrient cycling and waste decomposition
5. Nutrient imbalance resulting in salinity, acidity, alkalinity, and accumulation of toxic metal ions such as aluminum and iron, particularly in ultisols

Such factors, operating singly or in combination with others, prevent the soil from performing to its maximum potential. The twin considerations of land use and soil management are shown in Fig. 15.2.

15.11.1 Soil Resilience and Sustainable Soil Management

The management of soil for sustainable production should focus on preventing degradation of soil resources or implementing timely remedial measures to ameliorate any deficiency as and when it occurs. Although prevention of degradation is the best

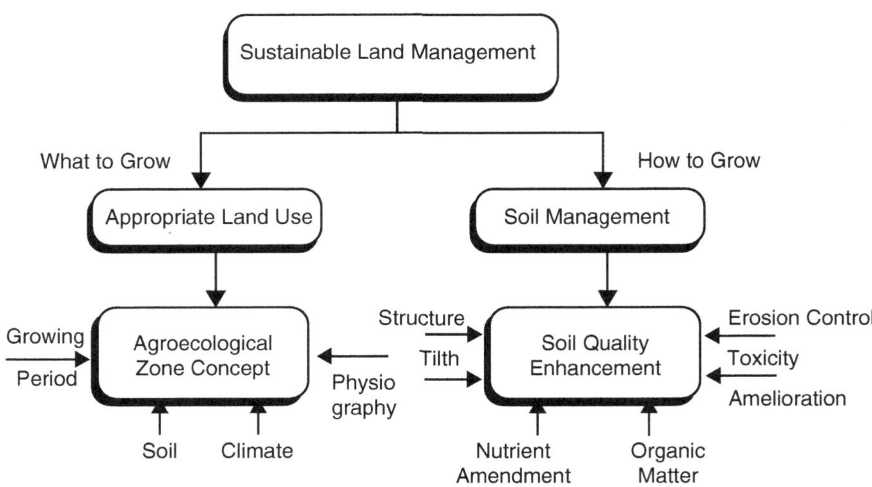

Fig. 15.2 Components of sustainable land management

course, it may be an ideal concept rather than a practical possibility, because, some degree of soil and land degradation is unavoidable over one or more crop cycles, depending upon the management and the intrinsic qualities of the soil. Fortunately, not all soil management induced soil degradation is irreversible. It is in this context that "The Nutrient Buffer Power Concept" comes into active play. Often, it is temporary and can be ameliorated with appropriate quantities and types of inputs and management initiatives, as has been amply demonstrated in practical terms, through the success stories of the author catalogued in this book. The very high estimates of arable land degradation by GLASOD initially sent shock waves through the agricultural and scientific communities. Soon, however, comfort was derived from the fact that soils are resilient and have an inherent capacity to recover from such stresses if suitable amendments or ameliorative measures are carried out. The concept of "soil resilience" is very important in this context, and the subject was discussed in detail at the International conference on "Soil Resilience and Sustainable Land Use" (1994) to understand and deal with different types and degrees of land/soil degradation and the capacity for partial or full recovery.

Any form of agriculture disturbs the natural equilibrium among the biophysical components of the soil ecosystem (soil nutrients, organic matter, soil biota etc.). This disturbance is greater with intensive and conventional agriculture than with traditional systems of crop production. This can lead to a temporary decline in soil productivity and ecological functions after one or more crop cycles. Resilience refers to the feasibility of improving the performance of the soil to its original level, or to an acceptable level, with proper technological or management inputs. The extent of recovery depends on the intensity of the disturbance and the amount of inputs provided.

The concept of soil resilience helps bring a balanced perspective to the issue of land degradation and sustainable agriculture. Only 6% of the overall 38% of human-induced crop land degradation indicated by GLASOD project is irreversible. The remainder is amenable to full or partial restoration with appropriate inputs.

Resilience can be interpreted as resistance to change or as recovery to original performance state (Eswaran 1994). The former interpretation refers to the stability of the system in response to external forces and is suitable for dealing with changes in attributes such as the structural stability of soils, which changes very slowly over thousands of years, only under drastic conditions, and, takes equally long to recover. In the context of soil management, it is more appropriate to use resilience to refer to the ability of the soil system to revert to its original or near-original performance or state. Depending on the nature and magnitude of disturbance and the manner of amelioration, the time for recovery can be as short as a few days to a few months.

15.11.2 Factors Affecting Soil Resilience

Soil is a dynamic and living entity. This gives it the ability to bounce back and restore most of its ecological functions after being disturbed by anthropogenic interventions, provided the disturbances caused are not irreversible and sufficient time is allowed. Thus, soil restoration depends on the balance between the soil-degradation and soil-resilience processes and the ameliorative measures applied. Soil biota, especially soil microorganisms, are responsible for the soil recovery process.

Soil resilience arises from both endogenous and exogenous factors. Endogenous factors are inherent soil properties, and climate and include soil fluxes, such as, weathering, erosion, soil formation, synthesis and decomposition of organic compounds, redox properties, and release and fixation of nutrients. It is in the context of the last property that "The Nutrient Buffer Power Concept" assumes great significance. A number of attributes such as rooting depth, good soil structure, structurally active clays, and good drainage enhance soil resilience, as do topography and climatic features. High soil organic matter content is among the most important characteristic of a productive and resilient soil. It also lends sufficient positive impact to nutrient buffering. Such soils can withstand reasonable deficiencies in management without undergoing degradation or a decrease in productivity.

Exogenous factors that influence soil resilience include mass and energy fluxes imposed by management practices, such as, tillage, application of chemicals, drainage, irrigation, contour interventions, removal of crop yields and residues, contamination by industrial effluents, and sludge application. Exogenous and endogenous fluxes are closely related and often interact and overlap (Szabolcs 1994).

15.11.3 Soil Resilience and Degradation

Soil resilience can be looked upon as the buffering capacity of the soil against physical, chemical, mechanical and biological impacts. The true scientific significance of the buffering capacity of a soil has been clearly explained by this author through the revolutionary soil management technique, now globally known as "The Nutrient Buffer Power Concept," which is also the central theme of this book. Soil resilience in the context of agriculture includes all the processes that enable it to counteract stresses and alterations imposed by cultivation processes. Soils and soil processes are generally conditionally renewable, provided degradation has not exceeded threshold limits, beyond which it may not be possible to renew their productivity or other ecological functions. For example, removal of soil by erosion is a serious stress and it takes a long time to regenerate the lost soil, but, other forms of degradation may be more easily amended with external ameliorates and practices.

Neither degradation nor resilience is soil specific, and, degradation and recovery do not proceed at the same rate. Knowledge of the rates and patterns of impacts serves to contain degradation and also to ascertain restorative capacity. Whereas structural changes are difficult, expensive, and take very long periods to restore, agricultural productivity performance is much simpler to restore and is amenable to rehabilitation through appropriate management intervention. It is in this connection that "The Nutrient Buffer Power Concept" plays a very crucial role, as has been amply demonstrated by the field success stories from South and Central Asia, as well as Africa. These are the regions where farmers need the best practical advice

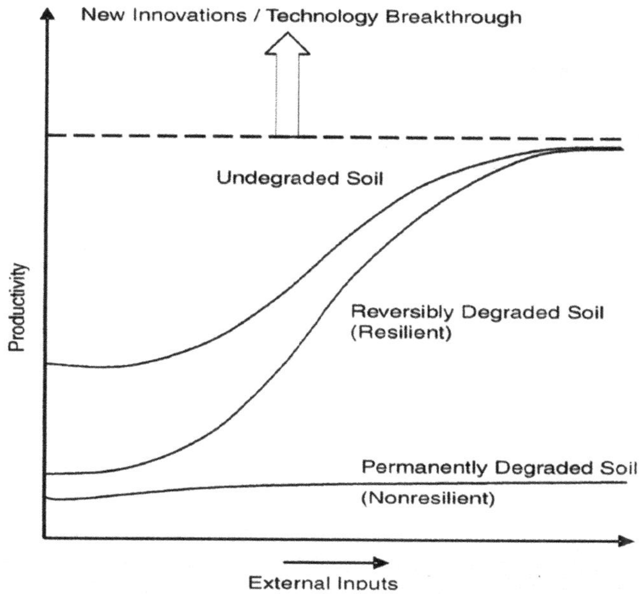

Fig. 15.3 Soil resilience and management. (Source: Adapted from Lal 1994)

and the concept is fulfilling that objective very remarkably. The best way to retain resilience is to adopt benign technologies to minimize the development of serious disturbances. Cautious monitoring and timely intervention are ideal.

The prevalence of large-scale global degradation of land poses a plethora of challenges to sustaining per capita production at comfortable levels. Maintaining or enhancing the productivity of the good and prime lands and preventing them from developing deficiencies is crucial to sustaining food security. Much human-induced degradation is of such a degree that it can be corrected or partially restored with judicious planning and implementation of appropriate scientific technologies, land use, and restorative measures. The extent to which the productivity of a soil can be boosted by technological inputs depends on the type and degree of stress, the resilience of the soil, and the effectiveness of the technology as shown in Fig. 15.3.

Full or partial restoration or rehabilitation of degraded lands to their original level of fertility and productivity forms a major component of any sustainable land management program.

15.12 Core Sustainable Soil Management Strategies

Soil physical, chemical and biological properties and soil processes are at the heart of land management for sustainable agriculture, because, along with climate, they determine the intrinsic productivity potential of any soil. High-quality soils, such as alluvial soils and mollisols with high organic matter content and fertility and inceptisols with high base content and rooting depth, support high and sustained productivity over long periods. On the other hand, alfisols, aridisols, vertisols, and the like are more problematic and suffer from poor attributes, such as, poor structure, low nutrient content, and low permeability and require more careful management to keep them productive. Appropriate land use, technology, and management practices can compensate for inherent deficiencies and enhance or sustain productivity even on deficient and fragile soils in ecologically sensitive regions.

Anderson (1998) used the phrase "growing the soil to grow crops" in relation to the following parameters:

1. Maintaining good soil depth and other favorable physical properties such as soil structure by preventing erosion
2. Feeding the soil by providing chemical nutrition to plants
3. Encouraging deep and prolific root systems to promote nutrient uptake from sub soil horizons, an attribute particularly important in shallow and low fertility soils such as ultisols and oxisols. This is relevant to all soils because a healthy soil bed and prolific root systems stimulate biological activity, which is vital for enhancing soil resilience and promoting sustainability. Optimum water management is essential for sustainable soil management because soil and water can destroy each other if they are not managed synergistically. In addition, it is also necessary to enhance above and below ground bio diversity and prevent the occurrence of certain soil-specific toxicity factors such as aluminum and iron toxicity.

The core strategies for sustainable soil management can be categorized broadly as follows:

1. Preventing soil erosion to maintain good soil depth and tilth
2. Maintaining favorable fertility and nutrient status
3. Enhancing soil organic matter status
4. Facilitating good soil structure
5. Preventing performance-debilitating problems such as salinity, alkalinity, sodicity, and aluminum and iron toxicity in specific soils
6. Enhancing soil bio diversity through crop rotations, intercropping and mixed cropping
7. Practicing sustainable water management

These strategies are often interlinked and contribute synergistically to sustainability. For instance, good soil structure improves several other attributes, including water retention and movement, habitat for soil biota, soil strength, porosity, and pore distribution. Similarly, soil organic matter amendment brings about improvements in soil structure, erosion control, nutrient bio availability, and soil biota. A brief account of the rationale of some of these strategies concludes this section.

References

Adams, C. R., & Eswaran, H. (2000). Global land resources in the context of food and environmental security. In *Advances in land resources management for the 21st century. Proceeding of International conference on land resource management for food, employment and environmental security (ICLRM), soil conservation society of India, November 9–13, 2000* (pp. 35–50). New Delhi: New Delhi Angkor Publishers.

Anderson, G. D. (1998). Grow the soils to grow the crops in Africa. In R. Lal (Ed.), *Soil quality and agricultural sustainability* (pp. 237–250). Chelsea: Ann Arbor Press.

Berger, G. W. (2004). Photonic dating of prehistoric irrigation canals of Phoenix, Arizona, U. S. A. *Geoarchaeology, 19*, 381.

Brown, L. R. (1997). *The agricultural link: How environmental deterioration could disrupt economic progress* (World Watch Paper 136). Washington, DC: World Watch Institute.

Clement, R. M., & Horn, S. P. (2001). Pre-Columbian land-use history in Costa Rica: A 3000 – year record of forest clearance, agriculture and fires from Laguna Zoncho. *Holocene, 11*, 419–426.

Crosson, P. (1995). Soil erosion costs and estimates. *Science, 269*, 461–464.

Dent, D., & Young, A. (1981). *Soil survey and land evaluation*. Wallingford: CAB International.

Dumanski, J. (1993). *Sustainable land management for the 21st century* (Vol. 1). Bangkok/Ottawa: International Board for Soil Research and Management and Agriculture Canada.

Erickson, C. L. (1992). Prehistoric landscape management in the African Highlands – Raised field agriculture and its environmental impact. *Population and Environment, 13*, 285–300.

Eswaran, H. (1994). Soil resilience and sustainable land management in the context of agenda 21. In D. J. Greenland & C. Szabolcs (Eds.), *Soil resilience and sustainable land use* (pp. 21–32). Wallingford: CAB International.

Eswaran, H., Beinroth, R., & Reich, P. (1999). Global land resources and population supporting capacity. *American Journal of Alternative Agriculture, 14*, 129–135.

FAO. (1976). *A framework for land evaluation, FAO Soils Bulletin* (Vol. 32). Rome: Food and Agriculture Organization.

FAO (1978–1991) *Agro-ecological zone project*. Rome: Food and Agriculture Organization.

FAO. (1990). *FAO yearbook 1989: Production* (FAO statistical series 94) (Vol. 43). Rome: Food and Agriculture Organization.

FAO. (1994). *Land degradation in South Asia: Its severity, causes and effects on People*. Rome: Food and Agriculture Organization.

FAO. (2000). *Agriculture Towards 2015/30 technical interim report, Global Perspectives Unit*. Rome: Food and Agriculture Organization.

Hartshorn, A. S., Chadwick, O. A., Vitousek, P. M., & Kirch, P. V. (2006). Prehistoric agricultural depletion of soil nutrients in Hawaii. *Proceedings of the National Academy of Sciences of the United States of America, 103*, 11092–11097.

Highman, C. F. W., Kijngam, A., Manly, B. F. J., & Moore, S. J. E. (1981). The bovid third phalanx and prehistoric ploughing. *Journal of Archaeological Science, 8*, 353–365.

Horrocks, M., Shane, P. A., Barber, I. G., D'Costa, D. M., & Nichol, S. L. (2004). Microbotanical remains reveal Polynesian agriculture and mixed cropping in early New Zealand. *Review of Palaeobotany and Palynology, 131*, 147–157.

Karlen, D. L., Mausbach, M. J., Doran, R. G., Cline, R. F., & Harris and Schuman G.E. (1997). Soil quality: A concept, definition and framework for evaluation. *Soil Science Society of America Journal, 61*, 4–10.

Knorzer, K. N. (2000). 3000 years of agriculture in a valley of the high Himalayas. *Vegetation History and Archaeobotany, 9*, 219–222.

Koohafkan, A. P. (2000). Land resources potential and sustainable land management: An overview. In *Advances in land resources management for the 21st Century. Proceeding of international conference on land resource management for food, employment and environmental security (ICLRM), November 9–13, 2000* (Soil conservation society of India, New Delhi) (pp. 1–22). New Delhi: Angkor Publishers.

Lal, R. (1994). Sustainable land use systems and soil resilience. In D. J. Greenland & I. Szabolcs (Eds.), *Soil resilience and sustainable land use* (pp. 41–67). Wallingford: CAB International.

Lal, R., & Stewart, B. A. (1992). *Soil restoration vol 17 advances in soil science*. New York: Springer.

Lal, R., Kimble, J. M., Levine, E., & Whitman, C. (1995). World soils and green house effect: An overview. In R. Lal, J. M. Kimble, E. Levine, & B. A. Stewart (Eds.), *Soils and global change* (pp. 1–7). Boca Raton: Lewis Publishers.

Lal, R., Reicosky, D. C., & Hanson, J. (2007). History of plowing over 10,000 years. *Soil and Tillage Research, 93*, 1–12.

Lange, C. H. (1992a). Canal irrigation in prehistoric Mexico – The sequence of technological change. Doolittle, WE New Mexico. *Historic Review, 67*, 194–195.

Lange, L. (1992b). Microbes and microbial products in plant protection. *Progress in Botany, 53*, 252–270.

Latham, M., & Syers, J. K. (1994). Collaborative research networks to promote sustainable land use. In D. J. Greenland & I. Szabolcs (Eds.), *Soil resilience and sustainable land use* (pp. 513–520). Wallingford: CAB International.

Lepofsky, D. (1995). A radio carbon chronology of pre-historic agriculture in the Society Islands. *French Polynesia, Radio-carbon, 37*, 917–930.

Logan, T. J. (1990). Chemical degradation of soil. In R. Lal & B. A. Stewart (Eds.), *Soil Degradation* (Vol II of advances in soil science) (pp. 187–216). New York: Springer.

Mann, C. C. (2002). The real dirt on rainforest fertility. *Science, 297*, 920–923.

Masse, W. B. (1981). Prehistoric irrigation systems in the Salt River Valley. *Arizona Science USA, 214*, 408–415.

McCoy, M. D., & Hartshorn, A. S. (2007). Wind erosion and intensive pre-historic agriculture: A case study from the Kalaupapa field systems. Molokai Island, Hawaii. *Geoarchaeology, 22*, 511–532.

Meadows, D. H., Meadows, D. C., Randers, J., & Behrens, W. W. (1972). *The limits to growth*. New York: Universe.

Nair, K. P. P. (2013). The buffer power concept and its relevance in African and Asian soils. *Advances in Agronomy, 121*, 447–529.

Nordt, L. F., Hayashida, H. T., & Crawford, C. (2004). Late prehistoric soil fertility, irrigation management, and agricultural production in northwest coastal Peru. *Geoarchaelogy, 19*, 21–46.

Oldeman, L. R. (1994). The global extent of soil degradation. In D. J. Greenland & I. Szabolcs (Eds.), *Soil resilience and sustainable land use* (pp. 99–118). Wallingford: CAB International.

Oldeman, I. R., Hakkeling, R. T. A., & Sombroek, W. G. (1990). *World map of the status of human-induced soil degradation: An Explnatory note*. Wageningen: International Soil Reference Centre.

Palacios-Fest, M. R., Mabry, J. B., Nialis, J. F., Holmlund, J. P., Miksa, E., & Davis, O. K. (2001). Early irrigation systems in southeastern Arizona: The ostracode perspective. *Journal of South American Earth Sciences, 14*, 541–545.

Park, C. C. (1983). Paleohydrologic reconstruction from stratigraphic evidence – A case study from prehistoric irrigation canals in Peru. *Journal of the Geological Society, 140*, 321.

Pimental, D. J., Allen, J., & Beers, A. (1993). Soil erosion and agricultural productivity. In D. J. Pimental (Ed.), *World soil Erosion and conservation*. Cambridge, M.A: Cambridge University Press.

Pinstrup, A. P., & Pandya-Lorsch, R. (1994). *Alleviating poverty and effectively managing natural resources* (FAO Discussion Paper I). Washington, DC: International Food Policy Research Institute.

Premathilake, R. (2006). Relationship of environmental changes in central Sri Lanka to possible pre-historic land-use and climate changes. *Palaeogeography, 240*, 468–496.

Raman, K. V. (2003). Multiple livelihood opportunities towards sustainable human development. In *Reshaping our earth view: Creative thoughts and alternative*. New York: Tata McGraw-Hill Pub. Co.

Raman, S. (2006). *Agricultural sustainability: Principles, processes and prospects* (pp. 13904–11580). Birmingham: The Hawoth Press Inc.
Scherer, S. J., & Yadav, S. (1996). *Land degradation in the developing world: Implications for food, agriculture and environment to 2020* (Food and Agriculture and the Environment Discussion Paper 14). Washington, DC: International Food Policy Research Institute.
Scherr, S. J. (1999). *Soil Degradation: A threat to developing-country food security by 2020?* (Vision 2020 Brief 58). Washington, DC: International Food Policy Research Institute.
Scholes, R. J. (1990). The influence of soil fertility on the ecology of south African dry savannas. *Journal of Biogeography, 17*, 415–419.
Sims, G. K. (1990). Biological degradation of soil. In R. Lal & B. A. Stewart (Eds.), *Soil degradation vol II advances in soil science* (pp. 289–320). New York: Springer.
Smaling, E. M. A., Stoonvogel, J. J., & Windmeijer, N. P. (1993). Calculating soil nutrient balances in Africa at different scales. II. District Scale. *Fertilizer Research, 35*, 237–250.
Stevenson, C. M., Jackson, T. L., Meith, A., Bork, H. R., & Ladeffaged, T. N. (2006). Pre-historic and early historic agriculture at Maunga Orito, Easter Island (Rapa Nui). *Chile, Antiquity, 80*, 919–936.
Syers, J. K., & Latham, M. (1994). Sustainable land management for mitigating stresses. In S. M. Virmani, J. C. Katyal, H. Eswaran, & I. P. Ibrol (Eds.), *Stressed ecosystems and sustainable agriculture* (pp. 332–338). New Delhi: Oxford and IBH Publishing Company.
Sys, C. (1976). *Land evaluation: Part I*. Ghent: ITC State University Ghent.
Szabolcs, I. (1994). The concept of soil resilience. In D. J. Greenland & I. Szabolcs (Eds.), *Soil resilience and sustainable land use* (pp. 33–39). Wallingford: CAB International.
UNCED. (1992). Chapter 24: Global action for women towards sustainable and equitable development. In *Agenda* (Vol. 21, pp. 349–354). Rio de Janeiro: United Nations Conference on Environment and Development.

Chapter 16
Erosion Control and Maintenance of Good Soil Tilth

Abstract Soil physical conditions control several factors, including water infiltration, retention, and uptake of plant nutrients, gaseous exchange, rooting environment, and erosion control, which are important in biological productivity and environmental integrity. This chapter discusses the significance of each in the context of sustainable agriculture.

Keywords Soil tilth · Erosion management

Soil physical condition controls several factors, including water infiltration, retention, and uptake of nutrients by plants, gaseous exchange, rooting environment, and erosion control, which are important for biological productivity and environmental quality. The maintenance of good physical conditions is necessary to provide an overall hospitable environment for plant growth and other beneficial organisms such as soil-micro –and mesoflora and fauna. Soil physical condition depends on a number of soil physical properties, such as soil texture, bulk density, aggregate stability, porosity, and organic matter content, all of which interact with each other. Soil tilth is a composite attribute which describes soil physical condition and its fitness as an environment for soil microbial populations and processes, seedling emergence, and root proliferation. It is a dynamic attribute that is impacted by natural processes as well as management practices (Karlen and Sharpley 1990). Tilth is quantified using the tilth index, which is computed from soil bulk density, core penetration index, uniformity coefficient, plasticity index, and organic matter content. The tilth index takes values from 0 to 1, with 0 representing unusable conditions for plant growth and higher values indicating progressively favorable to non limiting conditions (Singh et al. 1993).

Erosion is the dominant threat to soil productivity and the ability of humans to feed themselves (Dregne 1990). Soil erosion (soil loss) occurs when the rate of removal of soil by wind or water exceeds the rate of soil formation. The rate of soil formation is very low: 1 cm every 100–400 years, 0.01–0.0025 cm per year, or approximately 0.3–1.3 t/ha/year. Endogenous erosion (natural erosion), except on mountains, generally takes place at <1.0 t/ha/year and is comparable to rates of soil formation (Morgan 1986). Accelerated erosion through human intervention can

K. P. Nair, *Intelligent Soil Management for Sustainable Agriculture*,
https://doi.org/10.1007/978-3-030-15530-8_16

exceed 10 or even 100.0 t/ha/year, which is 10–100 times the soil formation rate. Very high erosion rates have been recorded in the Himalayan foothills of Nepal and in some areas of China, with values exceeding 200.0 t/ha/year. Texture, organic matter content, structure, and permeability influence the erodability of soil. Soils with very low clay and organic matter content and poor structural stability are highly erodible. Soils with intermediate texture are more erodible than coarse-textured sandy soils and fine-textured clayey soils.

Erosion leads to the displacement of fine soil particles and reduction in soil depth and soil tilth. This translates into several other deficiencies, such as, reduction in soil fertility, nutrient status, and recycling, soil organic matter content, available water, soil structure, and water infiltration. It also reduces soil chemical imbalances and toxicity (Stocking 1984). Erosion control is especially important in tropical soils, where harsh temperature and water regimes and poor structural attributes exacerbate management deficiencies.

16.1 Erosion Management

Soil renewal does not occur over the human life span; it takes place on a geological time scale. However, the issues of improving agricultural productivity and sustainability have to be faced over much shorter time scales of a decade or less. Thus, management methods should aim to prevent erosion rather than try to restore soil after erosion has occurred. Land use for crop cultivation influences soil erosion in two ways: (1): Many land-use practices leave the soil devoid of protective vegetation cover with only partial cover for significant periods of time. This leads to wind and water erosion. (2): mechanical disturbance of the soil by conventional tillage loosens up soil particles and leads to vulnerable displacement. This is aggravated by accelerated decomposition of soil organic matter. Intensive cultivation causes accelerated erosion through uninterrupted soil manipulation and loss of soil cover.

Agronomic practices that improve vegetative cover, ensure good plant root networks, and shorten the periods when the land is left bare and curtail soil erosion. Thus, strip cropping, agroforestry, retaining crop stubble, after harvest, on the soil, and crop cover control erosion. (Lal 1994) describes several examples of restorative practices used to ameliorate eroded lands in Africa. Soil structure enhancing measures, including addition of manures, mulching with organic residues, reduced or zero-tillage operations, such as, fertilizer drilling, and synthetic soil conditioning for fine-tilth seed beds, prevent erosion. Mechanical techniques such as contour bunding, shelter belts, and check dams reduce the energy of the eroding agent and prevent erosion. Conservation tillage is most effective in controlling erosion, in addition to bringing about its several other benefits.

The choice of management and technology for erosion-prone land depends on several factors such as fragility (slopes), nutritional status, and socio-economic conditions. In the case of prime agricultural lands under intensive cultivation, with high external inputs, erosion control measures will be advisable. Marginal and fragile lands need to be managed according to agroecological principles and are best left uncultivated if off-farm employment opportunities are available because the cost of labor and erosion control may not be commensurate with the benefits obtained. Conservation tillage methods using soil biological cover or vegetative barriers offer several advantages in terms of erosion control on slopy lands with poor structure.

References

Dregne, H. E. (1990). Erosion and soil productivity in Africa. *Journal of Soil and Water Conservation, 45*, 431–436.

Karlen, D. L., & Sharpley, A. N. (1990). Management strategies for sustainable soil fertility. In J. L. Hartfield & D. L. Karlen (Eds.), *Sustainable agricultural systems* (pp. 47–108). Boca Raton: Lewis Publications/CRC Press.

Lal, R. (1994). Sustainable land use systems and soil resilience. In D. J. Greenland & I. Szabolcs (Eds.), *Soil resilience and sustainable land use* (pp. 41–67). Wallingford: CAB International.

Morgan, R. P. C. (1986). *Soil Erosion and Conservation*. London: Longmans.

Singh, K. K., Colvin, T. S., Erbach, D. C., & Mughal, A. Q. (1993). Tilth index: An approach to quantifying soil tilth. *Transactions of the American Society of Agricultural Engineers., 35*(6), 1777–1785.

Stocking, M. A. (1984). *Erosion and soil productivity: A review*. Rome: Soil Conservation Program. Land and Water Development Division, Food and Agriculture Organization.

Chapter 17
Soil Fertility and Nutrient Management

Abstract The level of productivity of a cropping system is determined primarily by the interaction of crop genetic potential, environmental factors, and management options. Maintaining an adequate supply of mineral nutrients to crops is one of the most vital requirements for sustained crop growth and productivity. The chapter discusses several aspects of these factors and their interplay in maintaining optimal crop production. Specific importance in the chapter is also given to the role of organic matter in maintaining good soil fertility *vis-à-vis* ensuring good and healthy soil structure.

Keywords Soil fertility · Cropping system · Nutrient budgeting · Organic matter · Mineralization · Carbon · Nitrogen · Soil aggregates · Soil tilth · Structural resilience · High inputs · Need-based input application · Sustainable land management · Soil classes · Climate · Poverty · Lack of awareness · Unsupportive governmental policies · Qualitative descriptive indicators · Quantitative analytical indicators · Statistical limits · Scoring functions · Soil quality index

The level of productivity of a cropping system is determined primarily by the interaction of crop genetic potential, environmental factors, and management options. Maintaining an adequate supply of mineral nutrients to crops is one of the most vital requirements for sustained crop growth and productivity.

In natural ecosystems, sustainability is viewed in terms of the maintenance and stability of ecosystem productivity through tight nutrient cycling. Natural ecosystems maintain high productivity through a variety of above ground and below ground processes which serve to conserve nutrients, water, and soil organic matter. Soil biological activity generally maintains soil organic matter and fertility. Agroecosystems are, by their very nature, perturbed systems. They export large amounts of nutrients in crop biomass and therefore require large inputs, regardless of the amount of internal recycling (Hendrix et al. 1990). In addition, the plant growth rate is generally much higher and hence the decline in soil fertility is faster in agroecosystems. Large subsidies are required in the form of inorganic and organic fertilizers to recover exported nutrients. Nutrient depletion can occur through the removal of nutrients in crop harvesting and also as a result of erosion, runoff, leach-

K. P. Nair, *Intelligent Soil Management for Sustainable Agriculture*,
https://doi.org/10.1007/978-3-030-15530-8_17

ing, and so on. The more intensive the cropping, the greater is the depletion of soil nutrients. Ideally, nutrient status at every cropping cycle should be brought back to the initial level before crop removal, that is, inputs should be greater than, or at least equal to, outputs for sustainable use of land.

The level of production in a cropping system ultimately depends on the capacity to replace essential resources lost from the system during one crop cycle with inputs during the next. The inputs may be external (fertilizers) or internal (nitrogen fixation) or minerals derived (phosphorus). Sustainability is enhanced if internal mechanisms for renewal of lost nutrients are available. For example, atmospheric nitrogen fixation to augment nitrogen loss and phosphorus-mobilizing agents such as mycorrhiza to make up for phosphorus fixation in acid soils enhance fertility and sustainability. A general structure for resource balance is given in Fig. 17.1. The sources of renewal may be different in succeeding cycles, but, quantitatively they have the same effect on crop yield (Fig. 17.2).

The maintenance of soil fertility for sustainable crop productivity has faced considerable challenges, both in low-input resource-based agriculture and high-input intensive agricultural systems. Resource-extractive low-input cultivation results in serious depletion of soil fertility, which is exacerbated by soil erosion, drought, and desertification. These processes have resulted in agro ecological stagnation and perpetuation of hunger and malnutrition in sub-Saharan countries. During the 30 years ending 1996, on approximately 100 million hectares of cultivated land, a staggering loss of 700 kg of nitrogen, 100 kg of phosphorus and 450 kg of potassium net per hectare is estimated to have occurred (Sanchez et al. 1996). Hence, per capita food production has been declining approximately 2% per annum since the 1960s in sub-

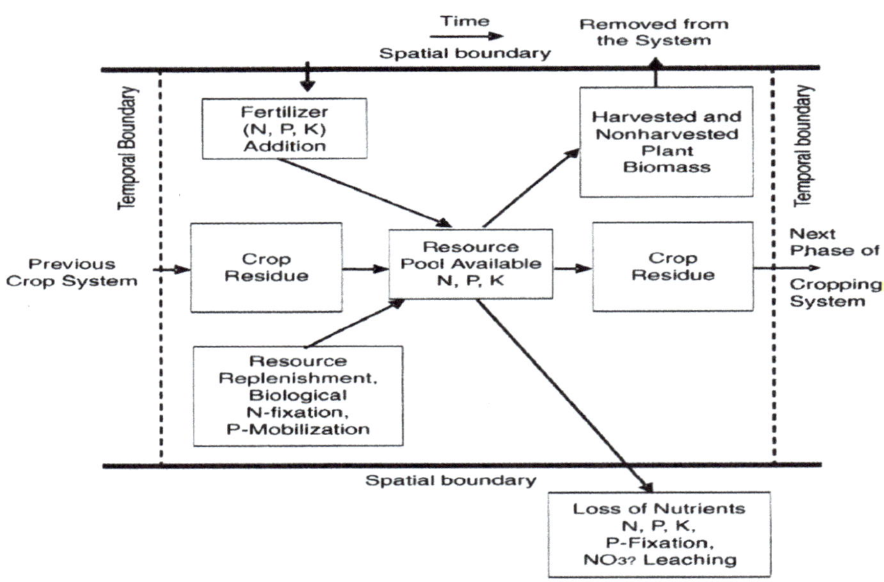

Fig. 17.1 Nutrient budget in a cropping system: sustainability = inputs>outputs

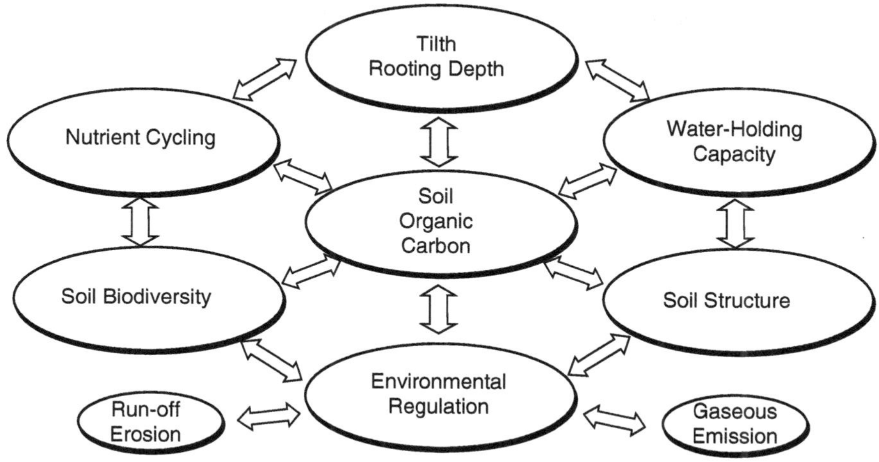

Fig. 17.2 Beneficial effects of soil organic matter

Saharan Africa, leaving almost 37% of the population chronically underfed, with the number almost doubling from 94 million in 1970 to 175 million in 1990 (Young 1997). This region will require approximately three times more fertilizer than it currently uses to maintain current levels of crop production without depleting soil nutrients according to Henao and Banaante (1999), where the nutrient requirements and fertilizer recommendations are made on the basis of routine soil testing. It is in such a context the utility of the "Buffer Power Concept," developed by the author, comes into crucial play, which will enable substantial reduction in fertilizer use by accurately quantifying the buffer power of each nutrient in these soils, following which precise fertilizer recommendations can be formulated, for which the author has developed accurate laboratory soil testing procedures.

Whereas the problems of extractive agriculture are relatively straightforward, the reasons for the threat to sustainability presented by high input intensive systems are more complex, although they are very real. The intensive long-term rice-wheat cropping systems in South Asia, such as, those of the Indo-Gangetic Plains, and the rice-rice system in Bangladesh and the Philippines are facing very serious threats to sustainability through declining and stagnating crop yields and low factor productivity, especially for nitrogen and phosphorus (Cassman et al. 1995; Pingali et al. 1995; Greenland 1997; Ladha et al. 2003). Several forms of nutrient-related deficiencies, including micronutrient deficiency, declining organic matter content, deleterious effects of excess fertilizers, and large energy inputs, have been posited as possible explanations. The integrated plant nutrient system has now been advocated as the best and most beneficial tool for maintaining soil fertility and effective nutrient management (Gupta et al. 2003). This will be discussed in one of the following chapters.

If a critical analysis is made of the above dangerous situation, one would conclude that it has much to do with the degradation of soil, which has resulted through various routes. And, of all, soil nutrient management is the most vital. It is very important to note in this context that of all the factors of crop production, the nutrient factor is the most crucial, influencing more than 50% of the crop performance, yet, it is least resilient to management (Fresco and Kroonenberg 1992). Declining or plateauing crop yields have their origin in improper soil nutrient management. This author has made critical field observations to this effect in the green revolution areas of India, notably Punjab State, popularly known as the "cradle" of Indian green revolution. Chemical fertilizer application over decades starting from mid sixties, when the high yielding and "miracle dwarf" rice and wheat varieties were introduced to the country, has been unbridled, in the hope of harvesting ever more crop yields. The intensive and extractive rice-wheat rotations supported by high chemical fertilizer input has taken a very big toll on soils' inherent productive capacity and soil organic matter has been depleted far beyond permissible levels. Ground water is loaded with excess nitrates, making it non potable. There are hundreds and hundreds of acres where even a blade of grass will not grow. Noting the very alarming soil situation, the State Government has enacted official governmental policies where the farmers have to compulsorily take a legume nitrogen fixing and soil organic matter enriching crop, like *Sesbania* or *Glyricidia,* and abandon continuous the rice-wheat monoculture, for some years. It is in these circumstances where "The Nutrient Buffer Power Concept" comes into active play in helping precisely quantify the nutrient buffer power of important soil nutrients, and help devise, reproducible and economic rates of fertilizer application, which enhances efficiency of their use and minimize soil pollution and environmental damage. Instances like these have also been cited in other parts of Asia, as discussed in the earlier chapters of this book.

Of late, widespread micronutrient deficiencies are seen in the tropical belt, of which, zinc deficiency is becoming rampant. In fact, in the Punjab State, due to unbridled use of nitrogenous, phosphatic and potassic fertilizers, much of inherent soil zinc has been mined out in the rice-wheat monoculture cropping system. There are areas where a mere application of 10 kg Zn SO_4 has been the determining factor in sustaining crop yields, with normal rates of N, P and K application. Yet, the question whether the prescribed rate is optimal, sub optimal or over optimal is left to doubt, because almost always the zinc fertilizer recommendations are based on the routine soil zinc test based on DTPA extraction. This has been found to be unsatisfactory. That a precise determination of the Zn buffer power leads to much better zinc fertilizer recommendations as, compared to routine DTPA test, has been reported by the author (Nair 2004).

17.1 Role of Soil Organic Mater

Soil organic matter is a key component of all terrestrial ecosystems, natural and managed. It is probably the dominant positive factor in terms of sustainable land management, because of its value-adding effects on almost all other key productivity- and sustainability enhancing attributes, such as, soil structure, water holding capacity, nutrient supply, and erosion control, all of which decline when soil organic matter status suffers.

Most of the beneficial effects of soil organic matter are attributable to the prolific microbial diversity and population that it generates and supports, which are conducive to soil resilience (Swift and Anderson 1994). Soil biodiversity and soil bio chemical processes have been recognized as vital to all ecological functions that impart resilience and sustainability to soil performance.

The quality and quantity of soil organic matter have an important impact on almost all ecosystem processes. Owing to its influence on soil physical properties and nutrient supply, soil organic matter is used as an index of soil fertility. In addition to the influence of natural factors, such as, soil type, climate, and place, the quality and quantity of organic matter in soil is profoundly influenced by management options, including crop selection, tillage practices, addition of organic matter (farm yard manures, green manures, mulching), and chemical fertilizers.

The soils of the tropics are especially low in soil organic matter as a result of the higher temperature and prevailing water regimes, which accelerate decomposition of soil organic matter. Soil disturbance through deep ploughing and removal of crop residues hastens soil organic matter decomposition. A decline in soil organic matter leads to a decline in crop productivity.

17.1.1 Organic Matter Fractions

On the basis of the individual components residence time, soil organic matter has been classified into decomposable plant material, resistant plant material, microbial biomass, and chemically and physically stabilized inner fractions. The lignin content has been implicated as an important factor at the rate of decomposition of organic residues. The Century model (Parton and Rasmussen 1994) proposes the existence of five different organic matter pools in the soil, based on their turnover time as depicted in the following figure (Fig. 17.3).

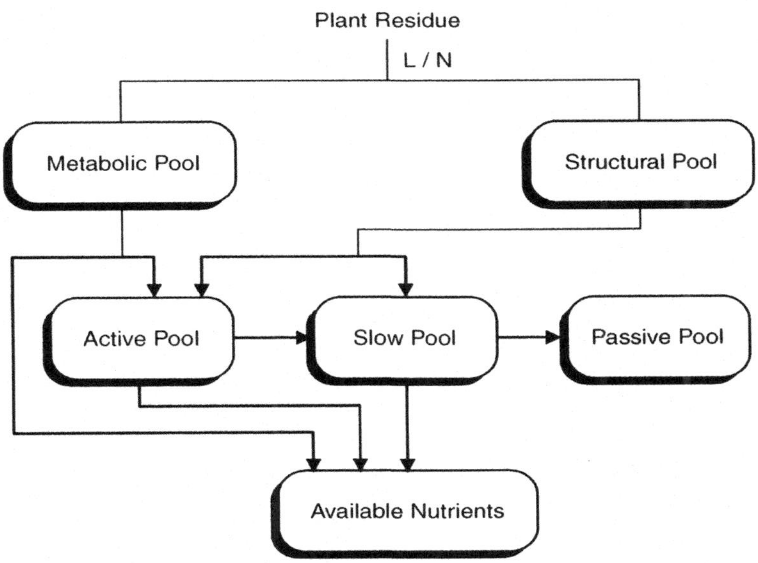

Fig. 17.3 Pathways for mineralization of carbon, nitrogen, and phosphorous in soil organic matter. (Source: Adapter from Parton and Rasmussen 1994)

Organic matter, such as, plant litter residues, root detritus, and animal droppings is divided into a structural pool and metabolic pool with 1–5 year or 0.1–1.3 year turnover times. These are transformed into active, slow, and passive pools. The active fraction has a short turnover time of 0.5–5 years and is called soil biomass. It is the most active pool and is active in the mineralization of carbon, nitrogen, phosphorus, and sulfur. The slow pool has a turnover time of 20–40 years. It contains these same elements, physically and chemically protected, and it decomposes slowly to liberate them. The passive fraction is the most inert, chemically stable, and physically protected and has a long turnover of 200–1500 years. This fraction plays a crucial role in the adsorption, release, and detoxification through immobilization of soil-applied pesticides (Raman 1987; Raman and Rao 1987).

Soil biomass is the home and elixir to millions of microorganisms, which are the prime controllers and mediators of all sustainability-enhancing features of soil. Soil biota plays a crucial role in the functioning of the ecosystem, its health, and its productivity. Sub groups of the soil biota are involved in a number of key functions responsible for plant growth and sustainability such as mineralization and recycling of important nutrients, maintenance of soil fertility and soil structure, plant protection and growth enhancement, detoxification, detoxification of naturally occurring and human-made pollutants, carbon sequestration, and synthesis and decomposition of organic matter. These functions are important for the long-term viability of the agricultural system. A host of herbivorous invertebrates, microbial pathogens, and rhizosphere organisms also, directly affect plant growth and plant health.

Maintenance and enhancement of soil fertility through biological control (that is, addition of organic matter) is an important pathway for sustaining soil fertility and resilience for crop production. The biological control of fertility through manipulation of soil biological processes is a complex and interactive process. Among the factors that independently determine the efficiency of use of added organic matter are:

1. The nature of organic matter added
2. The nature and size of the different pools of soil organic matter, which play critical roles in nutrient release, cation exchange, soil aggregation etc
3. The presence of soil microfauna, which work as regulators of decomposition, and microfauna, which act as soil engineers which improve soil physical properties, such as porosity and aeration
4. Surface litter and soil organisms, which have a decisive role in soil aggregation and water retention and availability and
5. Synchronization between plant growth demands and the release of plant nutrients from above-ground inputs and roots (Singh and Kushwaha 2000).

The Tropical Soil Biology and Fertility Program has conducted extensive research on all these aspects to develop appropriate and innovative approaches to sustain and improve tropical fertility through the management of biological processes and organic resources (Woomer and Swift 1994).

17.2 Soil Structure and Agricultural Sustainability

The structure of soil relates to the formation and arrangement of aggregates from soil particles to produce a continuous network of pores of different sizes which act as spaces for water and gases (Oades 1984). Soil aggregate is the basic component of soil structure and the process of aggregation and aggregate architecture play a key role in determining the overall structure of a soil.

Soil structure relates to the formation and arrangement of aggregates from soil particles to produce a continuous network of pores of different sizes which act as spaces for water and gases (Oades 1984). Soil aggregate is the basic component of soil structure and the process of aggregation and aggregate architecture play a key role in determining the overall structure of a soil.

Soil structure is affected by a number of intrinsic properties of the soil and also management options. The formation of aggregates and their stability depend on a number of factors, including soil organic matter status, presence of weathered oxide materials such as aluminum and iron oxide, organo mineral complexes, soil microorganisms, and the nature and amount of clay minerals. Illitic and smectitic soils are more prone to structural breakdown than kaolinitic soils. Various physical, chemical, and biological forces influence soil structure. Physical forces such as drying, shrink-swell, root growth, compaction, chemical forces involved in cation and polysaccharide-assisted binding of amorphous iron and aluminum oxides, and biological binding through fungal hyphae and microorganisms mould particles into aggregates.

Soil organic matter is among the most important factors in the formation of soil aggregates. Seventy percent of soil organic matter resides in the aggregates (Jastrow et al. 1996). Functional groups such as hydroxyl, carboxyl, and carbonyl strongly bind with clay minerals and oxides and act as active catalytic centers for binding and decomposing pesticides and other toxic molecules (Raman and Rao 1989). The stabilization of soil aggregates largely depends on different components in the organic matter. Based on the capacity to bind soil particles together, soil organic matter was divided by Tisdall and Oades (1982) into three categories: (1) The strong and persistent binding fraction consists of humic and fulvic acids, which form organomineral complexes with amorphous iron and aluminum oxides and with aluminum silicates. Between 52% and 98% total soil organic matter is involved in such complexes in the form of micro aggregates < 20 μm. (2) Roots and fungal hyphae constitute temporary binding agents and form macroaggregates >250 pm that are fairly long-lived, but, depend, heavily on soil management practices such as crop rotation. (3) Plant and microbial polysaccharides, gums, and more complex soil polysaccharides are transient binding agents and contribute to micro – 20–250 μm and macro >250 μm –aggregation. They are persistent only for short intervals, from weeks to a few months, and are affected by the soil and crop management practices.

Soil structure is among the most important factors in sustainable production because it has a major influence on the ability of the soil to support root development, to receive and store water and make it available to the plant, to recycle nutrients, to disperse agricultural chemicals, and to control erosion. Loss of soil structure can result in physical, chemical and biological impairments of the soil, which individually and collectively diminish soil functions and soil productivity.

The specific role of soil structure in crop yield has not been completely understood because different components within the concept of soil structure have themselves not been clearly defined or demarcated. Thus, soil structural form stands for total porosity and pore size distribution; structural stability represents the ability to withstand various forms of stress, such as, different forms of tillage, traffic by machines (tractors, and other ancillary agricultural machinery), and compaction; and structural resilience stands for the ability to recover from imposed stresses such as wetting and drying, root activity, and faunal and other biological activities. All these combined determine the structural vulnerability of a soil. The rate of changes of soil structural form is another parameter of interest, because, structural changes can occur over a wide range of time spans, from hours to years, to centuries. A reasonable time horizon compatible with the farmer's needs is optimal to monitor the way a soil structure responds to management and climatic changes (Kay et al. 1994).

The extent of aggregation within a soil controls soil porosity. A well-aggregated soil maintains a structure that allows a variety of pore sizes (Lynch and Bragg 1985). Pores of all sizes are important for root growth and proliferation, to accommodate soil microorganisms, such as, nematodes, to control infiltration and storage of water, and to facilitate gaseous exchange. Aggregates also provide microsites for the proliferation of soil microorganisms.

The breakdown of soil aggregates affects soil tilth, which is the most important holistic attribute important for plant growth and erosion control. Soil erosion is among the most common causes of soil degradation and consequent lowering of crop yield, and a threat to sustainable intensification of crop production. Soil erosivity depends quite significantly on aggregate stability, that is, the ability of soil aggregates to withstand the impact of raindrops and surface flow of water. Crusting of surface soil occurs when aggregates disintegrate under the impact of raindrops or flow of surface water and fill the pore spaces with fine soil particles, resulting in surface sealing. This impedes water infiltration and leads to surface runoff and erosion. Seedling emergence also becomes difficult. Soils with high organic matter content, and specifically those rich in carbohydrates and proteinaceous materials, resist crusting.

17.2.1 Soil Management and Soil Structure

Soil management options affect aggregate formation and stability. It is not possible to single out good or bad management practices in terms of maintaining soil structure. The technologies that farmers are constrained to adopt usually comprise of a mixed bag and exert both good and bad effects on soil structure. Specific options do not exist. In general, the system of technologies should be such that structural changes are balanced over time. Common agricultural practices do not generally have severe adverse effects on soil structure.

The addition of various forms of organic matter, such as crop residues and manure is particularly advantageous in terms of maintaining or improving soil aggregation. Residues also alter microbial community composition and numbers. Fungal biomass is important in aggregation (Gupta and Germida 1988) and is largely influenced by crop rotation.

Organic matter facilitates air encapsulation within the soil during water uptake and prevents aggregate breakdown and slaking. The breakdown of soil aggregates leads to oxidation of organic matter and disturbs atmospheric concentration of carbon dioxide. Small quantities of methane are released during their anaerobic decomposition (Sullivan 1990).

17.2.2 Management Constraints in Soils with Poor Soil Structure

The management of soils with poor structure, such as vertisols and alfisols, presents a challenge. The surface soil structure of vertisols can vary from a fine crumb to coarse blocks. The size of aggregates in the seedbed is important for germination and seedling emergence. Vertisols with geometric mean aggregate size of approximately 2 mm present no problem for seeding emergence, but, a coarser structure with bigger aggregates results in poor emergence because seed-soil contact is lower and light penetration to germinating seeds is higher.

Since surface structure influences ease of tillage, evaporation, erodibility, and infiltration, preparing seedbeds of vertisols with the proper tilth is a problem. The wetting and drying cycles of vertisols exert considerable influence on soil structure, and changes are not completely reversible. Successive wet and dry cycles decrease aggregate sizes and facilitate the production of fine tilth.

Natural development of tilth is important for small farmers without access to tillage equipment. Irrigation helps to manipulate wetting regimes to promote fine tilth. Capillary wetting of such soils is conducive to fine tilth, whereas flood wetting leads to coarse structure. Preparing a ridge-furrow system and applying water along every other furrow so that the intermediate ridges and furrows are wet by capillary action can facilitate the creation of fine tilth. In dry and semiarid regions, the tied -ridges system is used to conserve water and promote aggregation. Alternate furrow irrigation reduces problems related to the irrigation of coarsely structured vertisols.

17.3 Efficient Use of Inputs and Sustainable Agriculture

Previous sections of this chapter have stressed the importance of maintaining adequate levels of such inputs as mineral fertilizers, organic matter, and water for sustainable agriculture for optimal crop production. Perpetuating the potential productivity of land in an environmentally safe and economically viable manner requires that the magnitude, flux, and efficiency of use of inputs –in particular fertilizer, water and energy – are optimized. Improper management of these factors has been the cause of land degradation and environmental pollution in both low and high external input intensive (modern type industrial agriculture) agriculture.

It should be emphasized that the amount of external inputs needed is guided by the efficiency of their use. Nitrogen fertilizers are the most blatantly misused input in agriculture, especially among the developing countries, in particular, India. In India, since the last 3 years, price of nitrogen fertilizer (urea) has been decontrolled by the government, which means that there is no price fixation at high levels, in contrast to both phosphatic and potassic fertilizers, and, in consequence, millions of farmers go in for nitrogen fertilizer thinking that they can compensate for no or very meager application of both phosphatic and potassic fertilizers, whose prices have steeply increased, with additional urea application. This has led to a very great imbalance in nutrient status of crop fields and is creating a great damage to soil resources. And, the governmental machinery is not taking adequate measures to correct the imbalance. In fact, nitrogen fertilizer is the most commonly under used (in Africa) and over used (in Asia) fertilizer. Agriculture in sub Saharan Africa has suffered from inadequate fertilizer use, which has led to resource degradation, such as nutrient depletion and erosion, and also environmental pollution. High temperature and moisture regimes have exacerbated these conditions through accelerated oxidation of soil organic matter, which has left soils biologically dead and has led to serious erosion. The classic example is in the State of Punjab, the "cradle" of the so-called green revolution, where the most brazen fertilizer use was seen, especially nitrogenous, resulting in extensive soil degradation. In hundreds of acres even a blade of grass will not grow now. High temperatures and moisture regimes have exacerbated these conditions through accelerated oxidation of soil organic matter, which has left soils biologically dead and has led to serious erosion. The adverse effects are only too visible in the form of declining productivity and chronic hunger and malnutrition. India, in many parts, depicts this pathetic picture, despite the so-called green revolution. The International Institute For Tropical Agriculture, the Tropical Soil Biology and Fertility Institute and the International Centre for Agroforestry together will work on the problem of soil fertility degradation in Sub Saharan Africa under the SSA Challenge Program initiated by the Consultative Group for International Agricultural Research (CGIAR) to improve food security and sustainable livelihoods in Sub Saharan Africa. It is against the background of this dismal picture, the "Buffer Power Concept" aiming to rationalize fertilizer use targeting optimum crop yield assumes great global significance. And, the success stories earlier chronicled in this book lend credibility to the concept.

High-input agriculture has seen indiscriminate overuse of, especially, nitrogenous and phosphatic fertilizers, which has caused great damage to the production resources as well as to the environment. The excessive use of fertilizers often leads to wastage and run off, when there is especially heavy rainfall, causing pollution of surface and ground water and leading to excessive loading of nitrates in ground water making it non potable. The classic example of this phenomenon is in Punjab State in India, in the cradle of the so-called green revolution, characterized by unbridled use of these two fertilizers, especially nitrogenous. The choking of waterways with algal bloom, leading to anoxic environment, which is lethal to aquatic species, is a major problem. The availability of fairly cheap nitrogenous fertilizers (especially now in India following the order of the central government in New Delhi decontrolling nitrogenous fertilizers, where there is no ceiling price now as opposed to phosphatic and potassic fertilizers where the ceiling prices are quite high against the background of Indian economy, as the resource base for both is imported), and the availability of cheap nitrogenous fertilizers, and the prevalent policy of governments in providing subsidies to farmers to boost agricultural production, both in the developed and developing countries, especially in the latter, has led to farmers totally ignoring the efficiency factor, resulting in skewed use of these fertilizers leading to many nutrient related problems in the soil. Often blanket doses of fertilizers are recommended to the farmers by the extension (agricultural) staff, in countries like India, without concern for what quantities of nutrients are actually required during crop growth, and when, regardless of their availability to the roots in accessible forms and the pathways for their loss. The practical experience of this author in many instances in Africa and Asia is that such situations are more often than not encountered. This arises from the fact that in many cases only routine and "text book" oriented soil tests are performed, on the basis of which fertilizer recommendations are prescribed and the agronomists/soil scientists/extension specialists bank on these test results. "The Nutrient Buffer Power Concept" has very successfully ameliorated this anomaly in many field situations, some of which have been discussed in the earlier chapters of this book. Often, the most profitable rates of nitrogen, phosphatic and potassic fertilizer application are followed year after year by farmers across different agroecological zones, as in the case of rice-wheat crops production in South Asia. India and Pakistan are prime examples of this anomalous situation. Ironically, while the fertilizer industry is happy with this situation, the long-term damage done to soil resources far outweigh the boost in crop yields. Imbalance between the organic matter content and mineral content in the soil aggravates this situation further as has clearly happened in the Punjab State due to mistaken enthusiasm of the farmers and ill advice of the field staff on the high input technology, hallmark of the so-called green revolution. The disastrous environmental and social causes, even leading to the spread of dangerous diseases like cancer in the State, especially in districts like Gurdaspur, in Punjab State, is there for everyone to see.

Overuse of irrigation water, in irrigated agriculture, and its perils, of machinery for conventional tillage and harvesting, leading to excessive soil erosion, land degradation, and environmental pollution, are clearly dealt with, with specific reference to Punjab State (Prihar et al. 2000).

17.3.1 Need-Based Input Application

The best prescription to stall inefficient use of external inputs is to follow the golden rule of need-based supply of inputs, such as tillage, water and fertilizers and to check their wastage. Of the three, fertilizers, which assume crucial importance, as nutrients carried by them determine more than 50% of the crop's efficiency of production, nitrogenous ones need special care in handling. In addition to determining the precise quantity of nutrients and calibrating the quantity of fertilizer needed to supply it, split application of nitrogen and phosphorus based on site-specific research and the accuracy of the timing and location of application are important in enhancing efficiency of their use. Band application of phosphorus and nitrogen will reduce soil fixation of the former and gaseous loss of the latter through ammonia volatilization, as a consequence of urea hydrolysis, and also leaching of oxidized nitrates through leaching. The use of more efficient forms of fertilizers, such as urea granules and sulfur-coated and neem (*Azadiracha indica*)- coated urea, and the use of urease inhibitors in the tropics, help to slow down urea hydrolysis and loss due to ammonia volatilization. Nitrification inhibitors can control nitrate leaching and denitrification, especially in submerged soil conditions. Nair and Sharma (1979) have carried out some pioneering research on this aspect on maize nutrition with nitrogenous fertilizers.

Crop rotations provide an excellent means to reduce several adverse effects of continuous monoculture (as rice-wheat in Punjab State in India), especially pest and disease vulnerability and erosion, in addition to improving soil structure. The importance of sensible water management in irrigated agriculture can never be under estimated.

The basic goal of agricultural sustainability is to mimic natural processes and systems as much as possible by maximizing the internal recycling of nutrients. However, this is difficult and inadequate in agro ecosystems because of inevitable export of nutrients, which requires the input of chemical fertilizers to balance their removal from the soils. Our current knowledge of the necessity to retain natural and ecological processes not only to sustain agricultural production, but, more importantly, to safeguard our ecosystems has led to new integrated soil management techniques with multiple benefits to offer. Two of the most important techniques, that are being increasingly adopted globally, are conservation tillage practices and integrated plant nutrient management. "The Nutrient Buffer Power Concept" falls under the latter category.

17.4 Stressed Ecosystems and Context-Based Technologies

The techniques for soil management and land use described in this chapter serve as generic guidelines for sustainable land management. They need to be translated into appropriate technologies to match the desired productivity, environmental safety, and socioeconomic context. Modern science – and technology –based combinations of high yielding crop varieties, fertilizers, pesticides, and tillage have proved highly effective in boosting crop productivity in lands with enabling attributes, such as, good climatic conditions, low biotic stresses, high productive potential, and supportive socioeconomic and political environments, as in the case of the United States and Western Europe, where agriculture has become totally commercialized. In contrast, the poorer countries in the tropics have more than 90% of the global total of 5 billion ha of Class 1V, V, and V1 lands (43% in Africa, 28% in South America, and 20% in Asia) and support more than 50% of the global population. These are ecologically stressed conditions, which are not conducive to intensive cultivation and high productivity. The people are mostly poor and practice low-input-low output agriculture. Climate and soil constraints of various types dominate the tropics and preclude uniform agricultural practices of any kind. Climatic stresses can take the form of inadequate or excess rainfall or an unfavorable seasonal distribution, extremes of temperature, or unfavorable solar radiation distribution. Only 57% of land in the tropics is estimated to have high potential for crop production, because of good water availability. The remaining land is stress prone, with 21% experiencing low and uncertain rainfall and 22% experiencing excess rainfall, leading to soil erosion, leaching, soil acidification, and nutrient depletion (FAO 1989). In the former category falls a lot of land in central and south central India which only can grow crops like sorghum (*Sorghum bicolor*), *bajra* (*Pennisetum typhoides*), red gram (*Cajanus cajan*) and *ragi* (*Eleusine coracana*) (all of which are poor man's diet) and cotton, where the country's poor live, but, of late, cotton crop (the Bt varieties) have become the "killer crop" with thousands of poor farmers committing suicide as they become bankrupt owing to indebtedness, the cause being high seed cost, and for other external inputs like fertilizers and even supplemental insecticidal sprays (because despite the claim of the multinational seed agencies like Monsanto claiming Bt cotton to be pest free, invariably supplemental insecticidal sprays are needed), all of which escalate the cost of production and when the crop fails the farmers are led into penury and bankruptcy, which leads to their suicide, which has become a global scandal, putting India in such poor light).

17.4.1 Climate

In rainfed areas, climate imposes a number of limitations on productivity, soil conservation, and enhancement of bio diversity. The growing season can vary from place to place and from year to year, owing to variable patterns of distribution of rainfall which can be highly seasonal, extending from less than 2 months to more than 7 months. It is difficult to plan cropping systems under these constraints. Low fertility, low water holding capacity, surface-sealing soils, high evaporation rates, and tropical storms are additional problems which compound low-moisture regimes. Hence, the development of technologies for sustainable land use is far more challenging for the tropics than for temperate regions.

17.4.2 Soil Constraints

(i) *Texture*

Next to climate, soil-related constraints are the most common form of impediment to sustainable agriculture in the tropics. Soils in the tropics are mostly coarse textured alfisols and entisols with poor drainage owing to poor structure, which are quite difficult to manage. Most arid and semiarid regions are prone to accelerated water and wind erosion. This has a severe impact on crop yield and agronomic productivity. Past erosion has caused yield reductions of 10–20% in Africa (Dregne 1990).

(ii) *Nutrients*

Soils in the arid and semiarid areas of the tropics are inherently low in fertility because of low levels of organic matter and stored nutrients. Nutrient depletion and organic matter loss are alarmingly high in sub-Saharan Africa owing to oxidation of organic matter and nutrient leaching. The condition is relatively better in Asia and Latin America. In addition, specific nutrient constraints also exist which require special attention. Tropical soils are generally phosphorus deficient. In humid areas, where acid soils prevail, aluminum and manganese toxicity and phosphorus fixation are endemic. Similarly, nitrogen management is a problem in arid and semiarid regions. Low moisture level renders mineralization of nitrogen difficult, and excessive humid conditions accelerate nitrate leaching. Acidification also impairs nitrogen mineralization and uptake by plants. Potassium deficiency is also common, but, it is often masked by deficiencies of other nutrients. It occurs in semiarid regions, as a consequence of inherently low fertility or fixation by illitic minerals. In high-humidity areas, potassium is leached from soil because of the presence of low – activity clays, which cannot retain potassium (Vleck and Vielhauer 1994).

(iii) *Soil Organic Matter*

Organic matter is chronically low in alfisols and entisols in drylands when precipitation is < 500 mm annually. Organic matter is a function of soil, climate, management, and mean temperature and is influenced by the nature of the crop cultivated. Cultivation of newly cleared forests leads to a decrease in organic matter. The mineralization of organic matter in these soils is generally enough to support low productivity.

These constraints need to be addressed on a site-specific basis suitable for the needs and capabilities of the farmers in these regions. For example, water constraints in arid and semiarid regions can be overcome with the effective and ingenious use of rainfall. Different types of water management (water harvesting, external storage, *in situ* soil storage with crop cover and mulch management, tillage, bunding, ridge and contour cultivation, conjunctive use, and supplemental irrigation) are resorted to according to local requirements and possibilities. Similarly, liming, inoculation with mycorrhiza, and band application of phosphorus fertilizers tackle phosphorus deficiency and fixation. Nitrogen use efficiency is improved through alley cropping. Special land-use techniques such as agroforestry that help potassium cycling through the soil and vegetation system are needed to prevent potassium loss. Deep rooting vegetation helps nutrient pumping by the roots, of the large portion of the potassium residues in the litter and soil organic matter of the top layer. In the absence of this layer, potassium may leach below the root zone. Mixed cropping systems containing deeper rooting tree crops prevent losses by root interception, even under high rainfall conditions.

Slow-release potassium fertilizers are also beneficial under these conditions. Continuous replenishment of organic matter is necessary. The addition of crop residues is not very effective unless it is combined with farmyard manure or biogas slurry because synchronization of plant growth and mineralization is a problem. An exhaustive account of the nature and magnitude of the problems of the stressed ecosystems of the tropics and guidelines for the management of various constraints can be found in the proceedings of a conference on "Agroclimatology and Sustainable Agriculture in Stressed Environments" held at Hyderabad, India in 1993 (Virmani et al. 1994).

17.5 Socioeconomic Compatibility

Biophysical constraints apart, socioeconomic conditions such as poverty, lack of awareness, unsupportive governmental policies, lack of political will and commitment, small farm holdings, and technological constraints compound management problems in the tropics. On account of these, the pattern of agricultural practices has to be different and context specific. Land management strategies have to be tailored to the productive capacity of the land, socioeconomic conditions, long term environmental effects, ecosystem characteristics, population policy, marketing and infrastructure facilities, and cultural ethos of specific areas.

The countries of the tropics have a long history of traditional forms of agriculture suited to particular ecosystems and socioeconomic backdrops. Existing land systems have evolved through many complex interactions between several intrinsic and extrinsic forces such as soil physical, chemical, and biological properties on the one hand and human dominance and economic implications of various kinds on the other, especially in the decades following the so-called green revolution. Although the basic tenets of sustainability are the same, the priorities of traditional and high intensity commercial agriculture, euphemistically called the green revolution, differ on account of their very nature and goals. Whereas productivity, profit, efficiency of resource use, resource conservation and environmental safety, markets, and infrastructure are dominant issues in commercial agriculture, food security, stability, riskfree economic and social needs, and livelihood opportunities are paramount in traditional agriculture. These aspects, which are so very crucial to livelihood security of millions of poor farmers around the world, have not been addressed adequately and imaginatively by the high intensity input agriculture of the so-called green revolution, which assume a great importance in the context of this book. "The

Nutrient Buffer Power Concept," and its role in sustainable agriculture, hence, shows the right way forward.

17.6 Soil Quality and Sustainable Agriculture

The prime question regarding soil quality and sustainable agriculture, is, what do we understand by the term soil quality? The concept of soil quality has been around for a very long time in the limited context of the capacity of the soil for crop production. Larson and Pierce (1991) extended this definition of soil quality to include its environmental functions, such as, the capacity of the soil to regulate and partition water through the environment and to serve as an environmental filter and buffer. They defined it as the capacity of a soil to function within the ecosystem boundaries and to interact positively with the environment external to the ecosystem. The Conference on Defining Soil Quality for a Sustainable Environment (1992) endorsed a recommendation that the relationship of a soil to plant, animal and human health, be included as an additional component in the evaluation of soil quality. Accordingly, Doran and Parkin (1994) suggested that the quality of agricultural soils could be defined as "the capacity of a soil to function within biological productivity, maintain environmental quality and promote plant and animal health." The Soil Science Society of America adopted the general definition of soil quality as "the capacity of a specific kind of soil to function, within natural or managed ecosystem boundaries, to sustain plant and animal productivity, maintain or enhance water and air quality and support human health and habitation" (SSSA 1995). The terms "soil health" and "soil quality" are often used interchangeably. In general, though, the former is commonly used by producers and in popular literature, whereas the latter term is preferred in scientific and technical literature (SSSA 1995; Karlen et al. 1997).

Soil quality is a composite of inherent and dynamic components. Inherent soil quality depends on the various factors responsible for the formation of a soil, such as climate, topography, parent material, vegetation, and time (Jenny 1941). It is not drastically altered by management practices and varies slowly over time. Each soil has an innate functional capacity, a concept that is useful in comparing the suitability of soils for specific uses. The dynamic component of soil quality is influenced by soil management practices, such as, tillage, cropping systems, and use of fertilizers and chemicals used to control pests. These affect both the properties of the resource base and environmental quality, including water and air quality. The dynamic component of soil quality changes perceptibly over time and is important in evaluating soil quality in terms of agricultural sustainability as depicted in the following figure (Fig. 17.4).

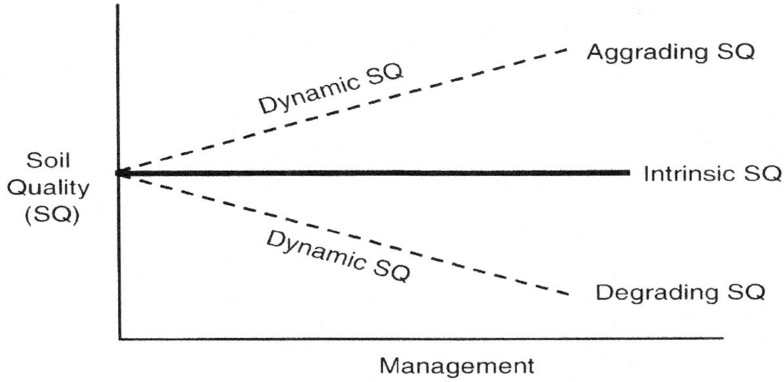

Fig. 17.4 Intrinsic and dynamic soil quality and management

17.6.1 Components of Soil Quality

Soil quality in relation to agriculture is a composite of the ability of a soil to be productive, ensure environmental integrity and health (both of plant and animal, including humans). Good conditions for the germination of seed, root growth, a balanced supply of plant nutrients and water to the growing plant, enabling biological conditions for nutrient storage and recycling, and the absence of adverse physical and chemical conditions, such as, soil salinity, soil alkalinity, acidity, toxicity, and pathogens which are conducive to proper plant growth and productivity. These attributes may be inherently present to different degrees in native soils and can be maintained or improved by cultural and management practices. Management to enhance soil attributes for sustainable crop production has already been discussed in an earlier chapter.

The environmental function of soil is well understood than the productivity function. The most important environmental functions of soil are infiltration and flow of water through the soil profile, nutrient retention and recycling, decomposition and mineralization of organic constituents and amendments, detoxification of organic molecules, such as, pesticides, and retention of deleterious metal ions, including mercury, cadmium, and aluminum. These are important in controlling *in situ* quality as well as the quality and extent of soil erosion, water run off, groundwater infiltration, and pollution.

The health component of soil quality is related to its impact on plant, animal, and human health. This is the least understood, also least investigated component, which includes the influence of soil deficiencies and the direct and indirect effect of toxins, pathogens, and other pollutants on the nutritional quality of food (Doran et al. 1996). There are multiple linkages between soil quality, the food chain, and health of plants, animals and humans.

Doran et al. (1996) listed three major pathways through which soil can affect the health of higher animals, including human beings as follows:

1. Direct poisoning from contaminants present in the soil
2. Direct and indirect poisoning through contaminants sequestered on soil organic matter being carried into water and air by erosion and transmitted directly or serially to plants, animals, and human beings and
3. Effect of soils with deficient or excessive quantities of certain elements on the quality of food grown on them

The first pathway pertains to direct health effects of soil. The effects are likely to be very local and occur as a result of high concentration of certain toxic metals, such as, mercury, cadmium, selenium, and lead, which may be naturally present or anthropically added through industrial waste and sludge. Similarly, pesticides and other organic chemical contaminants can enter directly into the food chain. Soils can reduce dietary risks from pesticides and other toxins by acting as filters (adsorption). Good-quality soils reduce the need for chemical pesticides by strengthening healthy soil microbial communities, which can suppress root diseases and also strengthen plant defense systems against pest attacks by mobilizing nutrients in the root zone.

The second pathway, through which soil contaminants are carried into the air and into the surface and groundwater and then transmitted to plants, animals, and humans, is the most common and is a very serious public health issue. Nitrate loading of surface and ground water, a common problem in intensive farming areas, leads to methemoglobinemia in children. Dissolved organic fractions are bioconcentrated in aquatic creatures and enter animal or human systems through the food chain. Similarly, toxic chemicals such as pesticides and harmful metals adsorbed on soil particles can enter streams and have direct and indirect effect on the health of animals and human beings. The serial transmission of ill effects from plants to animals, animals to humans, and plants to humans is also common. Good quality soils prevent water and air pollution, thus preventing health risks to plants, animals, and humans. Controlling air and water erosion and the run off of irrigation water can reduce the entry of pollutants such as nitrates, phosphates, and pesticides, into surface and ground water in particulate and dissolved form and can also reduce air pollution from wind erosion. Eating food grown on soils deficient in essential metals such as zinc, iodine, and iron can lead to their deficiency symptoms, including growth retardation and defective hormone and immune functioning. The topic of food quality has gained currency in recent years (Hurrel 2001).

17.6.2 Importance of Soil Quality

Soil quality is recognized as a critical component and a validatable parameter to monitor the sustainability of modern production systems and environmental quality in relation to management practices (Saroja Raman 2001). Soil quality is important to maintain healthy soil bio diversity, soil buffering against global climate changes, and sustaining ecosystem services and life-support systems critical to the biosphere. The National Research Council report Soil and Water Quality: An Agenda for Agriculture placed conservation and enhancement of soil quality as the first line of defense against air and water pollution (NRC 1993; Bezdicek et al. 1996). It advocated protection of soil quality, like protecting air and water quality, a fundamental goal for national environmental policy. Similar to air and water quality, the quality of soil has a profound effect on the environment and human health. However, unlike for water and air quality, it is difficult to set uniform and universal standards for soil quality because of the following reasons (1) soils are not uniform in composition and vary from place to place, (2) the body of physical, chemical, and biological properties that individually and collectively control various soil attributes and functions are far too complex and (3) the functions that soils perform are too diverse and are too closely inter-related to be singled out for cause and effect interpretations.

17.6.3 Measuring Soil Quality in Agriculture

The physical, chemical and biological properties of soil interact in a complex manner that determines a soil's capacity to function properly to support a crop. Soil quality cannot be measured directly; it has to be inferred from measuring the changes in certain properties or attributes of the soil. These are termed soil quality indicators, and both qualitative and quantitative indicators can be used.

17.6.4 Qualitative or Descriptive Indicators of Soil Quality

Simple visual and morphological indicators can be used to monitor soil quality in small fields. For example, soil crusting and surface sealing are indicative of the loss of soil organic matter, and rills and gulleys are indicative of water erosion. Ripple marks and sand against plant stems indicate wind erosion, and salt crust indicates salinization. Lack of plant response to fertilizer application is a possible indicator of soil chemical degradation (as is now commonly seen in Punjab, India, in the green revolution belt, water stagnation of poor drainage, and compact hardpan of structural deterioration). These are easily observed indicators and act as a timely early-warning signal of soil quality impairment. Harris and Bezdicek (1994) recommended that qualitative information should form an essential part of soil quality monitoring programs. Descriptive indicators are inherently qualitative and subjective, whereas analytical indicators are precise, quantitative, and without bias. Despite their "soft" nature, descriptive indicators can be of immense value in assessing the quality of a soil, especially for small, marginal and traditional farmers (Rooming et al. 1996).

17.6.5 Quantitative or Analytical Indicators of Soil Quality

Quantitative indicators of soil quality are based on measurable properties. As it is impossible to consider all the properties and attributes that control soil quality, Larson and Pierce (1991) proposed for the first time, a minimum data set (MDS) of the soil'sphysical, chemical and biological properties. Soil quality indicators and soil quality index are useful to assess the current state and act as a base from which to monitor changes over time, and, as a result of different management practices. Doran and Parkin (1994) suggested a specific set of criteria to make indicators easily measurable, reproducible, and indicative of changes resulting from anthropogenic factors. According to Doran and Parkin (1994), soil quality indicators should ensure the following:

1. Encompass ecosystem processes and relate to process-oriented modeling
2. Integrate soil's physical, chemical and biological properties and processes
3. Be accessible to many users and applicable to field conditions
4. Be sensitive to variations in management and climate where possible
5. Be components of existing databases

The commonly accepted MDS of the soil's physical, chemical and biological indicators individually and collectively relate to specific soil conditions or functions and can thus continuously reflect changes resulting from management interventions or other factors. This helps in adjusting the future course of action according to prevailing conditions. The MDS soil properties and their relationship to specific soil conditions and functions are given in the following table (Table 17.1).

There is no hard and fast rule for what an MDS should comprise. According to Kimble (1998), striving to create large data sets to solve all known and unknown problems is of no practical value. Often, the data needed are location-specific, soil-specific, land type-specific, (for example, rangeland, forest land, wet land or arable land), land-use specific, in particular, soil use function, cropping mix, crop and soil management, so on.

The relevance and integrity of MDS are what matter most, rather than the quantum of data generated. An MDS should be flexible, contextual, functional, and easily usable and is best formulated jointly by end users and researchers to suit requirements of a particular function. In this context, it is relevant to note that the soil organic matter by itself is quite useful as a composite indicator of soil quality. It interacts with other soil quality components and controls many other soil quality attributes, including pH, buffer capacity, cation exchange, aggregate formation, bulk density, mineralization of nutrients, persistence of pesticides, water infiltration and retention capacity, and activity and diversity of microbial populations. In this context, it should be noted that the chemically powered so-called green revolution in Punjab State in India and other parts of the country had the most adverse impact on soil organic matter, which led to not only crop yield plateauing, but, many other soil related properties enumerated above. The biologically active components of soil organic matter have been shown to be sensitive indicators of changes brought

Table 17.1 Minimum data set (MDS) of physical, chemical and biological indicators to screen the condition, quality and health of soil

Indicators of soil condition	Relationship to soil conditions and function
Physical	
Texture	Retention and transport of water and chemicals
Depth of soil,, topsoil, and rooting	Estimate of productivity potential and erosion
Infiltration and bulk density	Potential for leaching, productivity, and erosivity
Water-holding capacity	Related to water retention, transport and erosivity, available water
Chemical	
Soil organic matter	Defines soil fertility, stability, and extent of erosion
pH	Defines biological and chemical activity thresholds
Electrical conductivity	Defines plant and microbial activity thresholds
Extractable nitrogen, phosphorus and potassium	Plant available nutrients and potential for nitrogen loss; productivity and environmental quality indicators
Biological	
Microbial biomass, carbon and nitrogen	Microbial catalytic potential and repository for carbon and nitrogen
Potentially mineralizable nitrogen	Soil productivity and nitrogen supplying potential
Soil respiration, water content, and temperature	Microbial activity measure (in some cases plants), estimate of biomass activity

Source: Modified from Mausbach and Seybold (1998)

about by management practices (Kennedy and Papendick 1995). Similarly, simple production indicators, such as, crop yield, certain descriptive indicators, and ecosystem process indicators, such as, soil respiration, microbial biomass, organic matter decomposition (Visser and Parkinson 1992), and trends in canopy structure (Andiscott 1995) can provide a rapid early warning of the onset of changes in soil quality.

17.7 Reference Values for Soil Quality Indicators

17.7.1 Statistical Limits

Standard or reference values need to be established for soil quality indicators in relation to which current soil quality can be assessed. Pierce and Larson (1993) proposed defining limits to represent a desired level of quality for each indicator and then deciding whether the soil is at risk or out of control. They used statistical control concepts from manufacturing to establish upper and lower limits of changes based on natural variation. Reference values can be given as specific limits or ranges for each indicator for a particular soil or a group of similar soils. For intensively managed soils, as in production systems, reference values are established based on what is possibly sustainable for a specific land use under best management practices.

17.7.2 Scoring Functions

Karlen and Scott (1994) used a different approach to set standards for soil quality. They developed scoring functions for each indicator and normalized them to a scale of 0–1. The maximum value of 1 represents the best value; 0.5 represents the base, or the minimum acceptable value, values below 0.5 are not acceptable, and 0 is the least acceptable. Values between borderline d maximum potential (0.5–1) are the target values for good soil quality, and, management interventions are devised accordingly. Three different patterns exist depending on the nature of the soil quality indicator: (1) more is better, as in the case of soil organic matter content, microbial biomass, and rooting depth (2) less is better, as for bulk density and (3) a certain range is optimal, and values outside this range are undesirable. Soil pH and plant bio available nutrients fall in this category.

17.7.3 Soil Quality Index

The information obtained for an MDS of soil quality measurements needs to be integrated into an index to get a holistic, composite picture of soil quality with respect to its key functions. This index can be used to compare different soils for specific purposes, such as, productivity of a particular crop, to act as an early warning signal of soil quality deterioration, and to predict the effects of farming systems and management practices on soil quality and sustainability (Parr et al. 1992). Soil quality index should reflect both the potential for human use and the unique biophysical conditions specific to the location. Both qualitative and quantitative approaches have been used to evolve soil quality indexes.

17.7.4 Analytical or Quantitative Approach to Soil Quality Index

To assess a soil quality index, a framework is created through which soil quality indicators are linked to soil functions. The attributes of a soil pertinent to its effectiveness in terms of basic functions (for example, productivity, environmental quality, and health) are identified and are in turn linked to soil quality indicators. Larson and Pierce (1991) identified the following attributes as relevant to soil quality assessment:

1. Accepts, holds, and releases plant nutrients and other chemical constraints
2. Accepts and holds water and releases it to plant roots, streams, and ground water
3. Promotes and sustains plant root growth
4. Maintains suitable soil biotic habitat and
5. Responds to management and resists soil degradation

The comprehensive interactive framework of Izac and Swift (1994) is illustrative of this approach. Changes in soil quality indicators are linked to soil attributes and then to the corresponding soil function. Details of the computation of soil quality index for different purposes can be found in the literature (Doran and Parkin 1994; Harris et al. 1996). The use of soil quality indicators and soil quality indexes to evaluate the sustainability of agricultural management systems is discussed in one of the following chapters.

References

Andiscott, T. M. (1995). Entropy and sustainability. *European Journal of Soil Science, 46*, 161–168.

Bezdicek, D. F., Papendick, R. I., & Lal, R. (1996). Introduction: Importance of soil quality to health and sustainable land management. In J. W. Doran & A. J. Jones (Eds.), *Methods for assessing soil quality* (SSSA special publication) (Vol. 49, pp. 1–8). Madison: Soil Science Society of America.

Cassman, K. G., De Datta, S. K., Olk, D. C., Alcantara, J., Samson, M., Descalsota, J., & Dizon, M. (1995). Yield decline and nitrogen economy of long-term experiments on continuous irrigated rice systems in the tropics. In R. Lal & B. A. Stewart (Eds.), *Soil management : Exprimental basis for sustainability and environmental quality* (pp. 181–222). Boca Raton: CRC Press.

Doran, J. W., & Parkin, T. B. (1994). Defining and assessing soil quality. In J. W. Doran, D. C. Coleman, D. F. Bezdicek, & B. A. Stewart (Eds.), *Defining soil quality for a sustainable environment* (SSSA Special Publication 35) (pp. 3–21). Madison: American Society of Agronomy and Soil Science Society of America.

Doran, J. W., Sarrantonio, M., & Leibig, M. A. (1996). Soil health and sustainability. In D. L. Sparks (Ed.), *Advances in Agronomy* (Vol. 56, pp. 1–54). New York: Academic.

Dregne, H. E. (1990). Erosion and soil productivity in Africa. *Journal of Soil and Water Conservation, 45*, 431–436.

FAO. (1989). *World agriculture towards 2000*. Rome: Food and Agriculture Organization.

Fresco, L. O., & Kroonenberg, S. B. (1992, July). Time and spatial scales in ecological sustainability. *Land use policy, 9*, 155–167.

Greenland, D. J. (1997). *The sustainability of Rice farming*. Wallingford: CAB Inernational.

Gupta, V. V. S. R., & Germida, J. J. (1988). Distribution of microbial biomass and its activity in different soil aggregate size classes as affected by cultivation. *Soil Biology and Biochemistry, 20*, 777–786.

Gupta, R. K., Naresh, R. K., Hobbs, P. R., Jiaguo, J., & Ladha, J. K. (2003). Sustainability of post-Green Revolution agriculture: The rice-wheat systems of the Indo-Gangetic plains and China. In J. K. Ladha, J. Hill, R. K. Gupta, J. Duxbury, & R. J. Buresh (Eds.), *Improving the productivity and sustainability of the Rice-Wheat systems: Issues and impact* (ASA Special Publication 65) (pp. 1–26). Madison: American Society of Agronomy.

Harris, R. F., & Bezdicek, D. F. (1994). Descriptive aspects of soil quality/health. In J. W. Doran, D. C. Coleman, D. F. Bezdicek, & B. A. Stewart (Eds.), *Defining soil quality for a sustainable environment* (SSSA special publication 35) (pp. 23–35). Madison: American Society of Agronomy, Crop Science Society of America, and Soil Science Society of America.

Harris, R. F., Karlen, D. L., & Mulla, D. J. (1996). A conceptual framework for assessment and management of soil quality and health. In J. W. Doran & A. J. Jones (Eds.), *Methods for assessing soil quality* (SSSA special publication 49) (pp. 61–82). Madison: Soil Science Society of America.

Henao, J., & Banante, C. (1999). *Nutrient depletion in the agricultural soils of Africa* (Vision 2020 Brief 62). Washington, DC: International Food Policy Research Institute.

Hendrix, P. F., Crossley, D. A., Blair, J. M., & Coleman, D. C. (1990). Soil biota as components of sustainable ecosystems. In C. A. Edwards, R. Lal, P. Madden, R. H. Miller, & G. Haise (Eds.), *Sustainable Agricultural Systems* (pp. 637–654). Ankeny: Soil and Water Conservation Society.

Hurrel, R. F. (2001). Modifying the composition of plant foods for better human health. In J. Nosberger, H. H. Geiger, & P. C. Struik (Eds.), *Crop science: Progress and prospects* (pp. 53–64). Wallingford: CAB International.

Izac, A. M. M., & Swift, M. J. (1994). On agricultural sustainability and its measurement in small scale farming in sub Saharan Africa. *Ecological Economics, 11*, 105–125.

Jastrow, J. D., Boutton, T. W., & Miller, R. M. (1996). Carbon dynamics of aggregate-associated organic matter, estimated by C-13 natural abundance. *Soil Science Society of America Journal, 60*, 81–807.

Jenny, H. (1941). *Factors of soil formation*. New York: McGraw-Hill.

Karlen, D. L., & Stott, D. E. (1994). A framework for evaluating physical and chemical indicators of soil quality. In J. W. Doran, D. C. Coleman, E. D. Bezdicek, & B. A. Stewart (Eds.), *Defining soil quality for a sustainable environment* (SSSA special publication 35) (pp. 53–72). Madison: Soil Science Society of America.

Karlen, D. L., Mausbach, M. J., Doran, R. G., Cline, R. F., & Harris and Schuman G.E. (1997). Soil quality: A concept, definition and framework for evaluation. *Soil Science Society of America Journal, 61*, 4–10.

Kay, B. D., Rasiah, V., & Perfect, E. (1994). In D. J. Greenland & I. Szabolcs (Eds.), *Soil resilience and sustainable land use* (pp. 351–367). Wallingford: CAB International.

Kennedy, A. C., & Papendick, R. I. (1995). Microbial characteristics of soil quality. *Journal of Soil and Water Conservation, 50*, 243–248.

Kimble, J. M. (1998). Minimum data sets needed for soil and climatic characterization. In R. Lal (Ed.), *Soil quality and agricultural sustainability* (pp. 44–50). Chelsea: Ann Arbor Press.

Ladha, J. K., Pathak, H., Padre, A. T., Dave, D., & Gupta, R. K. (2003). Productivity trends in intensive rice-wheat cropping systems in Asia. In J. K. Ladha, J. Hill, R. K. Gupta, J. Duxbury, & R. J. Buresh (Eds.), *Improving the productivity and sustainability of the Rice-Wheat systems: Issues and impact* (ASA Special Publication 65) (pp. 45–76). Madison: American Society of Agronomy.

Larson, W. E., & Pierce, F. J. (1991). Conservation enhancement of soil quality. In *Evaluation for sustainable land management in the developing world* (Vol. 2.. IBSRAM Proc. 12 (2)). Bangkok: International Board for Soil Research and Management.

Lynch, J.M., Bragg, E. (1985). Microorganisms and soil aggregate stability. In: Advances in soil science. Vol 2, B.A. Stewart (Ed.). New York, Springer.

Mausbach, J. M., & Seybold, C. A. (1998). Assessment of soil quality. In R. Lal (Ed.), *Soil quality and agricultural sustainability* (pp. 33–43). Chelsea: Ann Arbor Press.

Nair, K. P. P. (2004). Sustaining crop production in the developing world through "the nutrient buffer power concept" – A case study with Black pepper (Piper nigrum) growing at low pH in the laterite soils of India. In *Proceeding of 16th international symposium on plant-soil interactions at low pH* (p. 392), Sendai, Japan, August 1–5.

Nair, K. P. P., & Sharma, P. B. (1979). Mineralization and field effectiveness of ordinary and coated urea, urea-aldehyde condensation product and urea treated with nitrification inhibitor. *The Journal of Agricultural Science, Cambridge, 93*(3), 623–627.

NRC. (1993). *Soil and water quality: An agenda for agriculture*. Washington, DC: National Academy Press.

Oades, J. M. (1984). Soil organic matter and structural stability: Mechanisms and implications for management. *Plant and Soil, 76*, 319–337.
Parr, J. F., Papendick, R. I., Hornick, S. B., & Meyer, R. E. (1992). Soil quality attributes and relationship to alternative sustainable agriculture. *American Journal of Alternative Agriculture, 7*, 5–11.
Parton, W. J., & Rasmussen, P. C. (1994). Long- term effects of residue management in wheat fallow: II CENTURY model simulations. *Soil Science Society of America Journal, 58*, 530–536.
Pierce, F. J., & Larson, W. E. (1993). Developing criteria to evaluate sustainable land management. In J. M. Kimble (Ed.), *Proceedings of the 8th international soil management workshop: Utilization of soil survey information for sustainable agriculture land use* (pp. 7–14). Lincoln: USDA-SCS National Soil Survey.
Pingali, P., Hossein, M., & Gerpacio, R. V. (1995). *Asian rice bowls: The returning crisis.* Wallingford: CAB International.
Prihar, S. S., Gajri, P. R., Benbi, D. K., & Arora, V. K. (Eds.). (2000). *Intensive cropping: Efficient use of water, nutrients and tillage*. New York: Food Products Press.
Raman, S. (1987). Mineralogy in relation to pesticide use and soil pollution. *Clay Research, 6*(2), 50–58.
Raman, S. (2001). Soil quality audit for resource conservation in Indian agriculture. *Social Change, 31*, 75–86.
Raman, S., & Rao, P. C. (1987). Kinetics of extraction of soil-applied metoxuron and its biological implications. *Water, Air, and Soil Pollution, 38*, 217–223.
Raman, S., & Rao, P. C. (1989). Adsorption of metoxuron and tebuthiuron on model clay-organo complexes. *Toxicological and Environmental Chemistry, 24*, 20–213.
Romig, D. E., Garlynd, M. J., & Harris, R. F. (1996). Farmer-based assessment of soil quality: A Soil health storecard. In J. W. Doran & A. J. Jones (Eds.), *Methods for assessing soil quality* (SSSA publication 49) (pp. 30–60). Madison: Soil Science Society of America.
Sanchez, P. A., Izac, A. M., Valencia, I., & Pieri, C. (1996). Soil fertility replenishment in Africa: A concept note. In S. A. Breth (Ed.), *Achieving greater impact from research Investments in Africa* (pp. 200–207). Nairobi: Inernational Center for Research in Agroforestry.
Singh, K. P., & Kushwaha, C. P. (2000). Ecological aspect of soil fertility management in Indian agro ecosystem. In P. K. Jha, S. B. Karmacharya, S. R. Baral, & P. Lacoul (Eds.), *Environment and agriculture: At the cross road of the new millennium* (Proc. International conference on environment and agriculture, November 1–3, 1998, Kathmandu) (pp. 205–222). Kathmandu: Ecological Society (ECOS).
SSSA. (1995, June 7). Soil quality: A conceptual definition. *Agronomy News.*
Sullivan, L. A. (1990). Soil organic matter air encapsulation and water – Stable aggregation. *Journal of Soil Science, 41*, 529–534.
Swift, M. J., & Anderson, J. M. (1994). Biodiversity and ecosystem functions in agricultural systems. In E. D. Schulze & H. Mooney (Eds.), *Biodiversity and ecosystem function* (pp. 15–42). New York: Springer.
Tisdall, J. M., & Qades, J. M. (1982). Organic matter and water-stable aggregates in soils. *Journal of Soil Science, 33*, 141–163.
Virmani, S. M., Katyal, J. C., Eswaran, H., & Abrol, I. P. (Eds.). (1994). *Stressed ecosystems and sustainable agriculture* (pp. 229–235). New Delhi: Oxford and IBH Publishing Company.
Visser, S., & Parkinson, D. (1992). Soil biological criteria as indicators of soil quality: Soil microorganisms. *American Journal of Alternative Agriculture, 7*, 33–37.
Vlek, P. L. G., & Vielhauer, K. (1994). Nutrient management strategies in stressed environments. In S. M. Virmani, J. C. Katyal, H. Eswaran, & I. P. Abrol (Eds.), *Stressed ecosystems and sustainable agriculture* (pp. 203–228). New Delhi: Oxford and IBH Publishing Company.
Woomer, P. L., & Swift, M. J. (Eds.). (1994). *The Biological management of tropical soil fertility.* Chicester: John Wiley.
Young, E. M. (1997). *World Hunger*. London: Routledge.

Chapter 18
How to Manage Water Use for Sustainable Agriculture?

Abstract Global fresh water resources are finite and non replenishable. The renewable fraction of fresh water constitutes less than 1% of the total fresh water pool. This chapter discusses, at length, several aspects of sustainable water use for agricultural production, *vis-à-vis* global needs for all other purposes, including domestic and societal needs, in short, for the very survival of human kind on planet earth with the implications on food security. The chapter also discusses several other aspects of the key element, Nitrogen, in sustainable agriculture, starting from the nitrogen cycle to global warming, contributed by the unbridled use of urea in green revolution leading to enormous nitrous oxide emission, on urea hydrolysis, leading to global warming.

Keywords Fresh water · Supply-Demand chain of fresh water · Ground water · Water availability · Water demand · Irrigated area · Irrigated agriculture · Salinization · Water logging · Overexploitation of water · Unsustainable Ground water use · Global climate assessment · Water · Ambient climate · Environment · Varietal improvement · Agronomic practices and crop choice · Need-based irrigation · Micro irrigation · Sprinkler irrigation · Water productivity · Rainfed agriculture · Brackish water · Agricultural drainage · Soil pollution · Innovative and low-cost technologies · Water sharing · Equitable distribution · Pricing water · Regularity mechanisms · Nitrogen fertilizer · Nitrogen use · Crop growth · Photosynthesis · N supply synchronization · Nitrogen demand

It is important to realize that global fresh water resources are finite and non replenishable. The renewable fraction of freshwater constitutes less than 1% of the freshwater pool. Freshwater cannot be replaced in any of its important functions, such as, for drinking, agriculture, hygiene, washing and sanitation, industrial purposes, survival of fish wealth, and other aquatic creatures, in fact, for almost all of the ecosystem functions and purposes. Only with this important perspective in mind can we approach water management for sustainable agriculture in a rational manner.

If one critically observes water use for agriculture globally it will be seen that agriculture makes use of water in the most wasteful manner. In fact, one of the principal contributing factors for the almost irreversible soil degradation, due to the so-

K. P. Nair, *Intelligent Soil Management for Sustainable Agriculture*,
https://doi.org/10.1007/978-3-030-15530-8_18

called green revolution in the State of Punjab, India, the "cradle" of green revolution in South Asia, has been through the prodigious use of irrigation water. For years on end, the wheat and rice fields of Punjab were irrigated by pumping water on soil surface (surface irrigation) –one of the most wasteful practices of water use. And, this contributed to a tremendous build up of soil salinity, which took a very heavy toll on soil fertility.

Globally, agriculture uses about 70% of freshwater. And in some regions it could be a lot more, and, the above cited case of Punjab is a classic example. The demand for fresh water will considerably increase in the coming two decades, especially in Asia and Africa, where most of the population increase will be concentrated. And a larger share will have to be allocated for civic use and industrial purposes. The number of people living in water-starved countries will increase six fold in the next two decades, which will prove a very big challenge to global and regional food security (IFPRI 1997a, b).

Prudent soil and water management is the bedrock of sustainable crop production. In terrestrial ecosystems, fresh water availability constitutes the most limiting factor for plant growth and crop yield. In any target environment, water along with crop choice is the key determinant of land productivity.

The second half of the twentieth century witnessed a pre-occupation with increasing land productivity through intensification of agriculture. The substantial increase in crop production in South Asia, especially in a country like India, is attributed to a 220% increase in irrigated land. In a situation whereas private ownership prevails over crop land, water, to a large extent, is held as a public good, ecologically and economically under priced, which leads to its overexploitation and wasteful use, especially in agriculture. To compound the situation, there are States in India, for example, Karnataka and Andhra, where such remarkable increases in irrigated land took place, it is important to note that in these States electricity is freely given to farmers as an incentive to pump water. This, invariably, leads to over exploitation of ground water resources and deep bored tube wells are a common sight in the country side in these states. There are instances, where, when water level shrinks, farmers dig up to 350 ft. below the soil surface, in search of water. And when electricity to pump water comes free of charges, the lure to overexploit cannot be resisted. This also leads to curtailing water for civic use, and metro cities like Bangalore in Karnataka State constantly face a "water crisis," especially when the monsoon fails, as did happen in 2008.

Despite serious shortages, the availability of water has often been taken for granted until recently, in calculations of future food security. The agricultural sector has become used to an abundance of water at low cost, and with minimal environmental responsibility. Various new technologies have, more often than not, overlooked the availability of water for their success. The classic example is the latest genetically modified crops, in particular, cotton in India. A careful evaluation of its performance in India shows, beyond a shade of doubt, that success has only been assured where irrigation water was aplenty. The total failure of the Bt cotton in *Vidharba* district of Maharashtra State in India, where thousands of Bt cotton farmers committed suicide when the crop failed is because of the rainfed nature of the region. When the rains failed to arrive the Bt cotton crop totally failed. The farmers' suicide rate became a global shame for India.

It has been predicted that freshwater will be the most limiting natural resource in the twenty-first century for sustainable development in general. This is especially critical in water-deficit developing countries with a large population in rural and agricultural areas. The World Summit On Sustainable Development in Johannesburg in South Africa in 2002 issued the highest red alert relating to water scarcity for regional and global sustainable development (WSSD 2002). With fragility of water availability in mind, "Water Resource for Food Security" was chosen as the theme for World Food Day 2002. The UN General Assembly designated 2003 as "International Year of Freshwater," with the resolve to improve the rate of productivity through water use. Fancy slogans like "More Crop Per Drop of Water" were bandied around, by those "arm chair scientists," without a serious understanding of the ground reality. Even after more than a decade, while colossal amounts of money was spent from public funds, like the UN kitty and other national treasuries, to host thousands of scientists in international conferences spread all over the world – the countries with political clout getting the benefit of "maximum photo opportunity" and media coverage – "arm chair scientists" and political big wigs only talked of water, while the ground reality has only deteriorated since then. Almost a parallel can be found in the climate talks. Almost everyone, save a microscopic minority, assume that water is available for agriculture and for all other purposes *ad infinitum.* How else can one explain the total apathy by most to this very precious and limited source for sustenance of plant, man and animal?

18.1 The Supply-Demand Chain of Fresh Water

Almost 75% of the earth's crust is covered with water, 97% of this is in the oceans, and unfit for agricultural use, human or animal consumption, and, even, for industrial purposes. Three-fourths of the fresh water is in the glaciers and icebergs and is not easily accessible. This is also the case for the one-fifth of water found deep in the soil. Less than 1% of fresh water is present in lakes or rivers. Freshwater is a finite resource that is widely available, but not everywhere in equitable proportions. Rainfall over land is the main source of fresh water fit for agricultural use and for human consumption.

The average annual rainfall over land is estimated to be 110,000 billion m^3. Of this, approximately 70, 000 billion m^3 evaporates back into the atmosphere, which is almost 64% of what has been received from the skies as rainfall. This portion of rainfall is called the "green water" and is the source of water for all non irrigated vegetation, including forests, woodlands, and rainfed crops. After allowing for evaporation from continental rainfall, approximately 40, 000 m^3 of fresh water enters lakes, reservoirs, streams, and aquifers in active exchange with surface water. This is called "blue water." The distribution of water among different sources is depicted in the following figure (Fig. 18.1).

Blue water is not only unevenly distributed, but also transient, as it evaporates or flows into unusable water sinks, such as, the sea or salt pans. While it remains on the surface, this water is not wholly in demand, and, it is also polluted during its flow as a result of anthropogenic activities. Approximately 20% of blue water is estimated

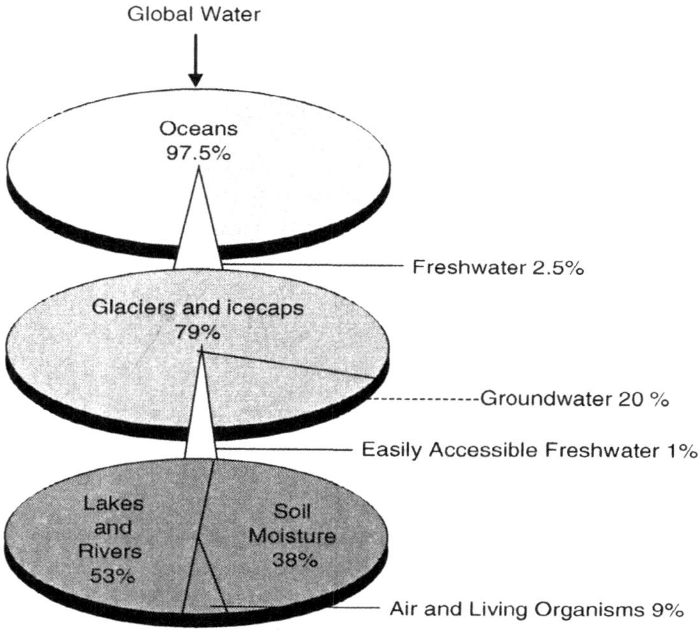

Fig. 18.1 Distribution of global water in various resources. (Source: Adapted from UNEP)

to be inaccessible. A large portion of the run off is not available when and where it is needed and is also wasted as flood water. Currently, an estimated 9000 billion m^3 is realistically accessible for human use. This, together with an estimated runoff of 3500 billion m^3 regulated by existing reservoirs, puts the available blue water at 12,500 m^3. Harnessing the remaining 27,500 m^3 is non useable owing to the cost factor, topography, remoteness from areas where it is required, and the social and environmental consequences of impoundment (FAO 1996a, b). These figures simply demonstrate the mammoth problems humanity faces with regard to water use.

18.2 Ground Water

Ground water accounts for approximately one-fifth of the freshwater resource. One-third of river flow emerges from ground water aquifers, and, this is the most stable component of surface flow. Ground water acts as a buffer against seasonal shortfalls in rainfall and plays a significant role in insulating agriculture from environmental and economic risks, more often than not, involved in rainfed agriculture, especially in the developing countries in their dry and semi arid regions. Ground water requires careful management because it may eventually run out if unbridled use exceeds the rate of natural recharge, as is happening in many parts of India, especially with farming, with unscrupulous elements who think of only to day, but, not tomorrow!. This is especially critical in the case of fossil groundwater, which cannot be renewed. In India, and many arid and semi arid regions of Asia, Africa, and the western U.S, considerable excess pumping of ground water takes place, resulting in severe depletion and lowering of the water table (Postel 1996). Ground water can also become salty and unfit for irrigation purposes. In coastal areas, ground water depletion results in the infusion of salt water from the sea into the aquifer, contaminating the underground reservoir itself.

18.3 Availability of Water Region-Wise

The following table (Table 18.1) depicts the distribution of water resources continent-wise. Although on a global aggregate basis per capita water availability appears quite comfortable, for the foreseeable future, the regional distribution of water resources raises concerns for future development.

On the basis of total water availability, as a rule of thumb, countries with fresh water resources of 1000–1600 m^3 per capita per year are considered water stressed. A total internal renewable water resource of less than 1000 m^3 per capita per year is classified as water scarcity and can lead to severe socioeconomic constraints, health problems, and environmental degradation. Currently, 23 countries face severe water shortage, globally, and 30 countries are water stressed, globally. The latter figure is predicted to rise to 50 by 2025, affecting three billion people in their life style. Water shortage will be due mostly to population increases in particular in the developing countries and also expansion in urban areas (in many countries due to rural-urban migration in search of livelihood, a tragic skewed development consequence in many countries like India). And this is expected to reach 59% of global population by 2025, up from 46% in 1996 (IFPRI 1997b).

Table 18.1 Distribution of water resources continent-wise

Region	Area (million km^2)	Volume (km^3)	Percentage of total	Potential water availability (10^3 m^3 $year^1$) Per km^2	Per capita
Africa	30.1	4, 050	9.6	134	5.72
Asia	43.5	13, 510	31.5	311	3.92
Australia and Oceania	8.95	2, 404	5.6	269	83.7
Europe	10.46	2, 900	6.8	277	4.23
North America	24.3	7, 890	18.4	324	17.4
South America	17.9	12, 030	28.1	672	38.2
World	135	42,785	100	317	7.6

Source: Adapted from Shiklomanov

18.4 Demand for Water

Broadly, the demand for available fresh water for human use falls into the following three categories:

1. Agriculture, Fisheries and Animal Husbandry
2. Civic Requirements in Rural and Urban Areas
3. Industrial Uses

Fresh water is also required to safeguard aquatic ecosystems, which are vital to sustain several ecosystem services, including water purification, maintenance of habitat for aquatic creatures, and flood control. The increasing demand for recreational facilities, as for example, fishing, rafting, snorkeling, and other sports activities, especially in the advanced countries, is likely to increase the use of fresh water.

The distribution of water withdrawal and water use by continent, and among the three major sectors of water is given in the following table (Tables 18.2 and 18.3).

Table 18.2 Distribution of water withdrawal and water use continent-wise

Continent	Water withdrawal (km^3 $year^1$)		Water use (km^3 $year^1$)		Water use per capita (m^3 $year^1$)	
	2000	2025	2000	2025	2000	2025
Africa	235	337	170	220	195.1	137.9
Asia	2,357	3,254	1,458	1876	387.6	382.4
North and Central America	705	786	243	269	499	452.1
South America	182	260	96	120	269.7	242.9
Australia and Oceania	32.5	39.5	18.7	22.3	595.5	576.2
Europe	463	559	197	256	285.1	374.3
World	3,973	5,235	2,182	2,764	353	350.9

Source: Adapted from Shiklomanov

Table 18.3 Continent-wise sectoral use of water

Continent	Agriculture (km^3 $year^1$)		Municipal (km^3 $year^1$)		Industry (km^3 $year^1$)	
	2000	2025	2000	2025	2000	2025
Africa	107	133	2.93	7.38	1.35	2.85
Asia	1322	1659	22.4	33.5	32.7	77.4
Europe	140	171	10.6	11.0	31.8	57.1
North and Central America	170	182	12.8	16.1	18.0	20.2
South America	70.8	80.9	3.7	5.6	3.4	9.9
Australia and Oceania	12.6	14.3	0.41	0.46	0.63	1.43
World	1,834	2,252	52.8	74.1	87.9	169

Source: Adapted from Shiklomanov

18.5 Water Needs of Agriculture

18.5.1 Current Picture

Of the all the aspects of water use, agriculture consumes almost 70% of total water withdrawal, which on average is 84% as a whole for Asia (IFPRI 1997b), of which India accounts 92%, while China comes next at 77%. As per the International Water Management Institute, globally this figure (current 84%), will still be as high as 65% in 2025. The largest fraction of agricultural water withdrawal goes for irrigation of field crops, which covers 276 million ha globally, constituting about 18% of arable land, which produces approximately 40% of the total food production and other agricultural commodities (WRI 2000). The continent wise irrigated area is shown in the following table (Table 18.4).

Irrigation uses approximately 2000–20,000 m^3 ha^{-1} annually, depending on climatic conditions, the nature of the field crop, the crop intensity, and the efficiency of water use. If the climate is conducive and water adequate, more than one crop can be grown annually in the tropics. In humid regions, irrigation merely supplements soil moisture provided by rains. In arid regions, all the moisture may have to come from irrigation.

Irrigated agriculture, along with copious fertilizer application, coupled with modern high yielding crop varieties (HYV) and chemical pesticides, have been the back bone of industrial type agriculture, euphemistically called the green revolution, in South Asia. The following table (Table 18.5) would prove that the exponential increase in food production, globally, in particular in South Asia, resulted from a parallel increase in irrigation (plus fertilizer use, though this table does not carry the relevant data on fertilizer use, but, the author knows it well), and, it is not correct to attribute this increase in crop production simply to the superior crop varieties alone. Thus, although the so-called green revolution left much of forest land and grassland from being converted to arable land, for crop production, it created a heavy demand for water.

Table 18.4 Global distribution of irrigated area

Continent	Irrigated area (Million ha)	Share of world total (%)	Share of Cropland which is irrigated (%)
Asia	190.5	70	34
Africa	12.7	5	6.2
North and Central America	31.3	11	11.7
South America	9.9	4	8
Europe	25.4	9	8.2
Oceania	2.7	1	5
World	272.5	100	17.9

Source: Adapted from FAOSTAT (1999)

Table 18.5 Decade wise sequential increase in food production and corresponding irrigated area

Year	Grain production (million metric tons)	Irrigated area (million ha)
1960	877	139.1
1970	1, 192	168.0
1980	1, 550	210.2
1990	1, 951	244.9
2000	2, 064	272.4

Source: Adapted from FAOSTAT

18.5.2 The Nature of Things to Come

It is grossly estimated that to feed the future increased population, an increase of 50–60% of current production (especially that of cereal grains, which are the most important staples of the world) is required between now and two decades hence (FAO 2000a). This will also depend on the increased productivity of crop land because rangelands and oceans have reached their productivity limits as a result of overgrazing and overfishing. In the past five decades, 20% of the world's pasture lands and rangelands have been degraded (Oldeman 1994) and per capita availability of fish decreased 8% in 1997 from a peak in 1988 (McGinn 1999). Serious constraints limit any increase in cropland area in the future on account of increases in population, urbanization, requirements for leisure activities and environmental services, and degradation of existing land. While these considerations for the requirement of land get priority attention in developed world, such considerations are totally lacking in the developing countries on the Asian and African continents, the worst situation being in the latter case. In a country like India, currently, a great debate is going on, on the acquisition of land by the State for industrial purposes. There are many vested interests in the industrial sector whose prime aim is to grab land, and this affects agriculture. It is here that the State has to have a visionary long range plan, which is more often than not, a victim of politicization. Agricultural specialists expect that most of the increase in food production in the future will have to come mainly from the developing countries and from the intensification of irrigated agriculture because the productivity of irrigated agriculture, even with lower external inputs like fertilizers and pesticides, is higher than that from rainfed dryland agriculture (FAO 2000b).

The crucial question is whether the currently available volume of water is adequate to meet future increases in food production to meet the demand of the increased population, along with increased demands from industrial sector and also other civic sectors (Postel 1998). Given the projected doubling of water needs for civic purposes in the developing world and the current appropriation of 87% for agriculture, a deep soul searching regarding current trends in water use must come about. For Asian countries, in particular India and China, which are home to almost 40% of global population, which are both predominantly agrarian, water availability for agriculture is very critical for future food security. Almost 30% of these countries' gross domestic product (GDP), which is a measure of economic prosperity, is contributed by the agricultural sector, and the proportion of the economically active population engaged in agricultural activities is as high as 60–70% of the total population. The global growth in irrigation has slowed down, and the irrigated area grew only 1.3% per annum between 1982 and 1994, compared with 2 per growth between 1970–1982, the active so-called green revolution phase, which fell on its face, dramatically starting the eighties. Worldwide irrigated agriculture is projected to grow at an average rate of 0.6% annually between 1995 and 2020. The rate in the developed countries will be minimal, at 0.2%, adding only about three million ha, compared to 0.7% increase in the developing world, amounting to 37 million ha increase. India is expected to add the largest share of 17 million ha of the above increase. Irrigation is not expected to result in any more significant increase in grain production in Sub Saharan Africa. Globally, per capita irrigated area has fallen by 5% since 1978 and is likely to be 17–28% lower by 2020, depending on whether the projected growth in irrigation occurs (Rosengrant and Ringler 1999).

18.5.3 What are the Future Constraints for Irrigated Agriculture?

Irrigated agriculture produces more than rainfed agriculture, even with low external inputs like fertilizer and pesticides. It can also save large areas of marginal and fragile lands. However, there is a price to be paid. "With irrigated agriculture, modern society may have inadvertently struck a Faustian Bargain with Nature" (Postel 1999), which seems to be extracting a price in many forms such as salinization of land and water (the most classic example being Punjab State, the "cradle" of the so-called green revolution in India), siltation, pollution by agricultural chemicals, especially pesticides (again in Gurudaspur district in Punjab with cancer spread, thanks to the green revolution), and declining biodiversity and ecosystem quality (both current problems in Punjab State). Exploitation of ground water beyond its recharging capacity, the constraints on creating new dams and diversions for irrigation, and the adverse impacts of global warming on hydrological cycles constitute additional challenges to expanding irrigated agriculture in future. Despite the grim prospects for the adequate availability of water for agriculture and other needs, rampant mismanagement of water is prevalent in agriculture (in many regions of irrigated agriculture in India, aided by the short-sighted governmental policies, such as free electricity for pumping out ground water). This represents a threat to not only future food security, but also, to the sustainability of ecosystems and the life-support services they provide.

18.5.4 The Consequences of Irrigated Agriculture –Pollution

The major cause for degradation of surface and ground water is irrigated agriculture where water, more often than not, is indiscriminately used. The US and Western Europe have become the largest non point source of pollution. Except for water lost through evapotranspiration, water used for agricultural purposes is recycled back to surface and ground water. The expansion of irrigation and the steadily increasing use of fertilizers and pesticides to sustain higher crop yields have had a great adverse impact on both the production resources and their environment. In order to conserve the precious resource of fresh water, it is necessary to control the pollution created by agriculture and other allied activities (FAO 1993). In addition to reducing the quality of water for drinking and other purposes, excessive pollution results in ecosystem dysfunction, loss of bio diversity, and in the contamination of ground water that is nearly impossible to clean up (FAO 1997).

Sedimentation of water resources as a result of cropland erosion is another major impact of improper land and water management. Soil and water, managed properly, have a symbiotic relationship in terms of boosting agricultural production and minimizing adverse environmental fallout. Without proper and effective management, they can be mutually destructive, resulting in serious losses for humans. Water erosion and sediment transport led to top soil loss, land degradation and sedimentation, and siltation of lakes and rivers. Further, the fine silt and clay particles of the sediment are the carriers of adsorbed chemicals, especially pesticides, toxic metal ions, and phosphate, all of which, when in excess, can seriously pollute the aquatic environment and lead to loss of aquatic wealth. Recently, huge fish swarms of dead fish were found floating in the hilly areas of the Western Ghats in Kerala State, India, and the cause was traced to unbridled use of Furadon, (red triangle highly toxic fungicide) used to control the wilt disease caused by *Phytophthora* in black pepper (*Piper nigrum*).

18.5.5 Salinization and Waterlogging of Irrigated Lands

Land degradation owing to irrigation-induced salinity and water logging are becoming the major threats to food security in developing countries. Irrigation of arid lands leads to salinization as a result of concentration of salts in the root zone. The use of excess water to wash down the salts results in a rise in the water table and water logging if proper drainage is not provided. In a similar manner, poorly drained soils undergo waterlogging and salinization when they are irrigated without drainage. It is estimated that 20% of total irrigated land in the world, covering 17.7 mha, is salt affected, leading to great loss of productivity as shown in the following table (Table 18.6). Vast areas of India, (especially in western India, in the States of Punjab –the "cradle" of the so-called green revolution, Haryana and Western Uttar Pradesh), China, Pakistan, central Asia (Egypt, Iran and Turkey in

Table 18.6 Extent of salinization in vulnerable countries

Country	Irrigated area (million ha)	Salt-affected area (million ha)	Percentage of salt-affected area (million ha)
China	44.88	6.70	15.0
India	42.10	7.00	16.6
CIS (Previously Soviet Union)	20.46	3.70	18.1
USA	16.10	4.16	23.0
Pakistan	16.08	4.22	26.2
Iran	5.74	1.72	30.0
Egypt	2.69	0.88	33.0
Argentina	1.72	0.58	33.7
World	227.11	45.40	20.0

Source: Modified from Ghassemi et al. (1995)

particular), and the US are deteriorating as a result of salinization. Approximately 36% of the affected land in India lies in the fertile plains of the Indus and Ganges rivers, including parts of the agriculturally important States of western India (Punjab, Haryana, Madhya Pradesh and Uttar Pradesh) as reported by Gupta and Abrol (2000).

Salinization affects 23% of irrigated land in the US overall, 67% in the lower Colarado basin, and 35% in California. The canal built to carry the salty farm drainage water from central California, which terminated in Keserton, turned out to be an ecological disaster. It was later dubbed the "Three Mile Island of Irrigated Agriculture," because, it became heavily contaminated not only with salt, but, also with toxic selenium, leading to severe damage to wildlife (Harris 1991).

The severe salt build up in almost 30% of the Aral Sea basin irrigated land in Central Asia is a monumental tragedy caused by over exploitation of the irrigation potential of the Amu Darya and Syr Darya. The consequent low flow of the rivers has led to a concentration of salts and toxic chemicals and has made the water supply hazardous to drink and damaging to crops, resulting in a 20–50% reduction in crop yield in 1994. In parts of Uzbekistan, which is the area most severely affected by the Aral Sea tragedy, 95% land is salt affected, with yields plummeting by 30%. The enormous quantities of water added during the non crop season to wash down the salt have resulted in the very rapid recharging of ground water, a rise in water table, and the mixing of the very deep, highly saline ground water with shallow ground water (UNDP 1995). It is estimated that the loss to farmers from soil salinization is approximately US $ 11 billion per annum by way of reduced income, and this figure is growing. Spreading at a rate of two million ha per year, soil salinity may be offsetting a good portion of the increased productivity achieved by expanding irrigation (Umali 1993). It thus seems, while irrigation *per se* might have a beneficial effect on crop yields, the consequences of salinization that follows, might offset this advantage in the long run, as has been clearly observed in Punjab, the "cradle" of the so-called green revolution in India.

18.5.6 Overexploitation of Water Resources

On account of the increasing demand, water sources in many countries are being very seriously and, in some cases to the extent of breaking point, exploited. Many rivers are being exploited beyond the environmental and economic threshold levels. Large rivers, such as, the Huang Ho in the People's Republic of China, the Colorado River in the southwest of the US, the great and legendary Ganges in India, the historical Nile in Egypt, the Amu Darya and Syr Darya in central Asia, and the Chao Phraya River in Thailand, are all so over exploited that they virtually have no output into the sea during certain times of the year. Excessive withdrawal of water for cotton irrigation and construction of flood storage reservoirs on the Amu Darya and Syr Darya in central Asia resulted in a colossal decline of inflow from 55,000 million m^3 of water in 1960 to 7000 million m^3 between 1989 and 1990. This led to the Aral Sea level falling by, 16 m in depth, resulting in the disappearance of 20 of 24 species of fish, consequently very adversely affecting the fish economy, and a staggering loss of 60,000 livelihoods (FAO 1998). The Yellow River in China did not have any flow into the sea for more than 7 months in 1997. Upstream and mid stream development, diversions, and excessive withdrawal caused the river to dry up before reaching the sea. Similar conditions prevail in the Indus and the Ganges delta region of India, spanning more than one million ha of irrigated cropland, that is home to very rich wildlife biodiversity, including several endangered species of birds and animals. Severe deterioration of the mangroves is occurring as a result of increasing salinity, declining nutrients, erosion, and land subsidence related to the significant curtailment of freshwater through development upstream (Postel 1999). A most striking adverse impact on the mangroves can be seen in the Arabian Sea coastal belt of Kerala State, in the Kannur district, where "misguided developmental" activities of the local governmental agencies are creating a severe havoc to the environment. Wasteful overuse of water in one area deprives users in others, leading to a fall in agricultural production and creating severe environmental problems. And the latest example of this totally "misguided developmental" activities in Kerala is the effort to establish an airport (a totally unneeded priority of the State) in the Aranmula wetlands in central Kerala, because of the vested interests and governmental collusion, destroying not only the Aranmula river, but posing a great threat to food security of the State, as it is one of the main paddy growing areas of the State. Despite the threat to paddy production and loss of livelihoods to ensue, the State and Central governments are apathetic to people's demands to stall the airport project. Fortunately, at the time of completion of writing this book, there is good news. Owing to the long and sustained agitation by the local people of Aranmula and the concerned ones all over Kerala, the vested interests of the State and the governmental collusion has been decreed to be stopped by the Supreme Court in New Delhi. This is a victory of the people of Kerala, which is reeling under a very corrupt government. The drying up and pollution of prime delta areas of potentially high productivity is a threat to any future increase in agricultural production.

18.5.7 Unsustainable Use of Ground Water

Most of the time, the deficits in surface water for agricultural purposes is made up through the exploitation of ground water. Many countries are drawing ground water at rates much higher than the recharge rate, which leads to its unsustainable use, bordering on over exploitation of the water resources. Excessive pumping of ground water is causing a steady drop in the ground water table, sometimes to dangerous levels and when the annual monsoon fails, the situation deteriorates further. This has been observed in several parts of India. Even in a monsoon plenty Kerala State, in India, the same has been observed. The situation is further compounded by the drying up of rivers, which are a perennial source of water supply, which has been caused by unlawful sand mining. This has been observed in the total drying up of *Bharatha Puzha* in northern Kerala State, which has been a great source of water for irrigation and livelihood since millennia, due to the activities of the sand mining mafia, which is in deep collusion with the unscrupulous and corrupt politicians who rule the State. The river which once over flowed with water is now a huge patch of dry sand covering large tracts in Northern Kerala.

It is estimated that the total ground water depletion globally is nearing 2000 billion m^3 per year of which a substantial portion goes for irrigation. This suggests that approximately 180 million tonnes of grain, equivalent to 10% of the global harvest, are being produced through unsustainable use of ground water. In the past two decades, ground water depletion has occurred in large areas of irrigated crop lands in China, India, Pakistan, Western US, North Africa, the Middle East, and the Arabian peninsula. India tops the list in terms of ground water deficit at 104 billion m^3, followed by China at 34 billion m^3, and the US at 13.6 billion m^3 (Postel 1999). It is estimated that 25% of India's grain harvest could be in jeopardy from ground water depletion (Seckler et al. 1998). Ground water pumping is causing social and economic problems for small and marginal farmers, especially in India, who cannot invest in powerful pumps and are forced to mortgage their lands to raise capital by borrowing from local unscrupulous money lenders at exorbitantly high interest rates. Similar threats are also found in Bangladesh, Pakistan and China, where the water table is falling by 1–1.5 m annually. California is overdrawing ground water equivalent to 15% of the US's net ground water use.

In North Africa and the Arabian Peninsula, where most of the ground water is from non renewable aquifers, the situation is very serious. Libya, Saudi Arabia, and the United Arab Emirates (UAE), the latter two of the Gulf Coordination Council grouping of countries, are the worst affected, followed by Kuwait, Bahrain, and Malta, as shown by data in the following table (Table 18.7). Given the water deficit incurred to support today's production level, it will have to be balanced in the future to safeguard our biosphere, given the current size of this deficit. With no alternative plan in place, it will be an extremely difficult task, if the current levels of ground water withdrawal is allowed to continue. This is where proactive and stringent political action at the highest level is called for. But, this is a cry in the wilderness, given the intransigence of policy makers and the suicidal thrust of some of these countries

Table 18.7 The mining of ground water in selected countries facing a future catastrophe

Country	Internal renewable water resources (km^3 year[1])	Actual withdrawal from renewable water resources (km^3 year[1])	Ground water mining as percentage of total water withdrawal
Kuwait	Negligible	2960	46.5
Bahrain	5, 981	206	40.2
Malta	359	359	32.2
United Arab Emirates	1, 405	1, 405	70.9
Oatar	559	538	14.9
Libya	767	767	90.0
Jordan	145	112	17.5
Saudi Arabia	709	709	79.7

Source: Based on FAO-AQUASTAT data (2005)

for "development," which, in the larger context of things become meaningless, when the future welfare of humanity is at stake. This author has noted, with utter dismay, the waste of water resources in some of the above-mentioned countries, where roads are paved on either side with flower beds and hedges, which are watered indiscriminately without strict monitoring, just to show to the visitors, how "green" are their countries, while in reality, years before the oil boom, which enabled these countries to harvest huge funds, they were large desert patches, thirsting for water, which, if left unchecked, will be the dire state of reality, in decades to come. Only sky scrapers and swanky buildings will be left over – nothing more. Iran is already heading towards this catastrophe.

18.5.8 Food Security Concerns Related To Water Shortages

Water shortage and improper use of water are twin problems posing challenges to self sufficiency in food production in many countries. Of the 30 odd countries which have per capita internal run off less than 1700 m^3 per year, almost all are net importers of more than 20% of their food grain requirements. This scenario is depicted in the following table (Table 18.8).

The net imports of these water deficit countries already exceed 25% of the global grain trade (Postel 1999). Considering that 3 billion people will face water shortages by 2025, most of whom will be from Asia and Africa, there is concern regarding procuring food from elsewhere, and what it would cost the national exchequer. Domestic competition for water will translate into international competition for food grain and escalation of cost, making it unaffordable to the poor, and adding to the number already hungry. If over populated countries such as China and India join the water deficit group in the next decades, the fiscal strain on food grain exporting countries, and the countries importing it, as well, will become unsustainable. Despite growing concern over water shortages and food security, irresponsible over-

Table 18.8 Food grain import pattern of selected water deficit countries in 1995[a]

Country	Internal runoff per Capita	Net grain imports as share of consumption in percentage
Jordon	249	91
Israel	309	87
Libya	115	85
South Korea	1, 473	77
Algeria	489	70
Yemen	189	66
Saudi Arabia	119	50
Morocco	1, 027	26
Somalia	645	26
Rwanda	808	20

Source: Adapted from Postel (1999)
[a]Less than 1, 700 m^3 annual runoff per person

use or misuse of water in agriculture continue unabated even in these water-stressed countries. This is particularly disturbing against the background of historical facts which have shown that when water scarcity loomed large, civilizations disappeared. On the contrary, abundance of water made civilizations to flourish. It is, also important in this context to note, that water shortage and soil mismanagement go hand in hand. When soils desertify civilizations are doomed. For instance, the Romans needed grain from the then North African continent (although geographical divisions as countries in North Africa did not occur at the time), and when the North African soils desertified, it gave way to the collapse of the Roman empire. In India, the history of the *Thar* desert (once a lush forest) is yet another classic example. One cannot discuss food security without water, much less without soil. Most unfortunately global discussions center around either water or soil, but not "Soil And Water" within one framework. A critical analysis of global developmental or research projects testify to this critical analysis.

The history of human civilizations is intricately linked to agriculture, and more specifically to the control of water for food production. It is replete with examples where once flourishing civilizations collapsed through inappropriate and unsustainable water use for crop production, which were exacerbated by other factors such as climate change, drought and human mismanagement (Evans 1998). There is indisputable indication that modern industrial type agriculture, euphemistically called the "green revolution," where unbridled use of water has been a hallmark, has fallen prey to such vulnerabilities a little too soon (in less than two decades 1960s–1980s), and, the twin challenges of providing more food for an exponentially increasing population, while at the same time containing the dwindling water resources due to mismanagement by human callousness, is the starkest gloom staring at humanity's face. The situation calls for all possible efforts to correct past deficiencies and launch a new paradigm of more efficient, economically, ecologically, and socially viable and equitable management of this most crucial, yet limited and scarce, resource. Sustainable balancing of its supply and demand should be at the top of the list of priorities for food security in water and food deficit countries at the least. India and China have not just the capacity, but, the moral responsibility as well, to lead the crusade. Whether the call would be taken up with the urgency it merits, or, these countries will get drowned in the pursuit for a so-called higher gross domestic product (GDP), in whose pursuit both industrial expansion and provision of utilities (services, for instance, the IT technology) catch the attention of those who govern these counties, in particular India, is the million dollar question.

18.5.9 Creating New Water Resources for Irrigated Agriculture

Since more than half of the global supply of fresh water is lost as run off in floods, there is enough to tap from this source for use in crop production. Approximately 40,000 large dams and 80,000 small dams have been built to block rivers and use the flood water. These combined have a storage capacity of 6600 billion m^3. However, constraints exist on building bigger dams in the future because the best sites for dams are already taken, the cost of building and maintaining these structures has become formidable and financially non viable owing to decreasing international food prices, and the storage capacity of the existing facilities is being reduced by 1% per year (66 billion m^3, Postel 1999). Tampering with the natural flow of a river by building dams, diversions, dikes, and so on destroys aquatic habitats and severs the river's connections with its flood plains, its delta and the sea into which it flows. This hampers the natural functions of the river. Serious environmental losses, including extinction of species and failure of fresh water ecosystem services are caused by extensive manipulation of rivers in the advanced countries. This has necessitated extreme measures, including breaching of some of the existing dams and diversions to re-create the pre-dam flow patterns (Postel and Carpenter 1997).

Displacement of the original inhabitants along the river coast is one of the most difficult problems to resolve, as has been shown by experience. The struggle of the local people along the river Narmada coast in Gujarat State, India, has taken a political color. It is called *"Narmada Bachao Andolan"* (Save the Narmada River Struggle in Hindi, India's common language, especially in northern India). The idea of the construction of Sardar Sarovar dam across the river Narmada is to provide irrigation water to the farmers and drinking water to the local inhabitants. The task of appropriating adequate volumes of water for agriculture, municipal and industrial use, and for environmental protection has triggered inter-and intra-regional conflicts, more often than not, involving not just the people concerned, but the local governments, as well. There are many such instances in India, which are festering. However, in developing countries, environmental claims have a lower priority than food production and personal use, and hence, environmental destruction through misuse of water continues unabated.

18.6 Water Supply and Global Warming

Global climate assessment report of the Inter Governmental Panel for Climate Change (IPCC 2001) and the one released in September 2013 predicts a rise of 1.4–5.8 °C over the next 100 years. This has enormous implications for the quantity and quality of water resources in general and for agriculture in particular in the future. An overall acceleration of the hydrological cycle, but, with more erratic patterns of precipitation, floods, and drought, raises concerns about the availability of water for irrigated and rainfed agriculture in different regions. Greater uncertainties in predictions at the regional level will further intensify the problem. Changing precipitation patterns will alter surface water availability, wetness, run off, and ground water recharge, all of which have profound implications for agriculture. Arid and semi-arid areas will be more sensitive to local hydrology and small changes in temperature and precipitation, which could cause large changes in run off. The fate of the 1.7 billion people who are already water-stressed and whose numbers are expected to rise to 5 billion by 2025 will be further aggravated by future stress on water supply. The tropics, especially sub-Saharan African countries, face a challenge because they already are threatened by land degradation and desertification. This, coupled with their low level of per capita income, fossil energy use, and technological primitiveness, will make it a lot harder for them to adapt to such changes. Rising sea levels as a result of melting glaciers and ice caps, a very dangerous and worrying consequence of global warming, will render coastal freshwater aquifers and estuaries saline, adversely affecting the fertile deltas in low lying areas such as Bangladesh and Mekong deltas (Nicholls and Leatherman 1995).

18.7 Conservation and Sustainable Use of Water in Agriculture

The precarious situation with regard to the supply-demand equation of water globally, and in particular, the developing countries, requires that measures for the conservation and sustainable use of water be strengthened. With the acceleration of demand in the civic and industrial sectors, it will be necessary to trim the disproportionate share of water use in agriculture. Agriculture today is faced with a formidable challenge to produce more food from lesser water than what has been hitherto the case. This, along with the environmental destruction and socioeconomic cost brought about by unsustainable water use in the past, requires that "more crop per drop of water" – as the catchy slogan goes –environmental safety, and social responsibility become the *mantra* of the current century. Just as increasing the productivity of the land – more yield per land space – was the agenda of the century past, increasing water productivity through thrifty and efficient use has to become the priority for both irrigated and rainfed agriculture, in particular the former, in the current century. It's importance needs no underscoring, especially for the developing countries, where exponential population increase and demand for water for civic purposes will continue to escalate making per capita water availability a very important parameter to be critically looked at. Formidable challenges lie ahead, because water management for agriculture is a lot more complex a problem, compared to crop varietal improvement, nutrient and pesticide use. For instance, while nutrient use will have its focus, primarily as the soil (of course in combination with water in the soil), water use will have to take into consideration soil and ambient climate. The following are some of the important constraints to be taken note of:

1. The qualitative and quantitative considerations applicable to the above mentioned inputs (like nutrient and/or pesticide use and crop variety) will have, albeit, an indirect and interactive effect, rather than a direct one
2. Water management is a more variable and contextual issue and has, seasonal, environmental, and strong socio-economic and political components, which are not easily definable, separable, quantifiable or actionable
3. The exact water requirement for a specific crop includes several variables, such as, nature of the crop, season, climatic and soil characteristics, surface run off and evapotranspiration
4. Unlike land for agriculture, which is immovable, and has clear ownership and price, water is generally in the public domain. The supply of water for agriculture in many countries is not strictly regulated or based on the precise cost incurred or the exact requirements. Complex socio-economic and political compulsions and populist policies of the government of the day (usually decided by the "vote bank politics"–as is seen in India)play a crucial role and lead to irrational and unsustainable supply and pricing
5. Most farmers in the developing countries are small holders and subsistence or marginal farmers. The accommodation of the socio-economic and political compulsions and affordable arrangements for equitable supply of water to small and marginal farmers are monumental challenges in the sustainable water use in agriculture.
6. Last, but not the least, the adverse impacts on cropland and the environment of wasteful and mismanagement in irrigation in the past need to be properly addressed and mitigated in the future. This is a very important lesson learnt from the impact of the so-called green revolution in India, in the "cradle" of the green revolution, Punjab State, where water was so very thoughtlessly used for more than two decades. The above stated multifaceted challenges demand multidimensional, supply-demand oriented strategies to enhance crop productivity through water use. Such strategies must be based on fool proof science, with a contextual and down to earth human touch. This author, to his great dismay, has observed that the people of the State of Kerala in India, including farmers, suffer severe water stress, both for agriculture and domestic needs, once summer sets in, despite receiving an annual rainfall of 350 cm plus, next to only Chirapunji in the State of Assam, which receives the highest at close to 500 cm. In the current state of affairs, the Kerala State is the classic example of this ironical situation on the waterfront! The anachronistic situation is, a hugely funded Center For Water

Development Research and Management (CWDRM) is situated in the State!!! The irony speaks volumes for the total mismanagement of the water resources of the state, and a clear lack of visionary leadership both at the technical, political and administrative levels is responsible for this mess, and, sadly, the poor is at the receiving end of this visionless handling of this scarce commodity.

18.8 How to Improve Productivity from Water Use in Crop Production?

Taking into serious account the gravity and urgency of the current water related problems, the Consultative Group on International Agricultural Research (CGIAR) set up a CGIAR Initiative as discussed below:

The initiative is a challenge program to facilitate improvement in water productivity for food security. The program involves five of the CGIAR's water related international institutes with the International Water Management Institute (IWMI) as the lead center (CGIAR 2002). The basic objective is to improve crop water productivity to a level at which it will be possible to meet internationally set targets to decrease malnutrition and rural poverty by 2015, the year of the "Millennium Goals" to reduce global hunger by 50%, – the year is past with no signs of the goal anywhere in sight, that is – while maintaining the global diversion of water for agriculture at the same level as that used in 2000. The immediate objectives are to provide food security to all at the household level, poverty alleviation and improved health through better nutrition, and less pollution and greater environmental safety with improved quality of water. A lead paper was prepared with reference to the opportunities and limitations of improving water productivity in crop production (Kijne et al. 2002). The sooner the affected communities become sensitized to the red alert and start working together toward these goals, the better will be their chances for a better future. But, this, most unfortunately, simply did not happen!

The term "water productivity" is used here to denote the average harvestable biomass per unit of freshwater withdrawal for crop production. It has components at the plant, field, ecosystem, and basin levels. Vigorous efforts are needed at all these levels and on a global basis for sustainable water management for agriculture. From the supply side and field perspective, water productivity can be defined as the ratio of total harvested biomass to water applied. This calculation is of interest to the farmer in terms of production and economics. Water supplied through irrigation is partitioned into beneficial and non beneficial depletion (Molden 1997). At the level of the plant, strictly speaking, transpiration is the only beneficial fraction which is directly related to crop yield. However, since it is almost impossible to separately quantify transpiration and evaporation from soil surface, crop water productivity is measured in terms of harvested biomass per unit volume of evapotranspiration and is termed "Water Use Efficiency" (WUE), (Sinclair et al. 1984), the same way in the efficiency of a specific important plant nutrient like nitrogen is expressed as

"Nitrogen (Nutrient) Use efficiency" (NUE). However, it is very important to point out in this context other very important parameters like the type of soil or climate in which a specific crop grows. For instance, from sandy soil the amount of harvestable biomass through the use of a specific amount of water will not be comparable to the same amount from the similar quantity of water used in a clayey or silty soil. Further, the ambient temperature would introduce a variable in the equation. Non beneficial depletion includes out flows such as evapotranspiration, seepage and percolation beyond the plant root zone, and run off, some of which are avoidable, but, some others unavoidable.

Logically there are two approaches to improve overall water productivity. First, by increasing the fraction of beneficially depleted, and, second, by limiting non beneficial depletion. The former is predominantly a crop and varietal attribute, (including the soil and climate influence), and the latter is environment and management dependent. An integrated genetic-environmentmanagement (G × E × M) approach combining genotype selection and management in a target environment is ideal for improving water productivity. The enhancement of water productivity through curtailment of non beneficial depletion has a bright future on account of technological advances and the greater awareness among farmers and other users of the need for thrift in water use. The main channels of non beneficial depletion of water applied during crop production are, over irrigation (as has almost always happened in the green revolution belt of Punjab State in India leading to many soil related problems), faulty or inefficient delivery systems, and deficient agronomic practices. Apart from wastage of water, these also lead to resource degradation such as water logging and salinization, which can be addressed through improved delivery systems and sound agronomic management.

Agriculture involves the consumptive water use. Hence, the use of additional water resources of marginal quality (for example naturally available brackish water), and using water more than once (for example reuse of agricultural drainage water, treated municipal and industrial waste water, and conjunctive water use) can expand the volume of water available for irrigation. The collection and storage of rain water for deferred and supplemental use is critical in rainfed agriculture. It is important that the means to improve water productivity are compatible with the socio-economic reality, perceptions, and capabilities of the farmers. This is essential in the developing countries, where a large number of farmers are small or marginal land holders, cultivate their lands which are biophysically stressed, and are burdened with several socioeconomic constraints. This is precisely where water productivity enhancement is acutely and urgently needed. Simple technologies for small farmers and institutions to ensure equitable distribution are also essential to get the most out of water used for irrigation.

There is ample scope for water conservation and its sustainable use in agriculture, leading to crop productivity, and equitable distribution among users by adopting a variety of measures such as:

1. Vartietal improvement of crops for higher WUE
2. Effective soil-water management

3. Appropriate agronomic management and choice of crops with a low water budget
4. Technological improvement in the water delivery systems
5. Need-based demand driven irrigation approaches
6. Using water of marginal quality and conjunctive water use
7. Creating affordable and appropriate technologies for small, poor and marginal farmers
8. Creating institutional arrangement for sustainable and equitable distribution of water

18.8.1 Varietal Improvement

Varietal improvement for higher crop productivity hinges on an increase in production per unit volume of water transpired without an increase in total water quantum depleted. Crops have different ranges for WUE, depending on environmental and agronomic conditions and plant genetic traits. Values up to 1 kg m^3 for rice and wheat, 8 kg m^3 for sorghum, 11 kg m^3 for potato as measured by yield/evapotranspiration can be found in scientific literature. Field-level water productivity is much lower. Hence, values such as 0.05–0.6 kg m^3 for rice (Tuong 1999) and 0.05–0.3 kg m^3 for sorghum (Rockstrom et al. 2002) have been reported.

The potential of genetic improvement to enhance crop water productivity was considered limited until recently. One important handicap is that beyond a certain limit, WUE tends to level off, irrespective of environment and agronomic practices. It is pertinent to note that the breeding of dwarf and short-duration varieties before the 1980s, aimed to predominantly increase the harvest index (HI)(partitioning plant biomass into harvestable yields), has led to simultaneous enhancement of WUE. However, the harvest index has almost reached its theoretical limit in most of the common field grown crops (Richarrd et al. 1993), and, the scope for further increase in HI in important staples is quite limited. However, recent advances in plant breeding and plant biotechnology have opened up new possibilities of taking the WUE to higher levels. One possibility is by way of introducing genes conveying higher photosynthesis per unit of water transpired, as in the case of C-4 plants. Some progress has been made in this direction (Bennet 2003). Search for other water productivity related genes holds promise, but, no spectacular results are worthy of note, as of date.

18.8.2 Soil and Water Management

At the heart of water conservation and its sustainable use is the question of soil and water management. Both soil and water have a symbiotic relationship in nature, affecting crop production. It is important to recognize that, without soil, there simply will be no water. Water cannot exist in its original form, except as vapor, in outer space. This is the reason why optimal soil management is key to ensuring optimal WUE. Agriculture thrives where soil is fertile, climate is favorable and adequate water is available. On the other hand, if the water is even temporarily insufficient, especially during critical periods of crop growth, agriculture fails.

Water productivity is dependent on the balance of water inputs and outputs. The total soil –water balance in cropped soil can be expressed as follows:

$$\mathrm{P} + \mathrm{I} = \mathrm{R} + \mathrm{ET} + \Delta S + \mathrm{D}$$

where

P = Water received as Precipitation
I = Water received as Irrigation
R = Runoff
ET = Evapotranspiration
Δ S = Water stored in the potential plant root zone of the soil
D = Seepage below the plant root zone

It is thus obvious that minimizing unproductive outflows through evaporation, run off, and seepage beyond the root zone, making efficient use of precipitation, and improving irrigation systems and schedules can reduce non beneficial depletion of water. The most comprehensive way to do this lies in providing demand driven or need based irrigation schedule, improving water delivery systems, and effectively harnessing rain water, especially in rainfed agriculture. Waterlogging and salinization are problems in irrigated agriculture. Drainage is a preventive option under such conditions. Hence, a number of measures are available to maximize WUE. A suitable choice has to be contextually made.

Good soil management and efficient water management are most often synergistic and complementary. Sustainable land management, such as, minimum soil tillage, contour cultivation, and soil mulching (a crop residue to mulch is preferable), will not only conserve water, but, also reduce wasteful run off and improve infiltration and water storage. Deep tillage reduces loss as a result of deep seepage.

Waterlogging, salinization, and high water tables are, as discussed earlier in this chapter, serious drawbacks of irresponsible irrigated agriculture, arising predominantly from over irrigation. The crisis in Uzbekistan and the surrounding areas is a stark example of such mismanagement. Providing artificial drainage helps overcome such adverse effects and improves crop productivity. In addition to providing an optimum air and salt environment in the plant root zone, drainage provides other advantages, such as less surface run off and soil erosion, less damage to soil texture, efficient disposal of agricultural wastewater, reduction in environmental pollution, and prevention of damage to biodiversity (Raman 1989, FAO 1997).

18.8.3 The Right Agronomic Practices and Crop Choice

Soil moisture is conserved by practices such as minimum tillage, contour cultivation and soil mulching, preferably with vegetative mulch. Cultural practices such as row spacing and the use of plant residues and mulches have great potential to increase water productivity.

Flooded rice cultivation on almost 40 million ha of irrigated land in South and Southeast Asia involves a colossal outflow of 30–50% of water through deep percolation and evaporation. Tremendous scope exists to save on water by switching over to upland, aerobic, and direct-seeded rice and new varieties with shorter growing season. Suitable cropping systems based on ecological factors and appropriate crop rotations to use up residual moisture (for example rice-legume rotation) go some way in improving the efficacy of water use.

In this connection a mention must be made of the System of Rice Intensification (SRI) developed by a Priest in Madagascar. This is a system where seedling number is increased and the rice field is not kept "field flooded," as conventionally practiced by farmers in the developing countries, especially Asia. In this manner much water is lost through field surface evaporation. On the other hand, in SRI, the field is only intermittently flooded and the water is drained, and a high seed rate is used. The SRI has shown remarkable success in many regions of India.

Shifting water away from water-thirsty crops such as sugar cane, cotton, and tobacco to nutritional crops such as wheat, millets and other small grains is a good option in areas where both food and water security are threatened. Reviving the practice of growing traditional and underutilized crops and varieties, for example fonio, teff and millets, which are resistant to dry conditions and have low level of water requirement is ideal for conserving water and improving food security in some water stressed areas (NRC 1999). Under rainfed conditions, inter cropping may be more beneficial than single crops in sequence. For example, intercrops of corn and pigeon pea and of sorghum and pigeon pea have been found suitable in vertisols with dependable rainfall in South India.

Advances in biotechnology have made it possible to introduce drought resistant genes into plants; this can be a safeguard for yield under water stress. CGIAR has initiated breeding programs for enhanced drought tolerance of nearly all of its recommended crops, with emphasis on farmer participation to improve security in the dry and drought prone areas of the world, which are home to the majority of the poor and unreached. Conventional breeding methods with wild relatives are currently used. Several drought tolerant crop varieties of rice, maize, barley, wheat, cowpea, groundnut, lentil, faba bean, and sweet potato have already been released for evaluation. Matching the duration of the vegetative and reproductive periods to the expected water availability and genetically modifying root systems for effective water utilization are useful methods in breeding for drought tolerance.

In rainfed agricultural systems, especially on poorly drained soils, periodic flooding and water logging lead to lower yields and salinization of the soil. Under these circumstances, and when using water of poor or marginal quality, it is beneficial to grow salt-tolerant varieties of crops. Although genes for salt tolerance have been identified in some crops, only little progress has been made in transferring them, because, generally a complex pyramid of genes is involved (Edmeades et al. 2001).

18.8.4 Need-Based Irrigation

The most practical and effective way to conserve water is to adopt a timely and need-based irrigation schedule to avoid wastage and resource damage. Although this seems simple and sensible on paper, many farmers have a tendency to over irrigate their crops when water is plentiful. Not only does this lead to wastage of water, but, it also results in several other agronomically deleterious effects such as nutrient leaching, waterlogging, and salinization, leading to lower productivity. Small and marginal farmers in developing countries generally adhere to a fixed schedule, which works quite well.

Recent advances in information, communication, and space technology have given rise to technologies which are being used to regulate water application to crops on large and mechanized farms in the developed countries. Precision agriculture is becoming increasingly popular among large-scale farmers in developed countries, where technology and commercial services are readily available. Newer technologies in weather monitoring and forecasting and satellite technology which can monitor the water needs of crops have been used in the US to deliver water at appropriate times in required quantities. The Department of Water Resources in California operates a network of more than 100 automated and computerized weather stations that collect a series of local weather data from which the evapotranspiration rate for each crop is calculated. This enables the farmer to know the exact time at which to irrigate. The result is an estimated 8–13% reduction in water use and economic benefits. Free weather data or more sophisticated, partially commercial services, are also available in several states, which farmers can access for precise information allowing them to save up to 15–20% of water (De Sena 1998). Such sophisticated arrangements are not available in poorer countries, but, computer programs are now available for most climates and crops to calculate evapotranspiration, soil moisture changes, and irrigation requirements using local data on fields, crop and so on, as inputs. A useful first step in this direction globally is the World Water and Climate Atlas prepared by the IWMI in Sri Lanka, which is a computerized tool for irrigation and crop planning that integrates agricultural weather data from 56,000 weather stations around the world for the period from 1961 to 1990. The atlas can help farmers and planners in both rich and poor countries to match crops and crop varieties to local conditions.

18.9 Technical Improvement of Water Delivery Systems

Despite the evolution of advanced methods for crop production during the past few decades, irrigation systems in most parts of the world remain quite primitive and inefficient. Most farmers continue to irrigate their fields the traditional way, mainly through gravity-led surface irrigation along open channels. This leads to a certain amount of wastage of precious water, as run off, infiltration past the plant root zone, evaporation, and so on. It also leads to salinization, leaching of nutrients, erosion, reduced productivity, and pollution of surface and ground water. Inefficient irrigation practices are estimated to cause a loss of 10–20% of water (Molden 1997). Globally, this amounts to a loss of 250–500 billion m^3 annually, which could instead add as many million tons of grains to global grain production. Simple measures such as lining canals and irrigation channels and avoiding over irrigation can reduce water losses and land degradation by lowering evaporation and seepage. Surge irrigation is a newly developed water delivery system which channels water along furrows in a series of pulses or surges, rather than in a continuous stream. It cuts water use by as much as 10–40%.

18.9.1 Micro Irrigation Techniques

Micro irrigation techniques of drip irrigation, sprinkler irrigation and underground watering of the root zone are more effective methods in conserving water through need-based supply and reduction in evaporation and seepage losses.

Drip irrigation systems deliver water directly to the root zone of individual plants through a network of plastic tubing installed above or below the soil surface. When it is combined with soil moisture monitoring or other methods of assessing crop water requirements, drip irrigation can achieve almost 95% efficiency. Fertilizers can also be applied through the drip irrigation system. In addition, to reducing evaporation and seepage, drip irrigation minimizes land degradation and pollution due to run off, and it avoids the detrimental effects of flooding and drying (Hillel 1994). Currently, approximately 2.8 million ha, which is approximately 1% of the total global irrigated area, is under drip irrigation (Bucks 1995). Low – cost drip systems are now available which are affordable to small and marginal farmers in the developing countries. Drip irrigation has a significant role to play in small farmers' fields in areas of water scarcity (Postel 2001). In India, an impressive increase in acreage covered has taken place in drip irrigation, from 71,000 ha in 1991 to 162,000 ha in 1999, which translates to an increase of more than 128% in 8 years, which works out to 16% per annum. The coverage is eventually expected to increase to ten million ha. Compared with conventional surface irrigation, a 30–60% of saving in water use and an increase in yield of 5–50% have been reported in the case of different crops, which amounts to a doubling of water productivity (Sivanappan 1994). Drip irrigating thirsty crops such as cotton and sugarcane can save a significant amount of water.

Sprinkler irrigation systems deliver water in sprays to target destinations on time-controlled cycles. To reduce drift and evaporative losses, low-energy precision application sprinklers are now available which deliver water in small doses though nozzles placed just above the soil surface. These sprinklers save an additional 15–35% of water and operate at 90–95% efficiency compared to drip irrigation. Given that evaporative losses account 11% in upland crops which can go up to as high as 53% in puddled rice (Burt et al. 2001), curtailing them has a tremendous economic value in rice cultivation.

Israel has been a pioneer in the development and use of micro irrigation technologies such as sprinklers and drip irrigation systems, coupled with appropriate measures for pricing, watering, and area-based allocation of water. Between 1951 and 1985, Israel expanded its irrigated area a five-fold, with only a three-fold increase in water use. In addition, the yields of most vegetables, fruits and other high-value crops increased almost three-fold (Postel 1992).

18.10 Water Management in Rainfed Agriculture

Improving water productivity is, a lot more important in rainfed agriculture, than in irrigated agriculture, which covers more than 80% of cropping area and produces more than 60% of total global food grain production (Harris 1991). More than two-thirds of India's crop production is confined to raninfed areas. Farmers living in these areas are among the poorest; they practice subsistence farming, or when landless, depend on production elsewhere. Increasing production on these marginal lands is crucial to ensure food security and livelihood opportunities. Increasing production in rainfed areas eases the pressure on irrigated agriculture, which is already constrained by water shortages, environmental pollution, and resource degradation.

Water and land productivity can be safeguarded and stabilized with supplemental irrigation (Rosckstrom et al. 2002). In rainfed agriculture, the problem is often more of the temporal and spatial variability of rain than low cumulative precipitation. In India, rainfall is highly seasonal and is concentrated in 3 or 4 months, which makes rainfed farming quite a risky venture, during drier months. The most classic and very tragic example is that of what happened in Vidharbha district of Maharashtra State in India, the "cotton belt" of India, where the failure of the Bt cotton, which needed plentiful supply of external inputs like chemical fertilizers and water, led to thousands of poor and marginal farmers taking their lives to escape from the clutches of unscrupulous money lenders, who lent money at usurious rates of interest to the poor and marginal farmers, when the crop failed due to scanty rainfall. This has become a national, and, perhaps, an international scandal. Intraseasonal dry spells can cause heavy losses of productivity, in particular, if they occur during crucial moisture-sensitive periods of crop growth. On most dryland farms in India, China, and Africa only 15–30% of rainfall directly aids crop production, with 20–25% being lost through run off, 20–25% through percolation, and 30–35% through evaporation. Channeling a greater share of rainfall into the soil to conserve soil moisture during droughts protects crops from wasting. Water harvesting and supplemental irrigation are invaluable in rainfed agriculture.

Water harvesting is an affordable and simple technology to capture water during excess rainfall to be used during dry spells for supplemental irrigation. This is a common practice in traditional agricultural systems and is among the ancient practices being revived in countries of Asia and Sub Saharan Africa and in drier areas of West Asia and North Africa with a good measure of success (Rockstrom et al. 2002). Assured water availability, coupled with adoption of better management and improved genotypes of crops, leads to higher yields and economic benefits.

Combining the traditional water harvesting measures with water-efficient low-cost and low-technology devices, such as, sprinklers and drip irrigation, can enhance water productivity and the food, economic, and livelihood security of small farmers. In Madhya Pradesh, northern India, the rainwater is ponded in large fields of 2–12 ha with earthen embankments during monsoon season (June through August) and drained in October to plant crops. The heavy black clayey soil is able to retain enough moisture to sustain the crops, until harvest. Recently sprinklers have also come into use. Similarly, Tamil Nadu, in southern India, has a tradition of storing water in tanks called "*eris*" for irrigation. This practice allows the harvesting of rainwater for use during the dry season and also prevents flooding and soil erosion, consequent to sudden heavy downpour, as can often happen in India, and promotes ground water recharge. Check dams and dams across gullies in watersheds can trap large amounts of run off that can be stored and used or channeled to recharge ground water, as seen in Rajasthan and Gujarat, two states in north western India.

18.11 Effective Use of Water of Marginal Quality

Agriculture does not always need water of a very high quality. Conjunctive use of canal water and pumped ground water, which is generally of a lower quality than canal water, is useful to expand available water for agricultural use. Conjunctive use of canal water and ground water is common in such places as the Indus and the Ganges valleys in India and Pakistan. However, over extraction has led to deterioration of water quality and to severe salinization in Pakistan and harmful levels of arsenic in Bangladesh.

Slightly brackish and minimally treated municipal and waste water can be used for re-irrigating crops which have better tolerance to salt. In the Jordan Valley, 98% of the total irrigated area of approximately 24,000 ha is being irrigated with water of salinity up to 3 dS. m^{-1}. Reusing municipal waste water after minimal treatment is a good option to improve productivity of water. As the volume of urban waste water is likely to double over the next three decades, treated waste water will be a reliably expanding resource. Israel is a pioneer in this area, using 65% of its domestic waste water for crop production. Treated waste water accounts for 30% of the country's agricultural water supply and is expected to reach 85% by 2025 (Postel 1992).

Agricultural drainage, which is most often treated as a waste and discharged into surface water ways, causing pollution and resource degradation, can be used as a valuable resource to grow crops tolerant to salt. In addition to improving water productivity, it protects natural resources from deterioration. In Israel, cotton is irrigated with water from agricultural drainage. Tomatoes grown for canning seem to benefit from salt water. In Malaysia's Muda rice irrigation system, a 17% increase in water productivity has been obtained by pumping used irrigation water. Reusing drainage water from agriculture to grow salt-tolerant crops and halophyte trees and shrubs for the restoration of wasteland has become the subject of extensive investigation. The quality of the drainage water is an important consideration in deciding which crops can be grown and whether it can be used for supplemental irrigation. Highly saline drainage water cannot be used to irrigate salt-sensitive crops, but, it can be used for salt-tolerant forages, in a saline-agriculture-forestry system, and tolerant trees and vegetables. One of the most important considerations in using drainage water, especially industrial effluents, for agriculture is that it may, more often than not, carry toxic heavy metals, such as lead, mercury, cadmium etc., which can pose great health hazards. This aspect is getting increasing attention from nutritionists, and health sector workers.

Water reuse must be balanced against both short-term and long-term needs, with consideration given to both local and off-site effects. For the safe use of agricultural drainage water, it is essential that the characteristics of the water used, soil in question, and intended crops, be known in advance, and managed with caution and appropriateness.

Reuse of agricultural drainage water has its disadvantages as well. Lower yields and the risk of long-term effects on soil productivity and soil degradation as a result of salinization, loss of structure, alkalinization, and sodication (when Na content is high) are distinct deterrents. Ameliorative measures include proper crop sequence, conjunctive use with good-quality canal water, water table management, and chemical treatment (Tyagi 2002). In the recently implemented saline-agriculture-forestry and solar evaporation in California, drainage water is sequentially used to irrigate salt-tolerant crops, trees, and halophytes and then discharged into rivers during high water flows or into solar evaporators for salt crystallization. This prevents the poisoning of wildlife, as had happened in the Kesterston Valley of California (FAO 1997).

18.12 Need-Based Technologies

One of the most important and tragic drawbacks of Green Revolution has been that it benefitted only irrigated agriculture on large farms, owned by wealthy farmers, typically characterized by farming in Punjab State, India. Poor and marginal farmers owning 2 ha or less than that, mostly in South Asia (in particular India and Pakistan) have generally been bypassed and have not had access to major irrigation projects. Tragically, the powers-that-be connived to provide infrastructural facilities like extensive irrigation facilities to facilitate the wealthy farmers, owning large tracts of land, more often than not, neglecting the poor and marginal farmers. This is also the reason why more than two-thirds of Indian agriculture is still rain dependent. Availability of water for crops is the most vexed question of poor and marginal farmers. And, sometimes, big business corners water for its own mercantile interests, with official connivance through bribes and kick backs. The classic example is what happened in Kerala State, India, in the recent past, in the water-starved district of Palakkad, in southern Kerala State, where the government of the day gave official permission for a global giant soft drink manufacturer to tap into the soil for deep-seated water. Agriculture suffered in the process. And, a legal issue followed, only which could resolve the plight of poor rice farmers of the district. It is, therefore, imperative, that poor farmers also be provided with affordable sources of irrigation to ensure local food security.

The general belief that bigger farmers with large tracts of land are more productive than marginal or poor farmers with far too smaller land holdings does not hold true in the Green Revolution regions of South Asia (India, Pakistan, Philippines and Sri Lanka). It is the small farmers who have been shown to have a higher proportion of net sown area and higher cropping intensity, to cultivate more diversified, higher value, and more labor-intensive crops, and to obtain higher yields per unit of land area. Given their share of cultivated lands in these countries, small holders contribute disproportionately to the production of major crops. Their share of production was considerably larger than their share of arable land in 39 out of the 55 developing countries investigated by the FAO (1998). Many are subsistence farmers who need to coax more out of their limited land holdings to ward off hunger, but, remain poor and under nourished. It is high time to make available to them better technologies which suit their needs. Unfortunately, the policy makers of a large agrarian economy like India, have chosen to bypass this basic shift in paradigm and continue to goad the rich and the competitive in the belief that large industrial enterprises will contribute more to the national Gross Domestic Product (GDP) than small-scale agriculture. The recent surge in food inflation has belied their calculations. Food inflation has shot up well past the double digit figure in the country. There is now a definite supply crunch, be that of staple cereals, vegetables or fruits and the policy makers have been caught in a difficult bind.

A number of innovative and low-cost technologies suitable for small and marginal farmers of South Asia and Sub-Saharan Africa have been developed to increase the productivity of millions of hectares of crop land and provide food security for marginal and poor farmers. In places such as Bangladesh, the northeastern states of India, and Sub-Saharan Africa, underground water is plentiful at low depths, but, small and marginal farmers cannot afford to use energy source like diesel or electricity to pump the water out to irrigate their parched fields. To overcome this problem, a simple treadle mill, resembling a step exerciser, has been devised to pump out water from shallow depths up to about 6 m below soil surface. This device is used extensively by poor and marginal farmers in Bangladesh, Pakistan and India, during the drought period preceding and following the monsoon (Orr et al. 1991). Such pumps have created a revolution in ground water use; they effected a 136% rise in irrigated area between 1975 and 1995, that is close to 7% increase per annum. This is a remarkable jump, when compared against the global scenario, where the increase in irrigated area was just 35% during this period. Farmers use the pumps extensively to grow vegetables, and high yielding rice varieties, without the fear of floods washing them out. The devices are indigenously produced, offer a three-fold net return on invested cost, and can irrigate about half an acre.

Africa is one continent which has failed to reap the benefits of the green revolution, in particular, because of the lack of extensive irrigational infrastructure, unlike what has happened in India, which has been compounded by the continents difficult terrain, unlike in South Asia, with extensive flatlands, hostile climate, socio economic constraints, and, over all, a lack of political stability and will. The 6% of Africa's cropland which is irrigated lies in the north of African continent. In Sub Saharan Africa only 6.3 million ha are irrigated. More than 80% of the poor have no access to irrigation. Africa is well endowed with water, but, not water conducive for conventional irrigation. More than 80% of the precipitation is lost through surface evaporation and run off and transpiration because of the hot and dry climate, compared to the global average of 63% loss. Rainfall is erratic and drought is endemic. Ground water is widespread, but, constitutes only approximately 15% of the region's renewable water supply. Traditional, indigenous system of small-scale irrigation and water management (which often registers a higher water productivity than compared to conventional systems, Postel 1999) and simple devices such as small-scale pumps, treadle pumps, and shallow tube wells have enhanced water use for irrigation by small farmers. This has improved the food and income security of small farmers who cultivate seasonally water logged wetlands and valley bottomlands. Modern technologies such as drought-resistant varieties and early-maturing crops can bring additional advantages to farming in these environments.

18.12.1 Equitable Distribution and Sharing of Water

The creation of the physical infrastructure of small and big dams and reservoirs to store water and efficient delivery systems improve irrigation efficiency, but, the actual distribution of water to farmers in an equitable, sustainable, and affordable manner is equally, if not more, important. This *soft* component of irrigation is trickier to manage than the *hard* component.

Water distribution systems comprise a heterogenous group of stakeholders, including big and small farmers, water use organizations, state and federal governments, non governmental agencies, local and national level politicians, and, finally the tax payers, who have, more often than not, complementary or contradictory interests. Farmers at the head of a canal invariably drain more than their requirements, leaving tail enders with chronic water deficit. For example, 25–40% of the declared irrigated area in India suffers from tail end water deprivation. In South Asia, salinization owing to overuse of water at the head end and deprivation and low productivity at the tail end, are quite common. Thus, the full potential of expensively created, gigantic irrigation structures is not often realized, resulting in loss of crop productivity, leading to food insecurity, degradation of land mass, and inevitably culminating in social inequities. As the creation of more dams appears to be economically, environmentally, and socially non viable, institutions need to be created to facilitate effective and equitable delivery and distribution from existing dams to improve water productivity.

The *warabandhi* system which regulates the water from the Bakra Nangal dam (in Punjab State, India) to irrigate almost 14 million ha of crop land primarily in the State of Punjab, northwest India, and Pakistan is a model for the equitable distribution of stored water. As the water available in the Punjab State is adequate for only one-third of the total cropped area, water is rationed in an organized and equitable manner. Each sub group of farmers receives water in proportion to the cropped area served and according to a strict schedule to start and stop irrigation. Farmers adjust their cropping pattern to water supply and supplement it with ground water. The success of food production in this area is testimony to the success of this water distribution system. Social capital has a large role to play in the operation of distribution systems in a fair, sustainable, and efficient manner, without which the Tragedy of the Commons ensues (Hardin 1968), the destruction and devaluation of a public good. In this entire endeavor, the most important factor for success is the human spirit, which believes in fairness, a commodity that is fast dwindling in the excessively and needlessly materialistic world.

18.12.2 Getting the Price Right

Pricing irrigation water is among the most unrealistic, most contentious, and weakest components in the efficient and thrifty use of this public good. Supplying irrigation water below its cost price through various kinds of subsidies appears to be a trend the world over. It has an underlying political connotation. In India, the experience has been, it is a "vote catching" exercise of the party in power. Irrigation welfare is widespread, all the way from small water-deficient countries such as Tunisia and Sri Lanka to the mixed economy of India to the free market giant USA. This often makes it economically non viable even to maintain and operate the existing storage capacities. With water reservoirs silting at the rate of 1% per year, even their maintenance becomes difficult. Myers (1997) estimates the total subsidy for water at US $ 33 billion per annum, worldwide. It is mainly the rich and big farmers who are the beneficiaries of this system rather than the poor and marginal farmers. Keeping water prices artificially low encourages wasteful and irresponsible water use by farmers, which not only lowers efficiency, but, also causes a grave ecological damage and adds to the hidden cost of water supply.

18.12.3 Trading Water

Markets for formal and informal trade in water are gaining great importance, which are becoming increasingly useful in reallocating water for different needs. Formal arrangements with legally enforced entitlements to water and restrictions on where it can be traded exist in such countries as the US. This helps to channel water to various sectors according to need and availability. Informal trading in ground water is becoming popular in South Asia and Mexico. In India, especially in the State of Bihar in eastern India, and Pakistan, informal sale of water by big farmers to small and marginal ones in return for a monetary, crop or labor equivalent are prevalent. The small and marginal farmers gain the advantage of reliable timing and sufficient water to cultivate their holdings, which the arrangement ensures (Shah 1993).

Extreme views surround treating water either as an economic good subject to market forces or as a public good that should be exempt from competitive pricing and allocation (Perry et al. 1997). A context-specific compromise, combining both options with an optimal amount of additional market management, is warranted to encourage the adoption of beneficial technologies and avoid waste of this very precious resource. Institutional and policy reforms such as phasing out of power subsidies, as is frequently seen in India, for pumping water for agriculture, secure water rights where possible, and the participation of users in management are conducive to more effective and efficient use of water, in particular, in the developing countries with large-scale production. However, this has to be realistic so that poor and marginal farmers can also afford water and agriculture does not become uneconomical. Incentives for adopting good practices such as conservation agriculture, may be better options and in the interests of both small, marginal and also big farmers.

18.12.4 Institutions and Regulatory Mechanisms

It is clear from the above that rules, regulations, and mechanisms for implementation need to be put in place through proper policies, flexible and specific to the situation in question, to enforce economic, social, and moral discipline among the water users. Indiscriminate use of ground water leading to severe depletion of ground water resources, is among the biggest threats to global irrigated agriculture and food security in the future. The poor and developing countries will be the greatest victims of this scenario. Yet, there are no formal regulations in many countries. Unfair monopolistic pricing of ground water which encourages greed and overexploitation of ground water by sellers, whose entitlement to ground water is not legally limited and depends on their ability to invest in high-powered pumps still exists in countries such as India, which needs to be stringently curbed. For example, in India, ownership of land ensures ownership of the water beneath, and there is no rule to regulate to limit the quantum pumped out. This has led to severe depletion of ground water resources, which will, without doubt, deprive future generations of their rightful share. Some efforts to regulate ground water pumping, for example, through licensing, credit, restricting electricity provision, setting minimum well-spacing requirements etc., have been introduced recently in India, and violations of these stringent conditions is the norm rather than their adherence. And enforcement of these regulations is very lax, given the corrupt nature of many involved. Serious and stringent control of the quantum of ground water pumped out must be in place without any further delay. This author is aware of so many instances where deep bore wells, going up to a depth of 150–160 ft., are dug indiscriminately, by bribing local authorities. Although India passed a model ground water Act in 1992, the different states do not adhere to this legislation. It is to be hoped that the recently introduced water policy will lead to more efficient, equitable, and sustainable use of water for agriculture. Scientists, farmers, and government agencies should strive to create conditions conducive to balancing pumping and recharge to minimize water mining. Pumping of fossil water from deep aquifers, which is common in the Middle Eastern countries and the western United States, represents a permanent loss of this precious natural endowment. The implications of this damage are more serious than just the threat to food security. It cuts into the very roots of the sustainability of planet earth, and, needs to be very stringently monitored.

18.13 The Role of Nitrogen in Sustainable Agriculture

In food crop systems, nitrogen plays a key role, because, it is the main yielddetermining nutrient. The rise in food production certainly paid a price, a very huge one at that: more external inputs resulting in correspondingly more environmental harm. Until 1960 N fertilizer was used at a relatively low rate; then, crop nitrogen uptake depended strongly on organic manure applications, biological N fixation through legume intercropping, sole cropping, and nitrogen mineralization through the decomposition of organic matter. Since the early 1960s, following the introduction of the "miracle dwarf" wheat varieties from CIMMYT, Mexico, especially into countries like India, the use of nitrogenous fertilizers has grown exponentially – approximately seven-fold – and nowadays 30–80% of nitrogen applied to farmland is lost to surface and ground water and to the atmosphere (Goulding et al. 2008). In the atmosphere, the non reactive form N_2 is already present abundantly; only the reactive N forms NH_3, NO_x are harmful. Gourdriaan et al. analyzed that in 1990 fertilizer N accounted for 60% of the net primary production (NPP) and since 1980 has exceeded the amount taken up by crops globally. The economic response in crop yield far outweighs the cost of fertilizer, which led to over applications when only yield improvement was considered without any serious thought to environmental sustainability. This is amply illustrated by the shift in focus from the so-called green revolution phase of the 1960s to mid 1970s, where application of increasing quantities of soil inputs such as nitrogen fertilizers and pesticides was emphasized, to the "sustainable agriculture" phase from the early 1980s to the present (probably to continue?, Nair 1996) phase.

The role of nitrogen in future world food production and environmental sustainability was explored by Eickhout et al. (2006). Based on FAO projections, they concluded that despite improvements in efficiency of nitrogen use in food production systems in the developed world, total reactive N loss will grow strongly towards 2030 because of the intensification of animal and crop production systems in developing countries. Herridge et al. (2008) estimated the global inputs of nitrogen fixed biologically in agricultural systems by pulses, oilseed legumes and other cropping systems at 50–70 Tg N (Tg: million tons). Soybean represents 50% of global crop legume area and contributes about 75% of the N fixed by crop legumes. The increasing global trade of commodities such as soybean is also accompanied by a flow of nutrients. The annual total N flow in traded cereals from exporting to importing countries was estimated to be 11.5 Tg in 2004, mainly from Brazil and the US to China and Europe (UNEP 2007). Hence, there is an urgent need for a balanced use of fertilizer N. Only economic optimization of fertilizer use does not control overuse; therefore, public awareness about the risks of excess N in agroecosystems is urgently called for and proper legislation should be implemented to bring to the knowledge of the farming community, in particular, and, the public, at large, of the environmental risks. Unfortunately, in many of the developing countries, fool proof environmental-safe legislations are still a far cry.

18.13.1 Nitrogen Use, Crop Growth and Yield

Long- term field experiments, such as those at Rothamstead in England, show that wheat yields with fertilizer N exceeded those without external N input by a factor of 2–3. It was found for on-farm and research station experiments that grain yields of maize increased from 3–14 t/ha with a rise in plant N accumulation from 50 to 300 kg N/ha, while rice yields increased from 2–8 t/ha with a rise in plant N accumulation from 25 to 200 kg N/ha respectively (Dobermann and Cassmann 2002). The increased use of fertilizers, especially N, substantially enhanced crop growth and yield, and, as a consequence, the associated resource-use efficiencies, such as light use efficiency (g MJ^{-1} and water productivity (kg dry matter produced per unit of evap otranspiration).

In complex cropping systems, such as multiple and relay cropping systems, the response of the component crops to N application is less than in the case of that obtained in mono cropping systems. This was clearly shown in the case of rice-wheat cropping system. The total N use efficiency of relay intercropping systems of wheat and cotton was assessed by comparing the relative N yield to the relative yield total. The relative N yield varied from 1.4 to 1.7, while the relative yield total ranged from 1.3 to 1.4, indicating that intercrops used more N per unit of produce than monocrops. Hence, component crops in intercropping systems should get less fertilizer N than in the case of a monocrop.

Higher yields of cereals, like rice and wheat, were derived from breeding high yielding and N responsive varieties. The dwarf wheat and rice are classic examples of this. The optimal N use for growth and yield maximization is determined by the plant traits, such as, physiological processes, environmental conditions and nutrient management. The most significant increases in nitrogen use efficiency (NUE) have come from improved plant genotypes and agronomic practices. It is important to remember in this context that of all the factors that determine the potential yield of a plant type, the nutrient factor is decisive to the extent of more than 50%.

Some opportunities for enhancing Nitrogen Use Efficiency (NUE) are the following:

1. Improved genotypes: Both traditional plant breeding and modern biotechnology offer the promise of higher yields in plant varieties, though the latter, especially with regard to genetically modified crops, has come under a cloud of suspicion of late
2. Improved resource use: A better timing of N and water supply by time and site-specific management can avoid stress at critical growth stages. Higher productivity per unit of water use will lead to enhanced yields and thus higher NUE (Peng and Bouman 2007).
3. Improved cropping systems: Farmers can vary the sowing time/and the choice of crops in a cropping sequence to make a better match of the genetic-make up of a crop and the growing conditions determined by climate, soil and pests (Hobbs et al. 2008)

Under high land pressure, the emphasis will be on growing high yielding genotypes and optimizing the management of external inputs, while under low input conditions, adopting cropping systems, including cultivar choice, to the variability and resource availability of contrasting agroecological conditions will assume far greater importance.

18.13.2 Nitrogen, Photosynthesis and Plant Growth

Nitrogen is the key element in plant nutrition, limiting plant growth and crop yields in any agroecosystem, rainfed as well as irrigated systems. Crop photosynthesis is closely associated with light capture by the canopy and leaf N content depending strongly on the availability of nitrogen (Lemaire et al. 2007). Leaf N content is strongly related to the rate of photosynthesis (Cabrera-Bosquet et al. 2007). An early canopy closure and a delay of canopy senescence will enhance the amount of light intercepted, while a very high leaf area index (LAI) increases mutual shading and therefore decreases light –use efficiency (Russell et al. 1989). In reproductive crops, such as cereals and pulses, the duration of canopy photosynthesis is also determined by the functional balance between sink strength and source capacity (Yin and Van Laar 2005), and by the ability of roots to capture N at the end of the growing season (Kichey et al. 2007) matter.

The quantum of N in the harvested produce (crop part) is determined by the sink strength of the storage organs and is expressed as N harvest index. This value is usually high in cereals and tuber crops, for example, 0.60–0.80 for wheat (Lopez-Bellido et al. 2008) and 0.70–0.80 for potato (Biemond and Vos 1992), but, somewhat lower in legumes (Chapman and Muchow 1985). Cereals do reallocate N from the leaves to the grains, while in root crops most of the N absorbed is retained in the crop residues. In general, dry matter and nitrogen partitioning in wheat differ between old and modern cultivars; however, both parameters are not always genetically associated with each other (Van Glinkel et al. 2001). Martre et al. (2007) found that variations in weather and N treatments also affected the nitrogen harvest index of wheat. Further improvement will require a good understanding of genotype x environment interactions.

18.13.3 Synchronization of N Supply and N Demand

To secure crop yields and avoid N losses, the N supply should match the crop N demand in dose and time. The concept of *synlocation* and *synchronization* in crop plant nutrition was first proposed by De Willigen and Van Noordwijk (1987). However, implementation of this concept seems to be quite difficult. A more generic approach to achieve a demand-based N supply to crops is based on the functional relationship between N uptake and carbon acquisition through canopy photosynthesis of a sole crop or a multiple of cropping system. This relationship can be illustrated by the following simple equations:

1. Total C acquisition = {LI × LUE, where {LI is the total amount of light intercepted by the canopy of a sole crop or intercrop MJ m^{-2} and LUE is the light-use efficiency g MJ^{-1}. The {LI is mainly determined by the amount of incoming radiation and the growth duration of a sole crop or a sequence of crops (Keating and Carberry 1993)
2. Total N uptake = f × C-acquisition, where f is a parameter determined by the maximum N content of the biomass. This parameter depends on species-or cultivar-specific plant traits (Lemaire et al. 2007)
3. Total N supply = a × (N soil reserve + manure + mineralization) + b × (N fertilizer) where a and b are parameters determined by plant traits, such as, root length, rooting depth etc., which affect the recovery of applied N (Van Delden 2001).

More sophisticated algorithms were developed to describe the relationship between crop dry weight and nitrogen uptake (Van Delden 2001). To get an optimal match of N demand and N supply, crop growth models can assist in predicting the yield potential under specific climatic conditions and take into account the risk of growth limiting, for example, drought or excessive ambient temperature, and growth reducing, for example weed infestation, pests and disease conditions.

In irrigated rice systems, the use of water and fertilizers, especially N, has increased dramatically; a major point of concern for these systems, is that of the efficiency of use for nutrients, especially N, and water. By using a simple rice-N model, MANAGE-N-, it was possible to enhance the efficiency of use of this important nutrient in rice production (Aggarwal et al. 1997), and it was possible to improve the timing of N dressing. Aggarwal et al. (1997) obtained yield up to 9–10 t/ha in high yielding rice varieties using the above model. For each user-defined fertilizer N dose, the model identifies the timing and amount of application, associated with maximum grain yield and maximum agronomic N efficiency (kg grain obtained per kg N applied). Improved timing of N application in irrigated rice resulted in yield increases ranging from 4 to 10%, at a fixed total dose. Changing the N dose to the predicted economic optimum rate resulted in additional increases up to 13%. Thus, in rice, it appeared essential to match the seasonal pattern of N supply to the N demand of the crop at each stage of development to achieve maximum yields, but also to minimize N losses to the environment. Integrated nutrient management (INM) and precision farming have shown to be effective tools to enhance NUE (Pierce and Nowak 1999). Smart farming technologies aiming at both productivity increase and efficiency gains have been promoted. A large number of experiments have been carried out with site-specific nutrient management (SSNM) in rice. But, this method did not change fertilizer use significantly, and therefore, N losses continue to harm the environment (Ladha et al. 2005). To facilitate site-specific decision making, long-term multiple crop yield map data sets have been transformed into profit maps which contain economic thresholds (Massey et al. 2008). However, environmental thresholds are still lacking, and, as a consequence, farmers are not informed about the risk of N losses. It is expected that "The Nutrient Buffer Power Concept," which has successfully addressed the questions of both enhancing P, K and Zn use efficiency and economizing their application rates in different soils of Africa and Asia, while at the same time keeping to the minimum environmental pollution, could open up a new chapter in rice production technology.

Generally, the emphasis in N management is too strong on tactical finetuning of the N dose and too weak on strategic choices to make the cropping system less leaky. The transformation of flooded to aerobic rice systems (Bouman 2007) in regions with water scarcity is one of the best examples of a strategic approach to achieve a more sustainable use of limited natural resources. Benchmarking of low and high-input cropping systems is needed to make a full assessment of economic benefits and environmental harm; an example for various rice ecosystems is presented in the following table (Table 18.9). Rice yields vary from 2000 kg to 12,000 kg/ha with an associated N input ranging from 50 to 260 kg/ha.

Surprisingly, the variation in parameter values for NUEs does not differ much between low and high input systems. Genetic improvements of rice are most effective in enhancing physiological NUEs (Peng and Bouman 2007), while applying best management practices can raise apparent N recoveries (Campbell et al. 1995).

Table 18.9 Estimated rice yields and N-use parameters for different rice cropping systems

Parameters	Irrigated lowland rice[a]	Irrigated rice-wheat systems[b]	Aerobic rice systems[c]	Rainfall rice systems[d]
Rice yield (kg ha^{-1})	9–12,000	5–9000	4–6000	2–4000
Grain N uptake (kg ha^{-1})	100–150	60–100	50–85	30–60
Apparent N recovery (kg ha^{-1})	0.30–0.40	0.25–0.40	0.30–0.50	0.40–0.60
Physiological NUE (kg kg^{-1})	50–80	50–70	40–60	30–50
Agronomical NUE (kg kg^{-1})	30–60	20–30	20–50	25–30
Crop N demand (kg ha^{-1})	150–200	100–150	80–125	40–80
Recommended fertilizer N supply (kg ha^{-1})	200–260	150–200	120–160	50–80

Sources:
NUE Nitrogen Use Efficiency
[a]Samonte et al. (2006), Belder et al. (2005[a]), Cassman et al. (1997)
[b]Becker et al. (2007), Ladha et al. (2005)
[c]Belder et al. (2005)
[d]Saito et al. (2007)

It is thus concluded that N supply should match N demand in time and space – not only for single crops, but, for a crop rotation as an integrated system – to achieve a higher agronomic NUE. Alleviating factors that limit growth, such as drought, flooding, pests and diseases, will most likely positively affect apparent N recovery.

References

Aggarwal, P. K., Knopff, M. J., Cassman, K. G., & ten Berge, H. F. M. (1997). Simulating genotypic strategies for increasing rice yield potential in irrigated, tropical environments. *Field Crops Research, 51*, 5–17.

Becker, M., Asch, F., Maskey, S. L., Pande, K. R., Shah, S. C., & Shrestha, S. (2007). Effects of transition season management on soil N dynamics and systems N balances in rice-wheat rotations of Nepal. *Field Crops Research, 103*, 98–108.

Belder, P., Bouman, B. A. M., Spiertz, J. H. J., Peng, S., Castaneda, A. R., & Visperas, R. M. (2005). Crop performance, nitrogen and water use in flooded and aerobic rice. *Plant and Soil, 273*, 167–182.

Bennett, J. (2003). Opportunities for increasing water productivity of CGIAR for tolerance of water defects for using crops through plant breeding and molecular techniques. In J. W. Kijne, R. Barker, & D. Molden (Eds.), *Water Productivity in Agriculture: Limits and Opportunities for Improvement* (pp. 103–126). Wallingford: CAB International.

Biemond, H., & Vos, J. (1992). Effects of nitrogen and the development and growth of the potato plant 2. The partitioning of dry matter, nitrogen and nitrate. *Annals of Botany, 70*, 37–45.

Bouman, B. A. M. (2007). A conceptual framework for the improvement of crop water productivity at different spatial scales. *Agricultural Systems, 93*, 43–60.

Bucks, D. A. (1995). Historical developments in micro irrigation. In F. R. Lamm (Ed.), *Micro irrigation for a changing world* (Proc. Fifth International Micro Irrigation Congress, April 2–6, 1996, Orlando, FL) (pp. 1–5). St Joseph: American Society of Agricultural Engineers.

Burt, C. M., Howes, D. J., & Mutziger, A. (2001). Evaporation estimates for irrigated agriculture in California. In *Paper presented at the irrigation association conference*, November 4–6, 2001, San Antonio, TX

Cabrera-Bosquet, L., Molero, G., Bort, J., Nogues, S., & Araus, J. L. (2007). The combined effect of constant water deficit and nitrogen supply on WUE, NUE and Δ^{13} C in durum wheat potted plants. *The Annals of Applied Biology, 151*, 277–289.

Campbell, C. A., Myers, R. J. K., & Curtin, D. (1995). Managing nitrogen for sustainable crop production. *Fertility Research, 42*, 277–296.

Cassman, K. G., Olk, D. C., & Dobermann, A. (1997). Scientific evidence of yield and productivity decline in irrigated rice systems of tropical Asia. *International Rice Commission Newsletter, 46*, 7–16.

CGIAR. (2002). *Challenge program for water and food*. Washington, DC: Consultative Group for International Agricultural Research.

Chapman, A. L., & Muchow, R. C. (1985). Nitrogen accumulated and partitioned at maturity by grain legumes under different water regimes in a semi-arid tropical environment. *Field Crops Research, 11*, 69–79.

De Sena, A.M. (1998). Satellites take away the guesswork out of irrigation. *US Water News* 15(8), August, 1–8.

De Willigen, P., & van Noordwijk, M. (1987). Roots for plant production and nutrient use efficiency. Ph.D thesis Agricultural University, Wageningen, The Netherlands 282 p.

Dobermann, A., & Cassman, K. G. (2002). Plant nutrient management for enhanced productivity in intensive grain production systems of the United States and Asia. *Plant and Soil, 247*, 153–175.

Edmeades, G. O., Cooper, M., Lafitte, R., Zinselmeir, C., Ribaut, J. M., Habben, J. E., Loffler, C., & Banziger, M. (2001). Abiotic stresses and staple crops. In J. Nosberger, H. H. Geiger, & P. C. Struik (Eds.), *Crop science: Progress and prospects* (pp. 137–154). Wallingford: CAB International.

Eickhout, B., Bouwman, A. F., & van Zeijts, H. (2006). The role of nitrogen in world food production and environmental sustainability. *Agriculture, Ecosystems & Environment, 116*, 4–14.

Evans, L. T. (1998). *Feeding the Ten Billion: Plants and population growth*. Cambridge: Cambridge University Press.

FAO. (1993). *Prevention of water pollution by agriculture and related activities*. Rome: Food and Agriculture Organization.

FAO. (1996a). *Control of water pollution from agriculture* (Irrigation and Drainage Paper 55). Rome: Food and Agriculture Organization.

FAO. (1996b). *Energy for sustainable development and food security in Africa*. Rome: Food and Agriculture Organization.

FAO. (1997). *Management of agricultural drainage water quality* (Water Reports 13). Rome: Food and Agriculture Organization.

FAO. (1998). *Crops and drops: Making the best use of water for agriculture*. Rome: Food and Agriculture Organization.

FAO. (2000a). *Agriculture Towards 2015/30 technical interim report, Global Perspectives Unit*. Rome: Food and Agriculture Organization.

FAO. (2000b). *The State of Food and Agriculture: Lessons from the past 50 years*. Rome: Food and Agriculture Organization.

FAOSTAT. (1999). *FAOSTAT Database*. Rome: Food and Agriculture Organization.

Ghassemi, F., Jackerman, A. J., & Nex, H. A. (1995). *Salinization of land and water resources: Human causes, extent, management and case studies*. Sydney: University of New South Wales Press.

Goulding, K., Jarvis, S., & Whitmore, S. (2008). Optimizing nutrient management for farm systems. *Philosophical Transactions of the Royal Society B, 363*, 667–680.
Gupta, R. K., & Abrol, I. P. (2000). Salinity buildup and changes in the rice-wheat systems of the Indo-Gangetic plains. *Experimental Agriculture, 37*, 99–113.
Hardin, G. (1968). Tragedy of the commons. *Science, 162*, 1243–1248.
Harris, H. C. (1991). Implications of climate variability. In H. C. Harris, P. J. M. Cooper, & M. Pela (Eds.), *Soil and crop management for improved water use efficiency in Rainfed areas* (Proceedings of International Workshop held in May 15–19, 1989) (pp. 179–198). Ankara/Turkey/Aleppo/Syria: International Center for Agricultural Research in Dry Areas s (ICARDA).
Herridge, D. F., Peoples, M. B., & Boddey, R. M. (2008). Global inputs of biological nitrogen fixation in agricultural systems. *Plant and Soil, 311*, 1–18.
Hillel, D. (1994). *Rivers of Eden* (pp. 221–225). New York: Oxford University Press.
Hobbs, P. R., Ken, S., & Raj, G. (2008). The role of conservation agriculture in sustainable agriculture. *Philosophical Transactions of the Royal Society B, 363*, 543–555.
IFPRI. (1997a). *Water resources in the twenty-first century: Challenges and implications for action*. Washington, DC: International Food Policy Research Institute.
IFPRI. (1997b). *The world food situation: Recent developments, emerging issues and long term prospects* (Vision 2020 Food Policy Report). Washington, DC: International Food Policy Research Institute.
IPCC. (2001). In R. T. Watson (Ed.), *Climate change 2001: Synthesis report*. Cambridge: Cambridge University Press.
Keating, B. A., & Carberry, P. S. (1993). Resource capture and use in intercropping-solar radiation. *Field Crops Research, 34*, 273–301.
Kichey, T., Hirel, B., Heumez, E., Dubois, F., & Le Gouis, J. (2007). In winter wheat (Triticum aestivum L.) post-anthesis nitrogen uptake and remobilization to the grain correlates with agronomic traits and nitrogen physiological markers. *Field Crops Research, 102*, 22–32.
Kijne, J. W., Tuong, T. R., Bennett, J., Bowman, B., & Oweis, T. (2002). Ensuring food security via improvement in crop water productivity. In *Background paper 1, challenge program for water and food*. Washington, DC: Consultative Group for International Agricultural Research.
Ladha, J. K., Pathak, H., Krupnik, T. J., Six, J., & van Kessel, C. (2005). Efficiency of fertilizer nitrogen in cereal production: Retrospects and prospects. *Advances in Agronomy, 87*, 85–156.
Lemaire, G., Oosterom, E., Sheehy, J., Jeuffroy, M. H., Massignam, A., & Rossato, L. (2007). Is crop N demand more closely related to dry matter accumulation or leaf area expansion during vegetative growth? *Field Crops Research, 100*, 91–106.
Lopez-Bellido, R. J., Castillo, J. E., & Lopez-Bellido, L. (2008). Comparative response of bread and durum wheat cultivars to nitrogen fertilizer in a rainfed Mediterranean environment: Soil nitrate and N uptake and efficiency. *Nutrient Cycling in Agroecosystems, 80*, 121–130.
Martre, P., Semenov, M. A., & Jamieson, P. D. (2007). Simulation analysis of physiological traits to improve yield, nitrogen use efficiency and grain protein concentration in wheat. In J. H. J. Spiertz, P. C. Struik, & H. H. Van Laar (Eds.), *Scale and complexity in plant systems research: Gene-plant-crop relations* (Gene-plant-crop relations, Wageningen UR Frontis Series) (pp. 181–201). Berlin: Springer.
Massey, R. E., Myers, D. B., Kitchen, N. R., & Sudduth, K. A. (2008). Profitability maps as an input for site-specific management decision making. *Agronomy Journal, 100*, 52–59.
McGinn, A. P. (1999). *Safeguarding the health of oceans* (Paper 145). Washington, DC: World Watch Institute.
Molden, D. J. (1997). *Accounting for water use and productivity* (SWIM Paper 1). Colombo: International Irrigation Management Institute.
Myers, N. (1997). *Perverse Subsidies: Their Nature, Scales and Impacts*. Chicago: McArthur Foundation.
Nair, K. P. P. (1996). The buffering power of plant nutrients and effects on availability. *Advances in Agronomy, 57*, 237–287.

Nicholls, R. J., & Leatherman, S. P. (1995). Global sea levels rise. In K. M. Strzepek & J. B. Smith (Eds.), *As climate changes: International impacts and implications* (pp. 92–123). Cambridge: Cambridge University Press.

NRC. (1999). *Our common journey: A transition towards sustainability*. Washington, DC: National Academy Press.

Oldeman, L. R. (1994). The global extent of soil degradation. In D. J. Greenland & I. Szabolcs (Eds.), *Soil resilience and sustainable land use* (pp. 99–118). Wallingford: CAB International.

Orr, A., Islam, A. S. M. N., & Barnes, G. (1991). *Treadle Pump: Manual irrigation for small farmers in Bangladesh*. Dhaka: Dinajpur Rural Services.

Peng, S., & Bouman, B. A. M. (2007). Prospects for genetic improvement to increase lowland rice yields with less water and nitrogen. In J. H. J. Spiertz, P. C. Struik, & H. H. Van Laar (Eds.), *Scale and complexity in plant systems research: Gene-Plant-Crop relations* (pp. 251–266). Berlin: Springer.

Perry, C. J., Rock, M., & Seckler, D. (1997). Water as economic good: A solution or a problem. In *Research report 14*. Colombo: International Irrigation Management Institute.

Pierce, F. P., & Nowak, P. (1999). Aspects of precision farming. *Advances in Agronomy, 67*, 1–85.

Postel, S. (1992). *The last oasis: Facing water scarcity*. London: Earthscan.

Postel, S. (1996). *Dividing the waters: Food security, ecosystem health and the new politics of scarcity*. World watch paper 132. Washington, DC: World watch Institute.

Postel, S. (1998). Water for food production: Will there be enough in 2025? *Bio Science, 48*, 629–637.

Postel, S. (1999). *Pillars of sand: Can the irrigation miracle last?* New York: W.W.Norton.

Postel, S. (2001). Drip irrigation for small farms: A new initiative to alleviate hunger and poverty. *Water International, 26*(1), 3–13.

Postel, S., & Carpenter, S. (1997). Freshwater ecosystem services. In G. C. Daily (Ed.), *Nature's services: Societal dependence on natural ecosystems* (pp. 195–214). Washington, DC: Island Press.

Raman, S. (1989). Adsorption of metoxuron and tebuthiuron on model clay-organo complexes. *Toxicological and Environmental Chemistry, 24*, 207–213.

Richards, R. A., Lopez-Castaneda, C., Gomez-Macpherson, H., & Condon, A. G. (1993). Improving the efficiency of water use by plant breeding and molecular biology Irrigation. *Science, 14*, 93–104.

Rockstrom, J., Barron, J., & Fox, P. (2002). Water productivity in rainfed agriculture: Challenges and opportunities for smallholder farmers in drought-prone tropical agroecosystems. In J. K. Kijne, R. B. Barker, & D. Molden (Eds.), *Water productivity in agriculture: Limits and opportunities for improvement* (pp. 145–162). Wallingford: CAB International.

Rosengrant, M. W., & Ringler, C. (1999). *Impact on food security and rural development of reallocating water from agriculture*. Washington, DC: International Food Policy Research Institute.

Russell, G., Jarvis, P. G., & Monteith, J. L. (1989). Absorption of radiation by canopies and stand growth. In G. Russell (Ed.), *Plant Canopy: Their growth, form and function* (pp. 21–39). Cambridge: Cambridge University Press.

Saito, K., Atlin, G. N., Linquist, B., Phanthaboon, K., Shiraiwa, T., & Horie, T. (2007). Performance of traditional and improved upland rice cultivars under nonfertilized and fertilized conditions in northern Laos. *Crop Science, 47*, 2473–2481.

Samonte, S. O. P. B., Wilson, L. T., Medley, J. C., Pinson, S. R. M., Clung, A. M., & Lales, J. S. (2006). Nitrogen utilization efficiency: Relationships with grain yield, grain protein and yield-related traits in rice. *Agronomy Journal, 98*, 168–176.

Seckler, D., Molden, D., & Barker, R. (1998). *Water scarcity in the twenty-first century* (Water Brief 1). Colombo: International Water Management Institute.

Shah, T. (1993). *Ground water markets and irrigation development*. Bombay: Oxford University Press.

Sinclair, T. R., Tanner, C. B., & Bennett, J. (1984). Water-use efficiency in crop production. *BioScience, 34*, 36–40.

Sivanappan, R. K. (1994). Prospects of micro irrigation in India. *Irrigation and Drainage Systems, 8*, 49–58.
Tuong, T. P. (1999). Productive water use in rice production: Opportunities and limitations. *Journal of Crop Production, 2*(2), 241–264.
Tyagi, N. K. (2002). Managing saline and alkaline water for higher productivity. In J. W. Kijne, R. Barker, & D. Molden (Eds.), *Water productivity in agriculture: Limits and opportunities for improvement* (pp. 69–87). Wallingfrord: CAB International.
Umali, D. L. (1993). *Irrigation induced salinity*. Washington, DC: World Bank.
UNDP. (1995). *The Aral in Crisis: United Nations development project, Tashkent, Russia.* New York: United Nations Publications Division.
UNEP. (2007). Reactive nitrogen in the environment. In *Too much or too little of a good thing* (Vol. 51). Paris: The Woods Hole Research Center, USA/UNEP DTIE Sustainable Consumption and Production Branch. ISBN 978 92 807 2783 8.
Van Delden, A. (2001). Yield and growth components of potato and wheat under organic nitrogen management. *Agronomy Journal, 93*, 1370–1385.
Van Ginkel, M., Ortiz-Monasterio, I., Trethowan, R., & Hernandez, E. (2001). Methodology for selecting segregating populations for improved N-use efficiency in bread wheat. *Euphytica, 119*, 223–230.
WRI. (2000). *World resources 2000–2001. People and ecosystems, the Fraying web of life.* Washington, DC: World Resources Institute.
WSSD. (2002). *World summit for sustainable development. United Nations conference on environment and development*, August 26 – September 4, Johannesburg, South Africa
Yin, X., & Van Laar, H. H. (2005). *Crop systems dynamics: An ecophysiological simulation model for genotype-by-environment interactions* (p. 153). Wageningen: Wageningen Academic Publishers. ISBN 907 69 985 82.

Chapter 19
Primary Productivity and Biodiversity

Abstract Net Primary Productivity (NPP) in terrestrial temperate ecosystems is generally limited to N availability. However, excessive levels of reactive N in the soil, water and atmosphere constitute a major threat to biodiversity in natural ecosystems. The challenge for the future is how can we combine a high level of productivity – the capacity of agricultural land to produce biomass on a sustainable long-term basis with the provision of ecosystem services and soil biodiversity. The chapter discusses the various aspects connected to this crucial need.

Keywords Net primary productivity · Nitrogen use · Arable cropping system · Mixed farming system

Net primary productivity (NPP) in terrestrial temperate ecosystems is generally limited by N availability. However, excessive levels of reactive N in the soil, water and atmosphere constitute a major threat to bio diversity in natural ecosystems (Sudding et al. 2005). There is increasing evidence that diversity of soil biota as well as plants contributes to ecosystem functioning and improved nutrient use efficiency (Brussard et al. 2007). Diversity and functional complementarity leads to greater soil C and N accumulation on agricultural degraded soils (Fornnara and Tilman 2008). Most of the research on improved ecosystem functioning by an increased diversity were carried out in vegetations with relatively low external inputs and biomass yield. However, Oeremans et al. 2007 studied the impact of long term nutrient supply on plant diversity in grassland, where they found that increased N fertilization reduced the number of species, site-independently. A unimodal relationship was observed between productivity and species number. The highest number of species was found when N and K were co-limited.

The challenge for the future is how can we combine a high land productivity – the capacity of agricultural land to produce biomass on a sustainable long-term basis – with the provision of ecosystem services and soil biodiversity (Barrios 2007). The use of crop genetic diversity in maintaining ecosystem services and soil biodiversity in maintaining ecosystem services was reviewed by Hajjar et al. (2008). They concluded that crop genetic diversity can enhance agroecosystem functioning and provide ecosystem services, especially by contributing to stability in crop and

K. P. Nair, *Intelligent Soil Management for Sustainable Agriculture*,
https://doi.org/10.1007/978-3-030-15530-8_19

soil health. As a consequence, crop productivity and resource use efficiencies will also become more stable. There are still many gaps related to methodology, experimental and conceptual approaches, which prevent quick progress in the guidance for policy – and decision – making on changes needed in developing highly productive, sustainable agricultural systems. Swift et al. (2004) concluded that maintaining ecosystem services and biodiversity outside conservation areas lies in promoting diversity of land use on the landscape and farm rather than field scale. Good examples are arable and grassland field margins (Sheridan et al. 2008). These field margins can provide multiple ecological services, such as biodiversity and pest control (Olson and Wackers 2007). Biodiversity effects can be managed (Storkey and Westbury 2007). It was reported that these effects increase linearly with biotope space (Dimitrakopoulos and Schmid 2004). By specially allocating land (2–5% of the total area cultivated) for ecosystem services, complementary to land used for crop production, biodiversity can be enhanced within intensive cropping systems without a severe loss of production potential. In areas with hilly or rolling land, strips or banks dedicated to developing biodiversity richness may also reduce run off of nutrients and soil erosion.

19.1 Nitrogen Use at the Farm and Global Levels

A high nitrogen-use efficiency (NUE) is no guarantee that N losses do not exceed critical environmental thresholds. The most important factor to determine the risk of potential N losses is the total amount of mineral N left over after harvest in crop residues and in the soil. To assess the environmental impacts, the N dynamics should not be studied merely in one crop, but, in a diversity of cropping systems and in mixed plant-animal copping systems as shown in the following table (Table 19.1).

19.1.1 Arable Cropping Systems

The principal objectives of ecologically and economically sustainable agriculture seek to maintain soil fertility and improve crop productivity and stability. Management options are: site and time specific nutrient and water management, crop production measures and the choice of adapted, high yielding cultivars. The effects of the various measures which are of importance for the maintenance and use of resources base cannot be easily assessed within one crop growing cycle (normally one crop season), but, should be evaluated over a sequence of crops seasons. Crop rotation is an important component of an integrated approach of sustainable agriculture and resource conservation. Short and long-term effects of a cropping sequence and related management practices can be expressed in terms of physical properties such as water holding capacity and bulk density, chemical properties such as pH, soil organic carbon and nutrient contents, and biological soil properties

Table 19.1 Sustainability parameters to benchmark contrasting food production systems in temperate, non- water-limited regions

Parameters	Technological high-input systems	Ecological low-input systems	Mixed animal-plant systems	Dairy-Grazing systems
A. Quantitative Parameters[a]				
1. Biomass – NPP ($t\ ha^{-1}$)[b]	12–20	8–12	10–18	12–20
2. Total N supply ($kg\ ha^{-1}$)[c]	150–300	100–150	200–350	200–400
3. N use (NUE) $(N^{output})/\ N(^{input})$[d]	0.30–0.60	0.40–0.70	0.25–0.50	0.15–0.35
B. Qualitative Parameters[e]				
1. Marketable yield ($kg\ ha^{-1}$)	High	Low	Moderate	High
2. Ecological stability/diversity	Low	Moderate	High	Moderate
3. Nutrient recycling	Low	Moderate	Moderate	High
4. Environmental N load	Moderate	Low	Moderate	High
5. Profitability (net returns)	High	Moderate	High	Moderate
6. Sustainability (planet issues)	Low	High	Moderate	Moderate
7. Ethical acceptance	Moderate	High	Moderate	High

[a]Best guesses of the author
[b]Above ground biomass or net primary productivity (NPP), expressed as ton dry matter per hectare
[c]N supply by manure and fertilizer use
[d]N use defined as overall system recovery, expressed as the ratio between output (N in harvestable crop parts, meat or milk) and input (N in manure and fertilizers)
[e]After Pretty (2008): Principles of Agricultural Sustainability

such as microbial activity. Growing special crops in a rotation can improve the sustainability of the cropping system (Struik and Bonciarelli 1997), examples of which are the following:

1. Legumes to improve N availability
2. Green manure crops to improve the physical and biological soil fertility
3. Cover crops to prevent soil erosion and to store nutrients prone to leaching or run-off

The potential legumes can be established by comparing yields and N uptake under the same agro ecological conditions. Sibma and Spiertz (1986) carried out field experiments with three forage crops, namely, grass (*Lolium perenne*), lucerne (*Medicago sativa*) and maize (*Zea mays*), over a 3 year period on a fertile clay soil, under temperate climatic conditions. It was found that the above ground dry matter (DM) yields varied from 13.4 t/ha to 19.8 t/ ha, from 13.4 t/ha to 18.1 t/ha and from

13.7 t/ha to 17.1 t/ha for grass, lucerne and maize, respectively. The associated N yields ranged from 413 to 452, from 392 to 577 and from 188 to 220 kg/ha respectively in the three crops mentioned above (Spiertz and Sibma 1986). Grass showed the highest productivity in the first year and Lucerne in the last year. N fixation by lucerne varied between 462 kg/ha and 507 kg/ha without N fertilizer application and 107 kg/ha–195 kg/ ha with N fertilizer application. The after effects from soil N reserves of 1, 2 and 3 years cropping sequence of Lucerne (no N fertilizer applied) and grass (300 kgN/ha as fertilizer), measured as N uptake by an unfertilized maize crop, amounted to 140–175 kg/ha and 110–140 kg/ha, respectively. For comparison, the N after effect of a preceding maize crop centered around 110 kg N/ha. Even higher yields of DM (Dry Matter) were reported by Lloveras et al. (2008) for irrigated lucerne under Mediterranean conditions. The authors found a range from 16 t/ ha to 21 t/ha averaged for 3 years. These data show the high fertility enriching potential of a legume crop in fixing atmospheric N and enhancing DM production under favorable crop growing conditions.

19.1.2 Mixed Farming Systems

In addition to crop rotation, integration of crop and animal production on the farm and regional scales may be an opportunity to increase "eco efficiency" (Wilkins 2008). Nitrogen is very mobile in soil –plant-animal system and with required N inputs for high crop yields, and intensive livestock production, the risk of N losses increases (Van Keulen et al. 2000). Traditionally, nutrient management has been concerned with optimizing the economic returns from nutrient input for crop production. The main emphasis was on the expected crop response from applying nutrients to the soil. In practice, however, nutrients, in particular, manure (organic) are not always applied to optimize plant nutrient use. Such practice or the improper or untimely application of organic manure and chemical fertilizers may release plant nutrients into the air (volatilization as in the case of N from urea application) and water (nitrate loading of ground water due to excessive urea application, as has happened in the so-called green revolution belt – Punjab State, in India – rendering ground water non potable). The problem is most acute in regions where there is an intensive animal production (Aarts et al. 1992). The excess N compounds in manure has become an issue of major concern in many European countries and will also assume as an increasing problematic issue in countries, such as, India and China, with high animal stocking rates. The problem of nutrient pollution is not one exclusively created by farmers, as such, but through wrong recommendatory advice from experts and farm extension specialists. The experience of this author is also the same in many developing countries, in particular, India. This arises from the fact that advisory practice is still rooted in "text book knowledge." The most classic examples of this is with reference to Zn nutrition of Black pepper (*Piper nigrum*)*with* regard to Zn fertilizer recommendations based on DTPA soil test (Nair 2004) and in Cardamom (*Elettaria cardamomum*) nutrition with regard to K fertilizer recommendation based on the routine NH_4OAc test Nair et al. (1997), both in India. In both cases, the "Buffer Power Concept" has shown the most precise and dependable way. Beegle et al. concluded that nutrient management strategies will not be the same for all farms. They classified farms on a nutrient balance basis into the following three categories:

(a) Nutrient- deficient farms where nutrient imports are less than nutrient exports. Thus, additional nutrients in the form of purchased fertilizer or other sources are required to achieve optimum crop yields. A well planned nutrient management program emphasizing economic and agronomic efficiency should reduce the need for purchased inputs and thus should improve farm productivity and profitability
(b) Nutrient-balanced farms where nutrient imports are approximately equal to exports. Because these farms are often the upper limit of being able to safely handle all the nutrients in the production system, nutrient management planning may offer potential environmental benefit.
(c) Nutrient-surplus farms where nutrient imports significantly exceed exports. The nutrients in the manure generally exceed those required for crop production on the farm. A significant component of a nutrient management program involves acceptable off-farm uses for the excess manure.

A few countries in the world, such as Denmark, has given priority to agro-environmental schemes, resulting in the more efficient use of fertilizers and manures. These countries, such as Denmark, have given priority to agro-environmental schemes, restricting the use of fertilizers and manures. The impact of these measures was investigated by evaluation of farm gate nutrient balances (Kyllingscbaek and Hansen 2007). It was shown that nutrient surpluses at the farm gate were reduced. However, the effects on water quality were minimal. It is still unclear if there is no direct relationship or that the lag time of reducing the N load is longer than considered in the study.

References

Aarts, H. F. M., Biewinga, E. E., & van Keulen, H. (1992). Dairy farming systems based on efficient nutrient management. *Netherlands Journal of Agricultural Science, 40*, 285–299.

Barrios, E. (2007). Soil biota, ecosystem services and land productivity. *Ecological Economics, 64*, 269–285.

Brussard, L., de Ruiter, P. C., & Brown, G. G. (2007). Soil biodiversity for agricultural sustainability. *Agriculture, Ecosystems & Environment, 121*, 233–244.

Dimitrakopoulos, P. G., & Schmid, B. (2004). Biodiversity effects increase linearly with biotope space. *Ecology Letters, 7*, 574–583.

Fornara, D. A., & Tilman, D. (2008). Plant functional composition influences rates of soil carbon and nitrogen accumulation. *Journal of Ecology, 96*, 314–322.

Hajjar, R., Jarvis, D. I., & Gemmill-Herren, B. (2008). The utility of crop genetic diversity in maintaining ecosystem services. *Agriculture, Ecosystems & Environment, 123*, 261–270.

Kyllingsbaek, A., & Hansen, J. F. (2007). Development in nutrient balances in Danish agriculture 1980–2004. *Nutrient Cycling in Agroecosystems, 79*, 267–280.

Lloveras, J., Chocarro, C., Freixes, O., Arque, E., Moreno, A., & Santiver, F. (2008). Yield, yield components, and forage nutritive value of alfalfa as affected by seeding rate under irrigated conditions. *Agronomy Journal, 100*, 191–197.

Nair, K. P. P. (2004). Sustaining crop production in the developing world through "the nutrient buffer power concept" – A case study with Black pepper (Piper nigrum) growing at low pH in the laterite soils of India. In *Proceeding of 16th international symposium on plant-soil interactions at low pH* (p. 392), Sendai, Japan, August 1–5.

Nair, K. P. P., Sadanandan, A. K., Hamza, S., & Abraham, J. (1997). The importance of Potassium buffer power in the growth and yield of Cardamom. *Journal of Plant Nutrition, 20*, 987–997.

Oerlemans, J., Von Oberfeld, W. O., & Wolf, D. (2007). Impact of long-term nutrient supply on plant species diversity in grassland: An experimental approach on conventionally used pastures. *Journal of Applied Botany And Food Quality, 81*, 151–157.

Olson, D. M., & Wackers, F. G. (2007). Management of field margins to maximize multiple ecological services. *Journal of Applied Ecology, 44*, 13–21.

Pretty, J. (2008). Agricultural sustainability: Concepts, principles and evidence. *Philosophical Transactions of the Royal Society B, 363*, 447–465.

Sheridan, H., Finn, J. A., Culleton, N., & O'Donovan, G. (2008). Plant and invertebrate diversity in grassland field margins. *Agriculture, Ecosystems & Environment, 123*, 225–232.

Sibma, L., & Spiertz, J. H. J. (1986). Dry matter production and nitrogen utilization in cropping systems with grass, lucerne and maize I: Comparison of crop characteristics, growth and production. *Netherlands Journal of Agricultural Science, 34*, 25–35.

Spiertz, J. H. J., & Sibma, L. (1986). Dry matter production and nitrogen utilization in cropping systems with grass, lucerne and maize 2. Nitrogen yield and utilization with various cropping systems and their after-effects. *Netherlands Journal of Agricultural Science, 34*, 37–47.

Storkey, J., & Westbury, D. B. (2007). Managing arable weeds for biodiversity. *Pest Management Science, 63*, 517–523.

Struik, P. C., & Bonciarelli, F. (1997). Resource use at the cropping system level. *European Journal of Agronomy, 7*, 133–143.

Sudding, K. N., Collins, S. L., Gough, L., Clark, C., Cleland, E. E., Gross, K. L., Milchunas, D. G., & Pennings, S. (2005). Functional-and abundance-based mechanisms explain diversity loss due to N fertilization. *PNAS, 102*, 4387–4392.

Swift, M. J., Izac, A., & van Noordwijk, M. (2004). Biodiversity and ecosystem services in agricultural landscapes. *Agriculture, Ecosystems & Environment, 104*, 113–134.

Van Keulen, H., Aarts, H. F. M., Habekotte, B., van der Meer, H. G., & Spiertz, J. H. J. (2000). Soil-plant-animal relations in nutrient cycling: The case of dairy farming system "De Marke". *European Journal of Agronomy, 13*, 245–261.

Wilkins, R. J. (2008). Eco-efficient approaches to land management: A case for increased integration of crop and animal production systems. *Philosophical Transactions of the Royal Society B, 363*, 517–525.

Chapter 20
Environment and Management

Abstract Concerns about the environmental impact of intensive agricultural systems require improvement in production technologies to maximize resource use efficiencies, and to minimize the adverse environmental impact. These aspects are discussed against the background of the "Plant-Soil-Atmosphere" connect.

Keywords Intensive agricultural system · Green House Gases · Nitrous oxide emission · Nitrogen cycling · Irrigated rice systems

Concerns about the environmental impact of intensive agricultural systems require an improvement in production technologies to maximize resource- use efficiencies, and to minimize the environmental impact. N fertilizers constitute almost 60% of the global reactive N load attributable to human activities, especially, in China (UNEP 2007). N use has major impact on ecosystem functioning and the well being of human beings. In Europe, agriculture is responsible for 40–80% of total N loading to surface water. N losses associated with the application of N fertilizers can result in nitrate contamination of water resources and increased emission of ammonia and nitrous oxide N_2O, the latter being an important "long lived" green house gas (GHG) leading to global warming. The impact of 1 km N_2O on global warming is 300 times more than that of the same amount of carbon dioxide, another important "long lived" GHG. N_2O molecules stay in the atmosphere for an average of 120 years before being removed or destroyed by chemical reactions. In 2011, N_2O emissions into the atmosphere accounted for about 5% of all GHGs in the US, originating from human activity, primarily agriculture, through improper soil management. Unfortunately, the environmental havoc created by N_2O has been underplayed or totally neglected, inasmuch as global warming is concerned, because of the political pressure of the fertilizer lobby. This has been the bane in the State of Punjab in India, where unbridled use of nitrogenous fertilizers, especially urea pumped up the so-called green revolution. The disastrous environmental effect is there for all to see now after more than three decades of this high input agriculture. All of these have had a vast negative effect on human health (Wilkinson et al. 2007). Today, the agronomic and economic requirements of nutrient management remain central, but, in addition we must consider the potential impact of these nutrients on environmental

K. P. Nair, *Intelligent Soil Management for Sustainable Agriculture*,
https://doi.org/10.1007/978-3-030-15530-8_20

quality. Total reactive loss will grow in the period explored until 2030, because of an increase in fertilizer consumption in developing countries to meet the food requirements of an exponentially burgeoning human population, and concurrently, a steep rise in dairy and meat products consumption in emerging industrial countries like China and India, despite improvements in overall system of N recovery in developed countries (Eickhout et al. 2006).

Since land is a finite and fragile resource, its sustainable management depends on the husbandry of its different components, of which soil fertility and water availability are key factors for agricultural production. At the moment, agronomic N-use efficiency is often very low (Spiertz and Oenema 2005). Agronomically, farmers should aim at the minimum input of each production resource required to allow maximum utilization of all other resources (de Wit 1992). Consequently, above a certain minimum, higher input of yield increasing factors such as water and nutrients result in higher yields per unit area and are associated with higher efficiencies of other resources. Higher efficiencies, expressed as output per unit of input, might coincide with larger emissions per unit area, leading to environmental pollution. This, indeed, is the reason why there is a justification to follow the "Buffer PowerConcept" which while maximizing nutrient use efficiency keeps the environmental pollution to the minimum.

20.1 Plant-Soil-Atmosphere

A more precise understanding of the soil-plant N cycle and the soil, plant and environmental factors which govern it, has been facilitated by mathematical models. Environmental concerns are focused on nitrogen losses from soil, which will pollute the environment as depicted in the following figure (Fig. 20.1).

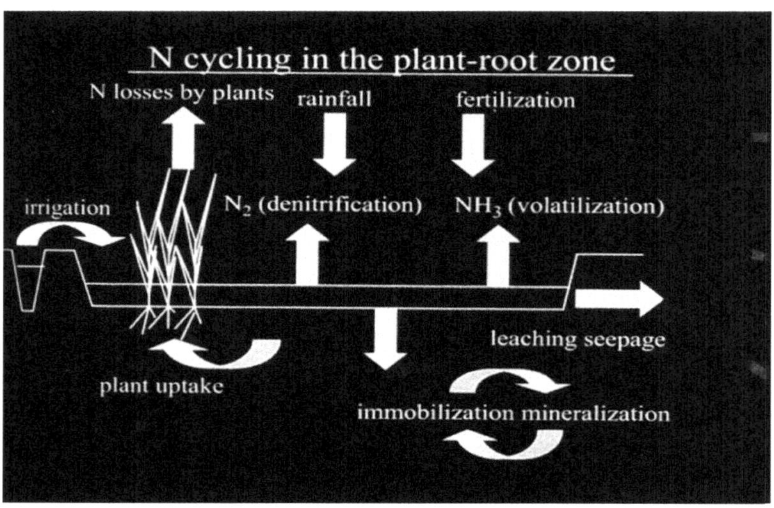

Fig. 20.1 N cycling schema for irrigated rice systems. (Modified after Bouman (2007))

Leaching is the major route by which nitrate enters the ground and surface water, while denitrification and nitrification are significant sources of N_2O a "long-lived," but, very toxic GHG, which stays up in the atmosphere for close to 120 years, before being detoxified or destroyed by other chemical reactions. Improved efficiency of N use on a field and farm scale, both increasing crop yield and quality and reducing losses, is dependent upon dynamic optimization to match supply of N to the crop and its needs on a field or farm scale. The optimization requires measurement and prediction of soil N supply, crop uptake and their variability (Stockdale et al. 1997). Models of crop growth, the soil-N cycle and plant-soil models have been developed (Bouman et al. 1996). However, these are little used in the current fertilizer and farm management recommendations. Farmers cannot wait for our understanding of soil-plant dynamics to be perfect, but, need researchers to put their current knowledge to maximum use. This great vacuum has been successfully filled by the "Buffer Power Concept."

20.1.1 Scale and Systems

A system approach can be used as a research tool, but also, as an instrument to train students and scientists. Research and training networks enable the necessary development of a common language of concepts, models and databases, and allow frequent interaction among actors (Ten Berge and Knoff 1995). System approaches were introduced in the 1970s and are used increasingly in research on food production studies, natural resource management, land use options and rural development. The system approach can be described as the systematic and quantitative analysis of agricultural systems. Agricultural systems are defined as well-delineated parts of the real world, consisting of many interacting elements (Neeteson et al. 2002). The system approach uses many specific techniques, such as simulation modeling, expert systems with databases, linear programming and geographic information systems. A system approach has been applied for studies of different aggregation levels, such as:

1. The plant level: Analyses and evaluation of genotype x environmental interactions in breeding programs
2. The crop level: Optimization of nutrient and water management
3. The farm level: Prototyping of integrated farming systems and analyzing the flows of nutrients on farms
4. The watershed and landscape level: Optimization of water use and saving water
5. The eco-regional level: Ex-ante assessment of the possible impacts of changes in technology or in the socio economic environment on agricultural development

Actually, in the last few decades, we have witnessed the integration of process-specific knowledge into very precise, widely accepted relationships between processes in the system and driving factors from outside the system (Van Keulen et al. 2000). This applies mainly for water management at the field and catchment levels (Bouman 2007). Mathematical quantification of N flows in space and time is more complex, because of the dynamic nature of N in the plant, soil, water and atmosphere (Erisman et al. 2007). The growing complexity of managing N in sustainable agricultural systems calls for problem-oriented, interdisciplinary research. Key disciplines are: agronomy, crop science, soil science (principally including physical chemistry -based understanding of processes within soil and plant cell with regard to ion uptake, and consequently plant nutrition), conservation biology, environmental sciences and systems modeling.

The current assessment of the impact of climate change on agriculture and the options for adaptation relies heavily upon both crop and climate modeling. The output of the models is greatly limited by the extent of our understanding of short- and long-term crop adaptation to changing environmental conditions, especially in soil traits and weather patterns. Howden et al. (2007) stated that complex problems require multi-disciplinary solutions, with a focus on integrated rather than disciplinary science.

References

Bouman, B. A. M. (2007). A conceptual framework for the improvement of crop water productivity at different spatial scales. *Agricultural Systems, 93*, 43–60.

Bouman, B. A. M., Van Keulen, H., Van Laar, H. H., & Rabbinge, R. (1996). The "School De Wit" crop growth simulation models: A pedigree and historical overview. *Agricultural Systems, 52*, 171–198.

DeWit, C. T. (1992). Resource use efficiency in agriculture. *Agricultural Systems, 40*, 125–151.

Eickhout, B., Bouwman, A. F., & van Zeijts, H. (2006). The role of nitrogen in world food production and environmental sustainability. *Agriculture, Ecosystems & Environment, 116*, 4–14.

Erisman, J. W., Bleeker, A., Galloway, J., & Sutton, M. S. (2007). Reduced nitrogen in ecology and the environment. *Environmental Pollution, 150*, 140–149.

Howden, S. M., Sousana, J. F., Tubiello, F. N., Chhetri, N., Dunlop, M., & Meinke, H. (2007). Adapting agriculture to climate change. *PNAS, 104*, 19691–19696.

Neeteson, J. J., Schroder, J. J., & Ten Berge, H. F. M. (2002). A multiscale system approach to nutrient management research in the Netherlands, NJAS Wageningen. *Journal of Life Sciences, 50*, 141–151.

Spiertz, J. H. J., & Oenema, O. (2005). Resource use efficiency and management of nutrients in agricultural systems. In *ERSEC Ecological Book Series – 1. On sustainable agricultural systems* (pp. 171–183). Berlin: Tsinghua University Press/Springer.

Stockdale, E. A., Gaunt, J. L., & Vos, J. (1997). Soil-plant nitrogen dynamics: What concepts are required? *European Journal of Agronomy, 7*, 145–159.

Ten Berge, H. F. M., & Kropff, M. J. (1995). Founding a systems research network for rice. In Bouma (Ed.), *Eco-regional approaches for sustainable land use and food production* (pp. 263–282). Dordrecht: Kluwer.

UNEP. (2007). Reactive nitrogen in the environment. In *Too much or too little of a good thing* (Vol. 51). Paris: The Woods Hole Research Center, USA/UNEP DTIE Sustainable Consumption and Production Branch. ISBN 978 92 807 2783 8.

Van Keulen, H., Aarts, H. F. M., Habekotte, B., van der Meer, H. G., & Spiertz, J. H. J. (2000). Soil-plant-animal relations in nutrient cycling: The case of dairy farming system "De Marke". *European Journal of Agronomy, 13*, 245–261.

Wilkinson, P., Smith, K. R., Joffe, M., & Haines, A. (2007). Energy and Health 1: A global perspective on energy: Health effects and injustices. *The Lancet, 370*, 965–978.

Chapter 21
Policy Making and Regulations

Abstract A more environmentally sensitive nutrient management on the field and farm levels can reduce N losses to a level that meets the standard. Decision making related to nutrient management occurs at the strategic, tactical and operational levels. The Chapter touches on these aspects.

Keywords Environmentally sensitive nutrient management · Agro environmental indicators

A more environmentally sensitive nutrient management on the field and farm levels can reduce nitrogen losses to a level that meets the standards (Goulding et al. 2008). Decision making related to nutrient management occurs at the strategic, tactical and operational levels. In Europe, society demands more and more accountability from farmers; therefore, more legislation has been implemented by the European Union (EU) and national governments. Legislation is inevitable whenever "profit" and "planet" goals are in conflict with each other. In Europe, especially in The Netherlands, there exists a set of regulations set up by the government to protect the environmental components – soil, water, and air – against nutrient losses from agricultural production. Legislation can only be successful when clear motivations and regulatory tools are provided to farmers (Schroeder et al. 2003). There are two approaches to nutrient management planning in a regulatory situation. One approach is to specify what should be done on all farms as a recipe for nutrient management. Another approach is to establish performance criteria on goals for farmers to meet as part of their nutrient management plan.

Langeveld et al. (2007) concluded that agroenvironmental indicators can be used for design, implementation and testing of farming systems, but, it should be kept in mind that indicators are not perfect, because of the complexity and highly variable processes involved in N cycling. Transitions occur at some cost; for instance, large savings on one limited resource, such as irrigation water, may have a trade off on yield. A methodology for an integrated analysis of trade-offs between economic and environmental indicators is available. Integrative modeling approaches to evaluate the impact of multifunctional agriculture have been developed. New approaches are needed that will integrate biophysical and ecological processes at the crop, farm and

K. P. Nair, *Intelligent Soil Management for Sustainable Agriculture*,
https://doi.org/10.1007/978-3-030-15530-8_21

landscape levels (Pretty 2008). A policy that will lead to N applications of manure and fertilizer balanced with crop N demand is urgently needed, not only in developed, but also, in developing countries. An intensive communication between all stake holders (environmental agencies, policy-makers, researchers and farmers) and a controlled implementation of indicators and guidelines may contribute to a balanced application of nutrients (Delgado et al. 2008). Such an approach will be needed to meet the "planet" and "profit" objectives and to prevent and stem the unbridled nitrate loading of ground water to conform to the established standards for safe human consumption, which is a "people" objective. This is especially relevant in the Indian context, in the green revolution belt, which was a short-term gain in terms of enhanced grain production, especially of rice and wheat, but, a "long-term environmental disaster."

References

Delgado, J. A., Shaffer, M., Hu, C., Lavado, R., Cueto-Wong, J., Joosse, P., Sotomayer, D., Colon, W., Follett, R., DelGrosso, S., Li, X., & Rimski-Korsakov, H. (2008). An index approach to assess nitrogen losses to the environment. *Ecological Engineering, 32*, 108–120.

Goulding, K., Jarvis, S., & Whitmore, S. (2008). Optimizing nutrient management for farm systems. *Philosophical Transactions of the Royal Society B, 363*, 667–680.

Langeveld, J. W. A., Verhagen, A., Neeteson, J. J., van Keulen, H., Conijin, J. G., Schils, R. L. M., & Oenema, J. (2007). Evaluating farm performance using agri-environmental indicators: Recent experiences for nitrogen management in The Netherlands. *Journal of Environmental Management, 82*, 363–376.

Pretty, J. (2008). Agricultural sustainability: Concepts, principles and evidence. *Philosophical Transactions of the Royal Society B, 363*, 447–465.

Schroder, J. J., Aarts, H. F. M., ten Berge, H. F. M., van Keulen, H., & Neeteson, J. J. (2003). An evaluation of whole-farm nitrogen balances and related indices for efficient nitrogen use. *European Journal of Agronomy, 20*, 33–44.

Chapter 22
Phosphate Solubilizing Microorganisms and Their Role in Sustainable Agriculture

Abstract The chapter discusses, at length, the role of Phosphate Solubilizing Microorganisms (Vesicular Arbuscular Mycorrhizae, VAM) and their benefits, especially with regard to economizing Phosphate fertilizer application to soil, in sustainable agriculture.

Keywords Phosphorus solubilization · Phosphorus solubilizing microorganisms · Phosphate fertilizers · Phosphorus solubilizing microorganisms · Inoculants · Mycorrhiza · Arbuscular Mycorrhizal fungi · Crop response

To meet the growing challenge of properly feeding a burgeoning global population, a great deal of effort needs to be placed on soil biological system and the agro-ecosystem, as a whole, to understand better the complex processes and interactions governing the stability of agricultural land. Chemical fertilizers, like, factory manufactured water soluble phosphatic fertilizers, have played a very significant role in the so-called green revolution, which are commonly recommended to correct soil phosphorus deficiencies. Most developing countries, however, import these fertilizers, which are often in limited supply and represent a major outlay for resource poor farmers. In addition, intensification of agricultural production in these countries necessitates the addition of phosphates to soil not only to increase crop production, but also, to improve phosphorus status of soils to prevent further soil degradation. Moreover, the use of chemical fertilizers is reaching the theoretical maximum use beyond which there will be no further increase in crop yields (Ahmed 1995). However, it is becoming increasingly evident that conventional agricultural practices can no more sustain crop production at the current level indefinitely.

Deficiency of phosphorus limits cop yields severely in tropical soils. Tropical and sub tropical soils are predominantly acidic, and often extremely phosphorus deficient (Gaume 2000), with very high phosphorus fixing capacity. On average, most mineral nutrients in soil solution are present in quantities approximating to millimolar, but, phosphorus is present in only quantities approximating to micromolar or even less than that (Ozanne 1980). The low levels of phosphorus are due to high reactivity of soluble phosphates with other elements, such as, iron, aluminum and/or manganese, mostly in acidic soils. In calcareous soils the reaction is with

K. P. Nair, *Intelligent Soil Management for Sustainable Agriculture*,
https://doi.org/10.1007/978-3-030-15530-8_22

calcium as precipitation as calcium phosphate. Organic phosphate also make up a large fraction of soluble phosphate, as much as 50% in soils with high organic matter (Barber 1984). Since the indiscriminate and excessive application of chemical fertilizers has led to health and environmental hazards, agronomists and soil scientists are desperate to find out alternate strategies which can ensure competitive yields, while preserving soil health. It is in this context that microbial inoculants, bio fertilizers, including phosphorus solubilizing microorganisms (PSM) in agriculture represents an environment friendly alternative to further application of high doses of chemical fertilizers. A continued exploration of the natural biodiversity of soil microorganisms and the optimization/manipulation of microbial interactions in the rhizosphere represents a pre-requisite step to develop more efficient microbial inoculants with phosphorus solubilizing ability.

22.1 Imminent Need for Phosphate Solubilizing microorganisms in Phosphate Nutrition of Crop Plants in Sustainable Agriculture

Since phosphate fertilizer production is highly cost intensive –close to US $ 4 billion per annum worth energy is required – to meet global need for phosphate needs of crop production (Goldstein et al. 1993), one has to critically look at means as to how a sustainable methodology could be formulated. The situation is further compounded by the fact that since almost 75–90% of added phosphate fertilizer is precipitated by metal cations (aluminum, iron, manganese and calcium), the problem assumes greater complexity. It has been suggested that the accumulated phosphates in agricultural soils is sufficient to sustain maximum crop yields world-wide for about one century (Goldstein et al. 1993). Thus, the dependence of fertilizer production based on fossil fuel energy, as a source of production, and the prospects of diminishing availability of costly input of fertilizer production in the future, obviously brought the subject of mineral phosphate solubilization to the forefront of research. It is imperative to source alternative paths to tackle this grave problem. Under diverse soil and agro-climatic conditions, the organisms with phosphate solubilizing (PS) abilities have proved to be an economically sound alternative to the more expensive superphosphates and possess a greater agronomic utility. The microbial system can siphon out appreciable amounts of soil nutrients from the natural soil reservoir. The crop-microbial ecosystem, thus, can be energized in sustainable agricultural systems, with considerable ecological stability and environmental integrity. The organisms with adequate phosphate solubilizing potential increase the soil availability of readily bio available phosphorus that can easily be taken up by plant root. The organisms can also enhance biological N fixation and further augment availability of iron, manganese, zinc, etc. They can also augment production of plant growth promoting regulators (Ponmurugan and Gopi 2006).

22.2 Nature of Phosphatic Fertilizers

There is a positive synergy between crop plant roots and rhizospheric microorganisms under phosphorus stressed conditions in soil. This association could result either in improved uptake of soil available phosphates or render unavailable phosphorus sources in soil accessible to the plant root. The vesicular arbuscular mycorrhizae (VAM) belong to the former category. The numerous bacteria and fungi in soil, capable of solubilizing insoluble phosphates in soil, belong to the latter category. This chapter will dwell on the theme of natural phosphate solubilizing organisms.

22.3 Phosphate Solubilizing Microorganisms

Soil microorganisms have both beneficial and hazardous effects. Rhizospheric microorganisms mediate soil dependent processes such as, organic matter composition, mineralization of soil nutrients and, consequently, nutrient mobilization, release of stored nutrients and water, fixation of nitrogen and denitirification. Phosphate solubilizing microorganisms have the faculty to solubilize fixed phosphates into readily abosorbable phosphates. The role of rhizospheric microorganisms in phosphate solubilization was known as early as 1903. Since then, there have been extensive studies on the mineral phosphate solubilization by naturally abundant rhizospheric microorganisms. Important genera of mineral phosphate solubilizers include *Bacillus* and *Pseudomonas,* while *Aspergillus and Penicillium* form the important fungal genera (Motsara et al. 1995). Very recently nematofungus *Arthrobotrys oligospora* was tested *in vitro* and *in vivo* for its ability to solubilize *kodjari* rock phosphate (KRO) from Burkina Faso, Togolese Rock Phosphate (TRP) from Senegal and *telimsi* rock phosphate (TIRP) from Mali all in central Africa, all of which were solubilized by the nematofungus, which also demonstrated the additional ability to solubilize phosphate from rock phosphates *in vivo* as well (Duponnois et al. 2006). Most of the impact with the use of microorganisms as biofertilizers has been directed towards understanding biological N fixation. In contrast, the fundamental research on phosphate solubilization by nodule bacteria has been substantially less, though it is known that phosphorus is the most limiting factor for N fixation by *Rhizobium-legume* symbiosis. There are but few reports on phosphate solubilization by *Rhizobium*(Chabot et al. 1996) and on non – symbiotic N fixer, *Azotobacter* (Kumar et al. 2001). Phosphate solubilizing microorganisms are ubiquitous whose numbers vary from soil to soil. In soil, phosphate solubilizing bacteria constitute 1–50% and fungi 0.1–0.5% of the total respective population.

Generally, the phosphate solubilizing fungi are 2–150 fold (Kucey 1983). The high proportion of PSM is concentrated in the rhizosphere and is known to be metabolically active than those isolated from sources other than rhizosphere(Vazquez et al. 2000). Conversely, the salt, pH and temperature tolerant phosphate solubilizing bacteria have been reported to be maximum in the rhizoplane followed by rhozosphere and root free soil in alkaline soils (Johri et al. 1999). The PSM strains with these stressed properties should, therefore, serve as an excellent model to study the physiological, biochemical and molecular mechanism(s) of phosphate solubilization under stressed ecosystems. Further, it has been observed that the phosphate solubilizing bacteria upon repeated sub culturing lose the phosphate solubilizing activity, but, such losses have not been observed in phosphate solubilizing fungus as well (Kucey 1983). In general, the phosphate solubilizing fungi produce more acids and consequently exhibit greater phosphate solubilizing activity than bacteria in both liquid and solid media (Venkateswaralu et al. 1984). The phosphate solubilizing ability of PSM also depends on the nature of the source of N used in the media, with greater solubilization in the presence of ammonium salts than when nitrate is used as the source of N. This has been attributed to the extrusion of protons to compensate for ammonium uptake, leading to a lowering of extra-cellular pH (Roos and Luckner 1984). In some cases, however, ammonium can lead to decrease in phosphorus solubilization (Reyes et al. 1999).

22.4 The Phosphorus Solubilizing Microorganisms- Trail of Search for It

To isolate the PSM, one has to use a serial dilution or enrichment culture technique in Pikovskaya medium (Pikovskaya 1948) from non – rhizosphere and rhizosphere soils, rhizoplane, and, also, from other environments, such as, rock phosphate deposit area and marine environment (Gaur 1990). Upon incubation of the organisms on to the solid plates containing insoluble phosphates, PSM are detected by the formation of clear halos around their colonies as in the following figure (Fig. 22.1).

Recently, a few other methods for the isolation and selection of PSM have been suggested. Since certain strains of phosphate solubilizing organisms exhibit many fold variation in phosphate solubilizing activity and instability with regard to their phosphate solubilizing activity (Illmer and Schinner 1992), they are repeatedly subcultured to test the persistence of phosphate solubilizing potential. Once the efficient phosphate solubilizing organisms are selected, they are tested for their ability to solubilize insoluble phosphates under liquid culture medium. Finally, the selected efficient phosphate solubilizing cultures are used for making the inoculants and their performance under pot/field conditions is tested on their efficiency in various field crops.

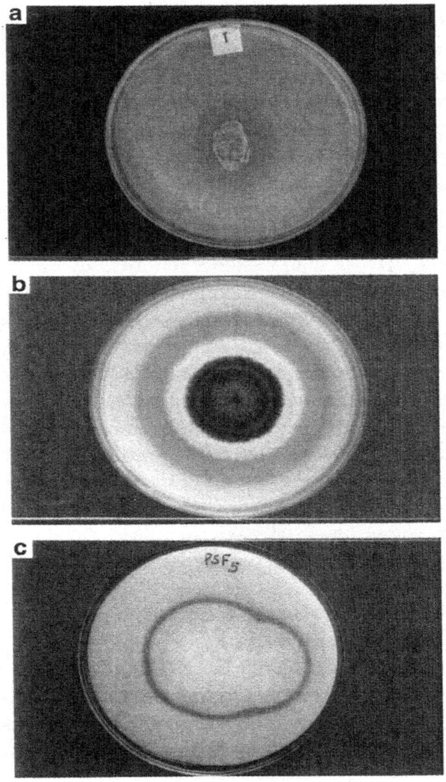

Fig. 22.1 Solubilization of insoluble P on Pikovskaya medium by phosphate solubilizing bacteria (**a**) and fungi (**b** and **c**)

22.5 Phosphorus Solubilization –How Does it Happen?

The phenomenon of phosphate solubilization by PSM in controlled liquid culture has been extensively investigated (Whitelaw 2000). The microbial solubilization of soil phosphorus in liquid medium has often been due to the excretion of organic acids as shown in the following table (Table 22.1).

For instance, oxalic, citric and lactic acids, in liquid culture filtrates were determined by paper chromatography, thin layer chromatography (TLC) or high performance liquid chromatography (HPLC)and specific enzymatic methods to allow more accurate identification of unknown organic acids (Gyaneshwar et al. 1998). Such organic acids can either directly dissolve the mineral phosphate as a result of anion exchange of PO_4^{2-} by acid anion or can chelate both iron and aluminum ions associated with phosphate (Omar 1998), while, in certain cases phosphate solubilization is induced by phosphate starvation (Gyaneshwar et al. 1999). However, no definite correlation between the acids produced by the PSM and amounts of phosphate solubilized is reported (Asea et al. 1988). The role of organic acids produced by PSM in solubilizing insoluble phosphate may be due to the lowering of pH,

Table 22.1 Details of the principal organic acid production by PSM

Organism – Phosphate solubilising fungi and actinomycetes	Predominant acids	References
Aspergillusniger, A. flavus, Penicilliumcanescens	Oxalic, citric, gluconic, succinic	Maliha et al. (2004)
A. niger	Succinic	Vasquez et al. (2000)
Penicillium regulosum	Gluconic	Reys et al. (1999)
Penicillium radicum	Gluconic	Whitelaw et al. (1999)
Penicillium variable	Gluconic	Vasilev et al. (1996)
A. niger	Citric, oxalic, gluconic	Illmer et al. (1995)
A. awamori, A. foetidus, A. terricola, A. amstelodemi, A. tamari, A. japonicus, A. foetidus	Oxalic, citric, gluconic, succinic, tartaric	Gupta et al. (1994) Singal et al. (1994)
Penicillium bilaji	Citric, oxalic	Cunningham and Kuiack (1992)
A. niger, P. simplicissimum	Citric	Burgstaller et al. (1992)
A. awamori, P. digitatum,	Succinic, citric, tartaric	Gaur (1990)
Scwaniomyces occidentalis	Succinic, fumaric, citric, tartaric, α-ketobutyric	Gaur (1990)
Penicillium sp.	Oxalic, itaconic	Parks et al. (1990)
A. niger	Succinic	Venkateswaralu et al. (1984)
Aspergillus sp., *Penicillium* sp., *Chaetomium nigricoler*	Oxalic, succinic, citric, 2-ketogluconic	Banik and Dey (1983)
Streptomyces	Lactic, 2-ketogluconic	Banik and Dey (1982)
A. fumigatus, A. candidus	Oxalic, tartaric, citric	Banik and Dey (1982)
Phosphate solubilising bacteria		
Enterobacter intermedium	2-ketogluconic	Hoon et al. (2003)
Bacillus amyloliquefaciens, B. licheniformis, B. atrophaeus, Penibacillusmacerans, Vibrioproteolyticus, Xanthobacteragilis, Enterobacteraerogenes, E. taylorae, E. asburiae, Kluyveracryocre-scens, Pseudomonas aero-genes, Chryseomonas luteola, Pseudomonas cepacia	Lactic, itaconic, isovaleric, isobutyric, acetic	Vasquez et al. (2000)
Bacillus polymyxa, B. licheniformis, Bacillus sp.	Oxalic, citric	Gupta et al. (1994)
Pseudomonas striata	Malic, glyoxalic, succinic, fumaric, tartaric, α-ketobutyric	Gaur (1990)
Arthrobacter sp.	Oxalic, malonic	Banik and Dey (1982)

(continued)

Table 22.1 (continued)

Organism – Phosphate solubilising fungi and actinomycetes	Predominant acids	References
Bacillus firmus	2-ketogluconic, succinic	Banik and Dey (1982)
Micrococcus sp.	Oxalic	Banik and Dey (1982)
Bacillus subtilis, Bacillus sp.	Oxalic, succinic, citric,2-ketogluconic	Banik and Dey (1983)

chelation of cations and by competing with phosphate for adsorption sites in soil (Nahas 1996). Inorganic acids, for example, HCl can also solubilize phosphate, but, they are less effective compared to the organic acids at the same pH (Kim et al. 1997). However, acidification does not seem to be the only mechanism of solubilization, as the ability to reduce the pH in some cases did not correlate with the ability to solubilize mineral phosphates (Subba Rao 1982). The chelating ability of the organic acids is also important, as it has been shown that the addition of 0.05 M EDTA to the medium has the same solubilizing effects as inoculation with *Penicillium bilaji* (Kucey 1988). Among nodule bacteria, for instance, *Rhizobium/ Bradyrhizobium*), the phosphate solubilizing activity of *Rhizobium* was associated with the production of 2-ketogluconic acid, which was abolished by the addition of NaOH, indicating that phosphate solubilizing activity of this organism was entirely due to its ability to reduce the pH of the medium (Halder and Chakrabarty 1993). However, the precise biochemical and molecular mechanisms of phosphate solubilization by symbiotic nodule bacteria are still unclear.

The word "Sustain" from Latin "Sustinere" ("Sus"- from below, "tinere"- to hold), to keep in existence or maintain, implies long-term support or permanence. As it pertains to agriculture, sustainable describes farming systems that are "capable of maintaining their productivity and usefulness to society indefinitely." Such systems must be resource conserving, socially supportive, commercially competitive, and environmentally sound. Thus the term sustainable agriculture means an integrated system of plant and animal production practices having site-specific application that will, over the longterm satisfy human food and fiber needs, enhance environmental quality and sustain the economic viability of farm operations. It is achieved through management strategies, which help the producer select hybrids and varieties, soil conserving cultural practices, soil fertility programs, and pest management programs. The goal of sustainable agriculture is, hence, to minimize adverse impacts to the immediate and off-farm environments while providing a sustained level of production and profit. However, the most important constraint limiting crop yields in developing nations, worldwide, and especially among resource –poor farmers, is soil fertility. Unless soil fertility is restored in these regions, farmers can gain but little benefit from the use of improved varieties and more productive cultural practices. Generally, the idea of Integrated Soil Fertility Management or Integrated Nutrient Management has attempted to address this problem. The crucial question is, these approaches are still rooted in classical "text book" knowledge. For instance, phosphate bio availability estimation is still determined either by the classical Olsen's or Bray's extraction methods, depending on soil pH, potassium availability is still rooted in normal neutral ammonium acetate extraction, and, zinc availability assessment continues to be the classical DTPA extraction, just to name a few examples. The idea of the "Buffer Power Concept" is, yet, to be followed throughout the world, though many regions, especially in the developing world, have very seriously taken to the new idea. Though the classical approaches would continue, as long as, farming is fundamentally energy dependent through fertilizers, the biochemical path is emerging as a challenging, yet, one with much promise, alternative, because, in this case, there is no expending of energy.

Most of the studies on phosphate solubilization by microbes center on the isolation of the microorganism from the rhizosphere, and subsequently evaluating its efficiency in phosphate solubilization in *in vitro* conditions. These studies have been followed in field conditions. The beneficial effects of PSM have been documented by Zaidi and Khan (2005). Nitrogen fixing bacteria are, perhaps, the most promising group of PSM on account of their ability to fix nitrogen symbiotically (in leguminous crops), or asymbiotically (in non leguminous crops), along with the ability of some strains to solubilize inorganic phosphatic compounds. A large number of published literature establish the phosphate-solubilizing capacity of strains of *Rhizobium, Bradyrhizobium* and *Azotobacter* which increase plant growth and enhance phosphorus content of both leguminous and non leguminous plants. An alternative approach for the use of PSM as microbial inoculants is either the use of mixed cultures or the co-inoculation with other microorganisms. In this regard, some experimental results suggest a synergistic interaction between vesicular arbuscular mycorrizhae (VAM) fungi and phosphate solubilizing bacteria. This complementary effect leads to better utilization of insoluble phosphate from problematic soil. Phosphate solubilizing bacteria have already been applied as effective inoculants in agronomic practices to raise crop productivity. For instance, in the former Soviet Union, a commercial bio fertilizer under the name "Phosphobacterin" was first made using *Bacillus megaterium* var. *phosphaticum* and later on was frequently applied in field conditions in Eastern Europe and also in India.

22.6 Application of PSM

Since the introduction of PSM inoculants, the application of PSM inoculants to seed surface prior to sowing is traditionally the most commonly used and easiest means of inoculation in the case of field crops. When properly applied, this method ensures that each seed receives the introduced microphos. Disadvantages include a limitation of the quantity that can adhere to the seed surface, direct contact with any external chemicals, movement of PSM away from rooting zone and exposure to environmental stresses after planting. However, the use of sticker solution, for example, gum Arabic, improves the adherence of the inoculants PSM to the seed surface. There are, however, certain situations where seed applications, for example, with seed-dressed with pesticide is incompatible with the PSM in question, which can restrict effectiveness of the PSM, under question. In such cases, soil application directly of the PSM is the better choice. Inoculants applied directly to the soil have the following advantages:

1. Greater population of PSM per unit surface area of soil
2. Direct contact with chemically treated seeds is minimized
3. Elimination of seed mixing
4. Ability to withstand low moisture content as compared to when the PSM is in powder form

In accordance with the above-mentioned conditions, two approaches can be applied for PSM inoculation, first, single culture approach (SCA), where phosphate solubilizing organisms can be used singly, and, secondly, the multiple or mixed culture approach (MCA), often termed "co-inoculation," where the PSMs are used along with the other beneficial rhizosphere microorganisms.

22.6.1 Factors Which Affect Survival of PSM

It is the ability of the PSM to survive and multiply in soil that determines its competitiveness. However, understanding this part of the use of PSM is the most limiting factor, and, it is quite difficult, if not impossible, to predict the behavior and efficacy of the inoculated PSM in a particular location. In general, the population size or density of the artificially introduced PSM decline rapidly upon the introduction in soils (Ho and Ko 1985). The survival of the inoculants strain, thus, depends upon various factors, such as, soil composition (Bashan et al. 1995) physiological status, temperature, pH, moisture content (Van Elsas et al. 1991) and the presence of recombinant plasmids (Van Veen et al. 1997). The biotic factors which affect the survival of the inoculated PSM include competition, predation and root growth that provide the substrates to the microbes.

22.6.2 Crop Response to Composite Inoculants

The soil is a habitat for a vast, complex and interactive community of naturally occurring soil organisms, whose activities largely determine the physicochemical properties of the soil and consequently promotes the growth of the crop plants. From seed germination until a plant reaches maturity, it lives in close association with soil organisms. The association is termed rhizocoenosis (Lynch 1983), and, hence, the rhizosphere that harbors heterogenous microbial communities with sufficient phosphate solubilizing abilities becomes an important soil habitat. Therefore, the focus of attention has now been shifted from plant microbe interactions to plant microbe-microbe interactions. Recently, some instances have been reported where plant growth has been markedly enhanced using two or three member association of rhizospheric organisms and such syntropic associations are of paramount agronomic utility.

22.6.3 Interaction Between PSM and Nitrogen Fixing Organisms

Nitrogen and Phosphorus are the two major plant nutrients and combined inoculation of nitrogen fixers and PSM may benefit the plants better than inoculating with either group of organisms alone. Both *in vivo* and *in vitro* investigations in this connection have been made (Sarojini et al. 1989). Phosphorus solubilization was observed by the mixed cultures (*Pseudomonas striata, Bradyrhizobium* sp., *Mesorhizobium ciceri*) suggesting that they could be used as mixed microbial inoculants, as depicted in the following figure (Fig. 22.2).

No antagonistic behavior of one organism towards others has been noted. Thus, the mixed cultures or co-inoculation with other microorganisms are generally preferred over single inoculation treatments. In this context, N fixers and PSM, when inoculated together colonized the rhizosphere and enhanced the growth of legumes by providing with Nitrogen and Phosphate respectively (Gull et al. 2004). The N fixing organisms not only provide N to the plants, but, also improve the N status of soil alone or in combination with PSM. Accordingly, the application of phosphate solubilizing bacteria (*Pseudomonas striata*) and nodule bacteria gave significantly

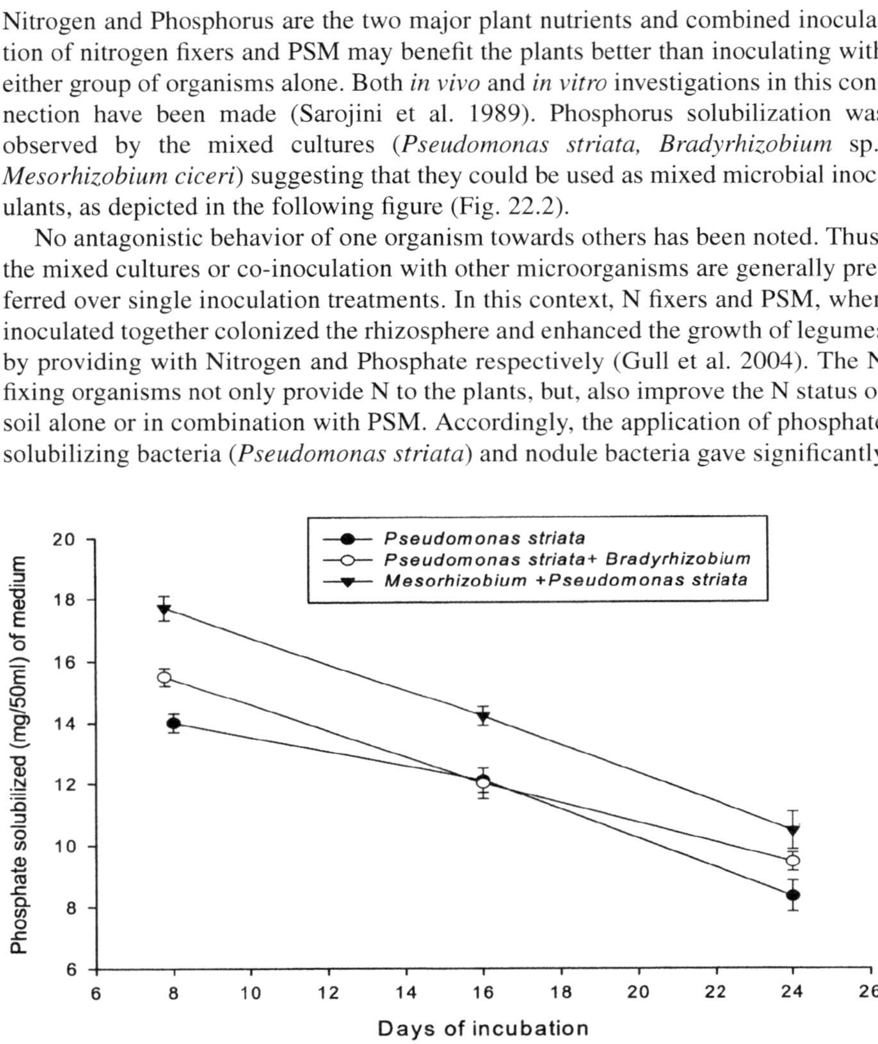

Fig. 22.2 In vitro solubilization of tricalciul phosphate by Pseudomonas striata alone and as influenced by Bradyrhizobium sp. (vigna) and Mesorhizobium cicero in liquid Pikovskaya medium at different incubation days

higher yield of green gram (Khan et al. 1997) and chickpea (Algawadi and Gaur 1988) than that obtained through the use of *Rhizobium* singly. Furthermore, *Rhizobium* and phosphate solubilizing fungi (*Aspergillus awamori*) when used as a seed inoculants, increased the grain yield of chickpea in field conditions. Similarly, the effect of interactions between three phosphate solubilizing fungi, namely, *Aspergillus niger, A. fumigatus* and *Penicillium pinophilum* and N fixing *Rhizobium leguminosarum biovar viciae* showed significantly higher positive growth, nutrients (N and P) uptake and, consequently, yield of *Vicia faba* under field conditions (Mehana and Wahid 2002). A decade-long trial using phosphate solubilizing bacteria and N fixing organism enhanced the seed production in soybean and was found more effective compared to superphosphate singly (Dube 2001). In a similar study, single or combined inoculation with PSM and N fixers had a positive effect on yield and nutrient uptake in cereals and legumes (Whitelaw 2000; Kumar et al. 2001).

A beneficial effect of phosphate solubilizing organisms alone and in combination with N fixer on cotton in the field and in wheat (Zaidi et al. 2004) has been reported in scientific literature. Co-inoculation studies using *Rhizobium* and phosphate solubilising bacteria have been carried out by many researchers (Kuacey 1987; Dowey and Van 1990). Similar results were obtained for French bean (*Phaseolus vrisulga*), when inoculated with *Agrobacterium,* a phosphate solubilizer. In contrast, beans grown in autoclaved soil inoculated with phosphate solubilizing *Penicillium bilaji* and *Rhizobium phaseoli* showed no significant increase in dry matter or total uptake of phosphate (Kuacey 1987). Further, a decrease in total N fixation in field peas due to dual inoculation of *Penicillium bilaji* and *Rhizobium leguminosarum* have also been reported (Downey and Van 1990). Further, a decrease in total nitrogen fixation in field peas due to dual inoculation of *Penicillium bilaji* and *Rhizobium leguminosarum* have also been reported (Downey and Van 1990). It was suggested that the adverse effect of *Penicillium bilaji* on N fixation might be explained by the mode of action of the fungus. The phosphate releasing fungi produce more of organic acids (Venkateswarlu et al. 1984), which enhance the solubilization of phosphate. However, most rhizobia prefer neutral or alkaline conditions during nodulation. These results suggest that before carrying out *in vitro* studies, the compatibility between the two associates must be checked in *in vivo.*

Combined inoculation of *Rhizobium* and phosphate solubilizing *Pseudomonas striata* or *Bacillus polymyxa* with or without added fertilizers on chick pea yield and nutrient content was studied under green house conditions. *Rhizobium* inoculation alone increased nodulation and nitrogenase activity, whereas the phosphate solubilizing organism increased the available phosphorus content in the soil. The combined inoculation increased nodulation, available phosphate of soil, as well as dry matter of the plants, grain yield and phosphorus and N uptake by plants. The inoculation effects, however, were more pronounced in the presence of added fertilizers (Algawadi and Gaur 1988). In a pot experiment, lentil seeds were inoculated with *Rhizobium leguminosarum,* along with increasing doses (50, 100, 200 and 400 kg/feddan, 1 feddan = 0.42 ha) of rock phosphate with or without a 1:1 mixture of elemental sulfur and rock phosphate in the presence or absence of phosphate solubilizing bacteria. Plant dry weight and N, P, Zn, Fe, Mn and Cu uptake increased with rock phosphate, sulfur and phosphate solubilizing bacteria compared with untreated control. Dry matter yield and nutrient uptake was slightly higher with sulfur application (Saber and Kalbesh 1990). A combination of *Azotobacter chroococcum* GA-1 and GA-3 with *Penicillium* HF-4 and HF-5 and *Aspergillus* GF-1 and GF-2 increased radicle and plumule length, but, the remaining culture combinations decreased radicle/plumule length. A significant increase in mungbean yield and groundnut yield was observed with the inoculation of *Rhizobium* spp., and phosphate solubilizing bacteria along with phosphatic fertilizers (Khan et al. 1998). Moreover, the microbes which are involved in phosphate solubilization as well as in better scavenging of soluble phosphate can enhance plant growth by improving the efficiency of biological N fixation, accelerating the availability of other trace elements and by production of phytohormones. Accordingly, increase in yield of various legumes have been observed following seed or soil inoculation with N fixing organisms and PSM or PSM(s) and arbuscular mycorrhizal fungus (Mukherjee and Rai 2000). It has been further suggested that about 50% of phosphatic fertilizer requirement could be saved by the combined inoculation of *Rhizobium* strain Tt 9 with *Bacillus megaterium* var. *phosphaticum* in groundnut. *Rhizoium* strain Tt 9 along with phosphobacteria at 75% phosphate level recorded higher nodule number, root length, shoot length and increased pod yield than the dual inoculation at 100% phosphorus level in groundnut (Natarajan and Subrammanian 1995). However, no significant increase in phosphate contents in pigeonpea plant inoculated with *Rhizobium* (CCI) with *Bacillus megaterium* var. *phosphaticum* was observed (Gunasekaran and Padiarajan 1995). Similarly, about 37% increase in the grain yield of blackgram was reported following the inoculation of *Rhizobium* and *Bacillus megaterium* (Prabhakaran et al. 1996). Dubey (1996) observed in a number of field trials carried out during 1991 and 1992, using soybean as test crop under rainfed conditions on a vertisol, a significantly higher yield and phosphorus content in the plants when seeds were bacterized with *Pseudomonas striata* and rock phosphate as compared to single super phosphate application. The mechanism of plant growth promotion by phosphate solubilizing microorganisms is depicted in the following figure (Fig. 22.3).

22.6.4 Symbiosis Between PSM and Arbuscular Mycorrhizal Fungi

Mycorrhizal fungi are ubiquitous soil inhabitants and form symbioses with terrestrial plants (Jeffries 1987). However, in association with N fixers, the arbuscular mycorhizal fungi increase nitrogen and phosphatic nutrients of the plants, especially in P deficient soils or soils having rock phosphate (Poi et al. 1989). The PSM can release some phosphatic ions from otherwise sparingly soluble P sources (Barea et al. 1983), which is trapped and translocated by the arbuscular mycorrhizal fungal

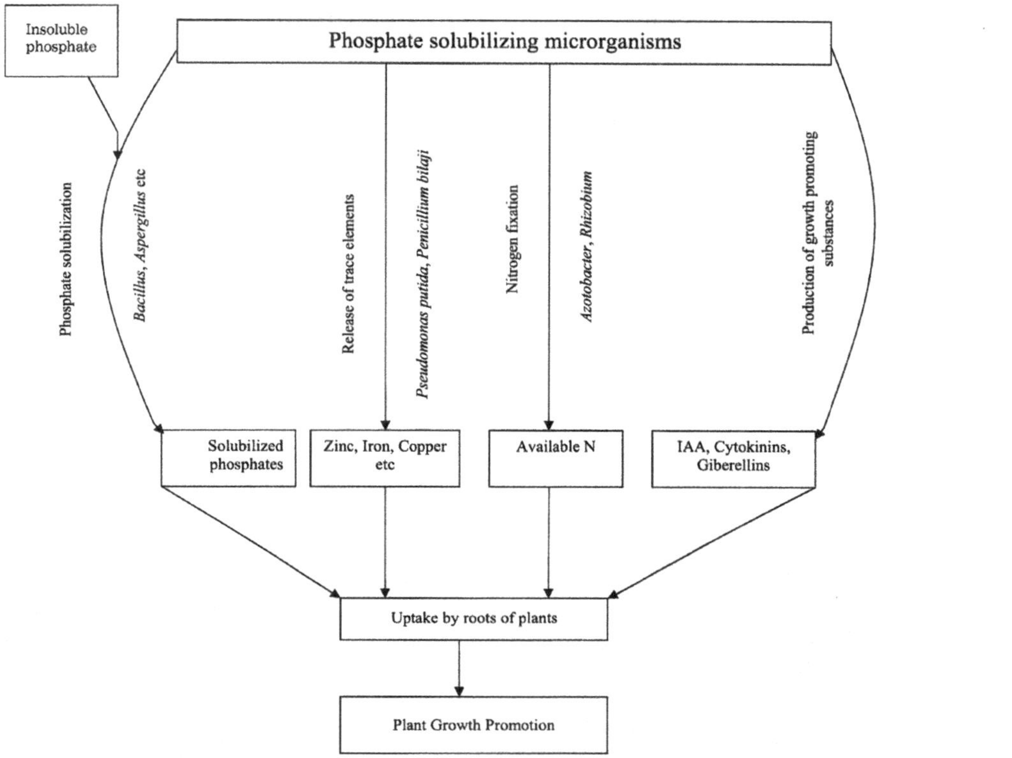

Fig. 22.3 Mechanism of plant growth promotion by phosphate solubilizing mirco organisms

hyphae to the plant (Azcon-Aguilar et al. 1986). Moreover, the PSM survives longer around mycorrhizal roots compared to non mycorrhizal roots and acts synergistically with the arbuscular mycorrhizal fungus leading to increased plant growth, especially where rock phosphate is applied to soil (Singh 1990). Simultaneous dual inoculation of arbuscular mycorrhizal fungi and PSM has been shown to stimulate plant growth more than inoculation of either microorganism alone in certain situations when the soil is P deficient (Piccini and Azcon 1987). The simplest interpretation of this fact is that mycorrhizal endophyte could be stimulated in quantity, efficiency, and longevity. The main effect of this mycorrizha in improving plant growth is through improved uptake of nutrients, especially P due to the exploration by the external hyphae of the soil beyond the root hair zone where P is depleted. Besides, root exudation and plasticity might change by PSM inoculation, which could also affect arbuscular mycorrhizal development. The arbuscular mycorrizhal fungi also produce plant hormones and increase the activity of N fixing organisms in the root zone (Bagyaraj 1984). However, a more thorough understanding of interactions between soil microorganisms is needed for an optimal utilization of these interactions with respect to growth and development of plants. When *Centrosema macrocarpum* plants were inoculated with *Rhizobium* strains and arbuscular mycorrhizal fungi *Glomus manihotis* or *Acaulospora longula,* a significantly higher dry matter production, mineral absorption, nodulation and infection by arbuscular mycorrhizal fungi was recorded(Satizabal and Saif 1987). However, a small amount of N fertilizer was suggested for application at the time of sowing. The dual inoculation of *Azotobacter chroococcum* and *Glomus fasciculatum* enhanced root infection of arbuscular mycorrhizal fungi, stimulated plant growth, and increased N, P and Zn contents in maize and wheat (Elgala et al. 1995).

22.6.5 Tripartite Symbioses Between Nitrogen Fixers, Phosphate Solubilizers and Arbuscular Mycorrhizal Fungi

Only limited references in literature are seen on the effect of combined inoculation of crop plants with N fixers and PSM in the presence of arbuscular mycorrhizal fungus. During the inter-generic interaction, N fixing microorganisms provide N to the plants and consequently improve the N status of the soil while PSM enhance plant growth by providing it with phosphates. Where N and P are limiting, arbuscular mycorrhizal fungi may improve phosphate uptake by plants, the higher P concentration in the plant benefiting the N fixers and the functioning of its nitrogenase activity leading to enhanced N fixation, which in turn promotes root and mycorrhizal development. Singh and Singh conducted an investigation on the associative effect of *Bradyrhizobium japonicum,* arbuscular mycorrhizal fungus and phosphate solubilizing microbes on soybean in a mollisol. Inoculation with endophyte alone resulted in 70% root colonization. Addition of rock phosphates or inoculation with phosphate solubilizing bacteria except *Bacillus polymyxa* stimulated root infection of native as well as introduced vesicualar arbuscular mycorrhizal endophytes. Application of rock phosphate with triple inoculation significantly increased grain yield, nodulation, N uptake and available soil phosphate. Further, Bethlenfalvay (1994) demonstrated the impact of *Glomus mosseae, Bacullus* sp., and *Rhizobium* sp., on plant growth and soil aggregation in *Pisum sativum* field. Complex interaction among the three microorganisms was found, which in turn, promoted dramatically plant growth and soil aggregation. A positive interaction, thus, appears to exist between root colonization, P uptake and plant growth promotion (Zaidi and Khan 2005).

22.6.6 The Reasons for Failure of Phosphate Solubilizing Microorganism

Doubts have been raised on the ability of PSM to liberate fixed phosphate under soil conditions (Tinker 1980). Growth promotion and crop yields following superphosphate supplementation despite the abundance of PSM in the rhizosphere zone and the bulk soil, raises important issues on the variations in the effectiveness of PSM inoculations. Why is P limited for plants when PSM (s) are abundant in soils? Many reasons have been attributed to account for the variations in the effectiveness of PSM inoculations on plant growth and crop yields (Kucey et al. 1989). Following are some:

1. Survival and colonization of inoculated PSM in the rhizosphere
2. Competition with native microorganisms
3. Nature and properties of soils and plant varieties
4. Insufficient nutrients in the rhizosphere to produce enough organic acids to solubilize phosphorus
5. Inability of PSM to solubilize soil P

It has been shown that at least two PSM showing phosphate solubilization ability in laboratory conditions could not release phosphate from alkaline vertisols, even when supplemented with other nutrients (Gyaneshwar et al. 1998). Similarly, no soluble phosphate was released when the non-sterilized bulk and rhizospheric soil samples were supplemented with carbon and nitrogen. The ability of the PSM to release phosphate from mineral phosphorus was significantly decreased possibly due to the high buffering capacity of the soils, coupled with the inability of PSM to secrete high quantities of organic acids. It is thus clear that the "Buffer Power Concept" can also exert an important impact on the effectiveness of PSM in solubilizing soil P. This is one area which deserves thorough investigation, in view of the significant role of the concept playing a very dominant role in the solubilization of important nutrients, such as, P, K and Zn from soils from different continents and location, as demonstrated by the research of this author (Nair 2013).

Despite promising results, PSM based bio fertilizers have not been used widely in agriculture, mainly because of the variable responses of crop species or genotypes to inoculation, depending on the bacterial strain used. Differential rhizosphere effect of crop species in harboring a target phosphate solubilizing bacterial strain or even the modulation of the bacterial phosphate solubilizing capacity by specific root exudates may account for the observed differences. On the other hand, good competitive ability and high saprophytic competence are the major factors determining the success of a bacterial strain as an inoculant. Therefore, investigations to unravel the competitiveness and persistence of specific microbial populations in complex environments, such as the rhizosphere, should be addressed in order to obtain efficient inoculants. Hence, efforts to obtain appropriate formulations of microbial inoculants, which protect the inoculant organism against environmental stresses and, at the same time, enhance and prolong its activity, may help in promoting the use of such beneficial bacteria in sustainable agriculture.

22.6.7 *Application of Genetic Engineering in developing Super phosphate Solubilizing Microbial Inoculants*

The performance of PSM inoculants depends largely on their ability to colonize a specific rhizosphere habitat. The investigation of colonization of phosphate solubilizer provides useful information to solve the deficiency of P in soils. Plate counting and most probable number techniques have been used for the study of PSM communities in rhizosphere. However, in general, only less than 1% of the microorganisms in the environment can usually be cultured by standard techniques, because these techniques fail to reproduce in artificial media. The niches of many microorganisms are found in high diversity environments such as the rhizosphere. However, using a strain specific DNA probes, one can decipher a considerable number of heterogenous group of microorganisms in the rhizosphere (Frederic et al. 2000). Recent developments in molecular biology, which do not rely on cultivation methods, allows microbial ecologists to reveal inhabitants of natural microbial communities, which have not yet been cultured (Hugenholtz et al. 1998). Consequently, these techniques are now widely applied to characterize microbial community structures in different environments (Holben et al. 1998).

Two of these techniques, cloning and sequencing, allows us to determine which microorganisms are present in the community, but, they are time consuming. Hybridization and probing are faster, but, requires a sufficient knowledge of the community to choose the appropriate target sequence (Aman 1995). However, the other molecular biology techniques such as amplified ribosomal DNA restriction analysis (ARDRA) or ribosomal intergenic spacer analysis (RISA) can be applied to study the PSM colonization or community structure. Even faster than hybridization and probing, ARDRA and RISA have been used in the analysis of mixed bacterial population from different environments (Martinez-Mercia et al. 1995). Although ARDRA gives little or no information about the type of microorganisms present in the sample, it can be used for a quick assessment of genotypic changes in the community over time, or to compare communities subject to different environmental conditions. In comparison, RISA is a method of microbial community analysis that provides a means of comparing differing environments or treatment impacts without the bias imposed by culture dependent approaches. In brief, RISA involves polymerase chain reaction amplication of a region of the rRNA gene operon between the small (16S) and large (23S) sub-units called intergenic spacer region (ISR). By using oligonucleotide primers targeted to conserved regions in the 16S and 23S genes, RISA fragments can be generated from most of the dominant bacteria in an environmental sample. While the majority of the rRNA operon serves as a structural function, portions of the 16S:23S intergenic regions can encode rRNAs depending on the bacterial species. However, the taxonomic value of the ISR lies in the significant heterogeneity in both length and nucleotide sequence. In RISA, thus, we attempt to exploit the length heterogeneity of the ISR. The use of such molecular techniques having greater quantitative efficiency can be extended to study the colonization and characterization of PSM inoculants under diverse environmental conditions.

The solubilization of mineral phosphate is brought about by the exuded organic acids from PSM, either by reduction in soil solution pH, or by the chelating action on cations associated with phosphorus. A better understanding of the generic basis of the release of organic acids could pave the way for transferring the mineral phosphate solubilizing (mps) ability to various bacteria which are competent to colonize a specific rhizosphere. Among the various factors, the competence of the rhizosphere is a focal point which decides the success or failure of the inoculants. Rhizosphere has tremendous amounts of carbon sources which are available to the heterogenous microbial communities in soil to produce different kinds of organic acids. In addition, the solubilized phosphates can be utilized by the plant before they get precipitated again. Among the various organic acids, gluconic acid mediated secretion seems to be the major mechanism of phosphate solubilization by Gram negative bacteria (Kim et al. 1998). Gluconic acid is produced by the oxidative metabolism of glucose by glucose dehydrogenase (GDH), which requires pyrroloquinoline quinine (PQQ) as a co-factor. Thus, the genes involved in the biosynthesis/ transport of PQQ can be cloned from various bacteria and be transferred to other bacteria (Babu-Khan et al. 1995). For instance, since the rhizosphere competent bacteria (RCB) like *Rhizobium* possesses apo-GDH, it would be interesting to transfer the genes involved in PQQ biosysnthesis to *Rhizobium* to make an effective and better PSM. The resulting *Rhizobium* strains will thus have phosphate solubilizing activity in addition to its natural N fixing ability. Another important rhizosphere competent bacteria (*Pseudomonas* spp.) can form gluconic acid through the oxidative glucose metabolism and over expression of PQQ biosynthesis and GDH genes could also make them better PSM. The alternative approach is to screen the mineral phosphate solubilizing (mps) genes directly in the target bacteria by over/under expression of genes followed by the selection of transformants with mineral phosphate solubilizing ability. Such an approach has been used to obtain mineral phosphate solubilizing genes from *Synecosystis* PCC 6803 in *E.coli* (Gyaneshwar et al. 1998). However, it remains to be seen if this will also be effective in other bacteria. Genetic engineering could also help in increasing the survival of the inoculant strains by incorporating the abilities to utilize certain nutrients better than the rest of the microbial populations (Glick and Bashan 1997). Also genes for utilization of salicylate were transferred to a growth promoting bacteria and the recombinant bacterium was able to survive and enhance plant growth better than the wild type (Colbert et al. 1993).

References

Ahmad, S. (1995). *Agriculture-fertilizer interface in Asia issues of growth and sustainability*. New Delhi: Oxford and IBH Publishing Co.

Algawadi, A. R., & Gaur, A. C. (1988). Associative effect of Rhizobium and phosphate solubilizing bacteria on the yield and nutrient uptake of chickpea. *Plant and Soil, 105*, 241–246.

Aman, R. I. (1995). Fluorescently labeled rRNA targeted nucleotide probes in the study of microbial ecology. *Microbial Ecology, 4*, 543–554.

Asea, P. E. A., Kucey, R. M. N., & Stewart, J. W. B. (1988). Inorganic phosphate solubilization by two Penicillium species in solution culture and soil. *Soil Biology and Biochemistry, 20*, 459–464.

Azcon-Aguilar, C., Diaz-Rodriguez, R., & Barea, J. M. (1986). Effect of soil microorganisms on spore germination and growth on the vesicular arbuscular mycorrhizal fungus (Glomus moseae). *Transactions of the British Mycological Society, 86*, 337–340.

Babu-Khan, S., Yeo, T. C., Martin, W. I., Duron, M. R., Rogers, R. D., & Goldstein, A. H. (1995). Cloning of a mineral phosphate solubilizing gene from Peudomonas cepacia. *Applied and Environmental Microbiology, 61*, 972–978.

Bagyaraj, D. J. (1984). Biological interaction with VA mycorrhizal fungi. In C. L. Powell & D. J. Bagyaraj (Eds.), *VA Mycorrhiza* (pp. 131–153). Boca Raton: CRC.

Banik, S., & Dey, B. K. (1982). Available phosphate content of an alluvial soil as influenced by inoculation of some isolated phosphate solubilizing microorganisms. *Plant and Soil, 69*, 353–364.

Banik, S., & Dey, B. K. (1983). Phosphate solubilizing potentiality of the microorganisms capable of utilizing aluminium phosphate as a sole phosphate source. *Zentralblatt fur Mikrobiologie, 138*, 17–23.

Barber, S. A. (1984). *Soil nutrient bioavailability: A mechanistic approach*. New York: Wiley.

Barea, J. M., Azcon, R., & Azcon-Aguilar, C. (1983). *Interaction between phosphate solubilizing bacteria and VA mycorrhiza to improve the utilization of rock phosphate by plants in non acidic soils* (pp. 127–152). Brussels: Third International Congress on Phosphorus Compounds.

Bashan, Y., Puente, M. E., Rodriquea, M. N., Toledo, G., Holguin, G., Ferrera-Cerrato, R., & Pedrin, R. (1995). Survival of Azorhizobium brasilense in the bulk soil and rhizosphere of 23 soil types. *Applied and Environmental Microbiology, 61*, 1938–1945.

Bethlenfalvay, G. J. (1994). Sustainability and rhizoorganisms in an ecosystem. *Sociedad Mexicana de la Ciencia del Suelo, 4*, 9–10.

Burgstaller, W., Straser, H., & Shinner, F. (1992). Solubilization of zinc oxide from filter dust with Penicillium simplicissimum: Bioreactor, leaching and stoichiometry. *Environmental Science & Technology, 26*, 340–346.

Chabot, R., Anton, H., & Cescas, M. P. (1996). Growth promotion of maize and lettuce by phosphate solubilizing Rhizobium leguminosarum biovar phaseoli. *Plant and Soil, 184*, 311–321.

Colbert, S. F., Hendson, M., Ferri, M., & Schroth, M. N. (1993). Enhanced growth and activity of a biocontrol bacterium genetically engineered to utilize salicylate. *Applied Microbiology, 59*, 2071–2076.

Cunningham, J. E., & Kuiack, C. (1992). Production of citric and oxalic acids and solubilization of calcium phosphate by Penicillium bilaji. *Applied and Environmental Microbiology, 58*, 1451–1458.

Downey, J., & Van Kessel, C. (1990). Dual inoculation of Pisum sativum with Rhizobium leguminosarum and Penicillium bilaji. *Biology and Fertility of Soils, 10*, 194–196.

Dubey, S. K. (1996). Response of soybean to rock phosphate applied with Pseudomonas striata in a typic chromustert. *Journal of the Indian Society of Soil Science, 44*, 252–255.

Dubey, S. K. (2001). Associative effect of nitrogen fixing and phosphate solubilizing bacteria in rainfed soybean (Glycine max) grown in vertisols. *Indian Journal of Agricultural Sciences, 71*, 476–479.

Duponnois, R., Kisa, M., & Plenchette, C. (2006). Phosphate solubilizing potential of the nematofungus Arthrobotrys oligospora. *Journal of Plant Nutrition and Soil Science, 169*, 280–282.

Elgala, H. M., Ishac, Y. Z., Abdel-Monem, M., El-Ghandour, I. A. I., Hang, P. M., Berthelin, J., Bollag, J. M., Mc Gill, W. B., & Page, A. I. (1995). Effect of single and combined inoculation with Azotobacter and VA mycorrhizal fungi on growth and nutrient content of maize and wheat plants. *Environmental Impacts of Soil Component Interactions, 2*, 109–116.

Frederic, B. G., Estefania, A., Jordi, B. F., Charles, A. A., Dolors, M., & Manel, P. (2000). Assessment of microbial community structure changes by amplified ribosomal DNA restriction analysis (ARDRA). *International Microbiology, 3*, 103–106.

Gaume, A. (2000). Low P tolerance of various maize cultivars: The contribution of the root exudation. Ph.D dissertation, Swiss Federal Institute of Technology, Zurich, Switzerland.

Gaur, A. C. (1990). *Phosphate solubilizing microorganisms as Biofertilizers* (p. 176). New Delhi: Omega Scientific.

Glick, B. R., & Bashan, Y. (1997). Genetic manipulation of plant growth promoting bacteria to enhance biocontrol of phytopathogens. *Biotechnology Advances, 15*, 353–378.

Goldstein, A. H., Rogers, R. D., & Mead, G. (1993). Mining by microbe. *Bio Technology, 11*, 1250–1254.

Gull, M., Hafeez, F. Y., Saleem, M., & Malik, K. A. (2004). Phosphorus uptake and growth promotion of chickpea by co-inoculation of mineral phosphate solubilizing bacteria and a mixed rhizobial culture. *Australian Journal of Experimental Agriculture, 44*, 623–628.

Gunasekaran, S., & Pandiarajan, P. (1995). Dual inoculation of rhizobium and phosphobacteria with two forms of phosphorus in pigeonpea. In *Microbiology abstracts XXXVI. Annual conference of the Association of Microbiologists of India* (p. 111). Hissar, Haryana State, India. November 8–10.

Gupta, R. R., Singal, R., Shanker, A., Kuhad, R. C., & Saxena, R. K. (1994). A modified plate assay for screening phosphate solubilizing microorganisms Gen. *Applied Microbiology, 40*, 255–260.

Gyaneshwar, P., Naresh, K. G., & Parekh, L. J. (1998). Effect of buffering on the phosphate solubilizing ability of microorganisms. *World Journal of Microbiology and Biotechnology, 14*, 669–673.

Gyaneshwar, P., Parekh, L. J., Archana, G., Podle, P. S., Collins, M. D., Hutson, R. A., & Naresh, K. G. (1999). Involvement of a phosphate starvation inducible glucose dehydrogenase in soil phosphate solubilization by Enterobacter asburiae. *FEMS Microbiology Letters, 171*, 223–229.

Halder, A. K., & Chakrabarty, P. K. (1993). Solubilization of inorganic phosphate by Rhizobium. *Folia Microbiologica, 38*, 325–330.

Ho, W. C., & Ko, W. H. (1985). Effect of environmental edaphic factors. *Soil Biology and Biochemistry, 17*, 167–170.

Holben, W. E., Noto, K., Sumino, T., & Suwa, Y. (1998). Molecular analysis of bacterial communities in a three compartment granular activated sludge system indicates community-level control by incompatible nitrification process. *Applied and Environmental Microbiology, 64*, 2528–2532.

Hoon, H., Park, R. D., Kim, Y. W., Rim, Y. S., Park, K. H., Kim, T. H., Such, J. S., & Kim, K. Y. (2003). 2-ketogluconic acid production and phosphate solubilization by Enterobacter intermedium. *Current Microbiology, 47*, 87–92.

Hugenholtz, P., Goebel, B. M., & Pace, N. R. (1998). Impact of culture independent studies on the emerging polygenetic view of bacterial diversity. *Journal of Bacteriology, 180*, 4765–4774.

Illmer, P. A., Barbato, A., & Schinner, F. (1995). Solubilization of hardly soluble $AlPO_4$ with P-solubilizing microorganisms. *Soil Biology and Biochemistry, 27*, 260–270.

Jeffries, P. (1987). Use of mycorrhizae in agriculture. *CRC Critical Reviews in Biotechnology, 5*, 319–357.

Johri, J. K., Surange, S., & Nautiyal, C. S. (1999). Occurrence of salt, pH and temperature tolerant phosphate solubilizing bacteria in alkaline soils. *Current Microbiology, 39*, 89–93.

Khan, M. S., Aamil, M., & Zaidi, A. (1997). Associative effect of Bradyrhizobium sp (vigna) and phosphate solubilizing bacteria on moongbean (Vigna radiata L.) wilczek. *Biojournal, 10*, 101–106.

Khan, M. S., Aamil, M., & Zaidi, A. (1998). Moogbean response to inoculation with nitrogen fixing and phosphate solubilizaing bacteria. In A. M. Deshmukh (Ed.), *Biofertilizers and biopesticides* (pp. 40–48). Jaipur: Technoscience Publications.

Kim, K. Y., Jordan, D., & McDonald, G. A. (1997). Solubilization of hydroxyapatite by Enterobacter agglomerans and cloned Escherichia coli in culture medium. *Biology and Fertility of Soils, 24*, 347–352.

Kim, K. Y., Jordan, D., & McDonald, G. A. (1998). Enterobacter aglomerans phosphate solubilizing bacteria and microbial activity in soil: Effect of carbon sources. *Soil Biology and Biochemistry, 30*, 995–1003.

Kucey, R. M. N. (1983). Phosphate solubilizing bacteria and fungi in various cultivated and virgin Alberta soils. *Canadian Journal of Soil Science, 63*, 671–678.

Kucey, R. M. N. (1987). Increased P uptake by wheat and field beans inoculated with a phosphorus solubilizing Penicillium bilaji strain and with vesicular arbuscular mycorrhizal fungi. *Applied and Environmental Microbiology, 53*, 2699–2703.

Kucey, R. M. N. (1988). Effect of Penicillium bilaji on the solubility and uptake of P and micronutrients from soil by wheat. *Canadian Journal of Soil Science, 68*, 261–270.

Kucey, R. M. N., Janzen, H. H., & Legget, M. E. (1989). Microbial mediated increases in plant available phosphorus. *Advances in Agronomy, 42*, 199–228.

Kumar, V., Behl, R. K., & Narula, N. (2001). Establishment of phosphate solubilizing strains of Azotbacter chrococcum in the rhizosphere and their effect on wheat cultivars under greenhouse conditions. *Microbiological Research, 156*, 87–93.

Lynch, J. M. (1983). *Soil Biotechnology: Microbiological factors in crop productivity*. Oxford: Blackwell.

Maliha, R., Samina, K., Najma, A., Sadia, A., & Farooq, L. (2004). Organic acids production and phosphate solubilization by phosphate solubilizing microorganisms under in vitro conditions. *Pakistan Journal of Biological Sciences, 7*, 187–196.

Martinez-Murcia, A. J., Acinas, S. G., & Rodriguez-Valera, F. (1995). Evaluation of prokaryotic diversity by restrictase digestion of 16S rDNA directly amplified from hipersaline environments. *FEMS Microbiology Ecology, 17*, 247–256.

Mehana, T. A., & Wahid, O. A. A. (2002). Associative effect of phosphate dissolving fungi, Rhizobium and phosphate fertilizer on some soil properties, yield components and the phosphorus and nitrogen concentration and uptake by Vicia fabia L., under field conditions. *Pakistan Journal of Biological Sciences, 5*, 1226–1231.

Motsara, M. R., Bhattacharya, P. B., & Srivastava, B. (1995). Biofertilizers, their description and characteristics. In *Biofertilizer technology, marketing and usage. A sourcebook – Cum Glossary* (pp. 9–18). New Delhi: Fertilizer Development and Consultation Organization.

Mukherjee, P. K., & Rai, R. K. (2000). Effect of vesicular arbuscular mycorrhizae and phosphate solubilizing bacteria on growth, yield and phosphorus uptake by wheat (Triticum aestivum) and chickpea (Cicer arietinum). *Indian Journal of Agronomy, 45*, 602–607.

Nahas, E. (1996). Factors determining rock phosphate solubilization by microorganism isolated from soil. *World Journal of Microbiology and Biotechnology, 12*, 18–23.

Nair, K. P. P. (2013). The buffer power concept and its relevance in African and Asian soils. *Advances in Agronomy, 121*, 447–529.

Natarajan, T., & Subrammanian, P. (1995). Response of phospho-bacteria along with Rhizobium at two levels of phosphorus on groundnut. In *Microbiology abstracts XXXVI annual conference of the association of microbiologists of India* (p. 111), Hissar, Haryana State, India November 8–10

Omar, S. A. (1998). The role of rock phosphate solubilizing fungi and vesicular arbuscular mycorrhiza (VAM) in growth of wheat plants fertilized with rock phosphate. *World Journal of Microbiology and Biotechnology, 14*, 211–219.

Ozanne, P. G. (1980). Phosphate nutrition of plants – General treatise. In F. E. Khasawneh, E. C. Sample, & E. J. Kamprath (Eds.), *The role of phosphorus in agriculture* (pp. 559–589). Madison: American Society of Agronomy, Crop Science Society of America.

Parks, E. J., Olson, G. J., Brinckman, F. E., & Baldi, F. (1990). Characterization by high performance liquid chromatography (HPLC) of the solubilization of phosphorus in iron ore by a fungus. *Journal of Industrial Microbiology & Biotechnology, 5*, 183–189.

Piccini, A., & Azcon, R. (1987). Effect of phosphate solubilizing bacteria and vesicular arbuuscular mycorrhizal fungi on the utilization of Bayovar rock phosphate by alfalfa plants using a sand vermiculite medium. *Plant and Soil, 101*, 45–50.

Pikovskaya, R. I. (1948). Mobilization of phosphorus in soil in connection with vital activity of some microbial species. *Microbiology, 17*, 362–370.

Poi, S. C., Ghosh, G., & Kabi, M. C. (1989). Response of chickpea (Cicer arietinum L.) to combined inoculation with Rhizobium, phosphobacteria and mycorrhizal organisms. *Zentral fur Microbiology, 114*, 249–253.

Ponmurugan, P., & Gopi, C. (2006). In vitro production of growth regulators and phosphatase activity by phosphate solubilizing bacteria. *African Journal of Biotechnology, 5*, 348–350.

Prabhakaran, J., Ravi, K. B., & Srinivasan, K. (1996). Response of Vamban-I Blackgram to N2 (2 subscript please) fixer and P mobilizers in acid soil. In *Microbiology abstracts XXXVII. Annual conference of the microbiologists of India* (p. 120), IIT, Chennai, Tamil Nadu, India, December 4–6.

Reyes, I., Bernier, L., Simard, R. R., & Antoun, H. (1999). Effect of nitrogen source on the solubilization of different inorganic phosphates by an isolate of Penicillium rugulosum and two UV induced mutants. *FEMS Microbiology Ecology, 28*, 281–290.

Roos, W., & Luckner, M. (1984). Relationships between proton extrusion and fluxes of ammonium ions and organic acid in Penicillium cyclopium. *Journal of General Microbiology, 130*, 1007–1014.

Saber, M. S. M., & Kabesh, M. O. (1990). Utilization of biofertilizers in field crop production II. A comparison study on the effect of biofertilization or sulphur application on yield and nutrient uptake by lentil plants. *Egyptian Journal of Soil Science, 30*, 415–422.

Sarojini, V., Verma, S., & Mathur, M. S. (1989). Biocoenotic association between nitrogen fixing and phosphate solubilizing microorganisms. *Current Science, 59*, 1099–1100.

Satizabal, E. J. H., & Saif, U. S. R. (1987). Interaction between vesicular arbuscular mycorrhiza and leguminous Rhizobium in an oxisol of the eastern plains of Colombia. *Acta Agronomica Journal, 7–21*.

Singal, R., Gupta, R., & Saxena, R. K. (1994). Rock phosphate solubilization under alkaline conditions by Aspergillus japonicus and A. foetidus. *Folia, 39*, 33–36.

Singh, H. P. (1990). Response of dual inoculation with Bradyrhizobium and VAM mycorrhiza or phosphate solubilizer on soybean in mollisol. In B. L. Jalali & H. Chand (Eds.), *Trends in mycorrhiza* (Research Proceedings of the National Conference on Mycorrhiza) (pp. 14–16). Hissar: Haryana Agricultural University.

Subba Rao, N. S. (1982). In N. S. S. Rao (Ed.), *Advances in agricultural microbiology* (pp. 229–305). New Delhi: Oxford and IBH Publ. Co.

Tinker, P. B. (1980). In F. E. Khasawneh, E. C. Sample, & E. J. Kamprath (Eds.), *The role of phosphorus in agriculture*. Madison: Soil Science Society of America.

Van Elsas, J. D., Van Overbeek, L. S., & Fouchier, R. (1991). A specific marker pat for studying the fate of introduced bacteria and their DNA in soil using a combination of detection techniques. *Plant and Soil, 138*, 49–60.

Van Veen, J. A., Leonard, S., Van Overbeek, L. S., & Van Ellsas, J. D. (1997). Fate and activity of microorganisms introduced into soil. *Microbiology and Molecular Biology Reviews, 61*, 121–135.

Vassilev, N., Fenice, M., & Federici, F. (1996). Rock phosphate solubilization with gluconic acid produced by immobilized Penicillium variable P16. *Biotechnology Techniques, 20*, 585–588.

Vazquez, P., Holguin, G., Puente, M., ELopez Cortes, A., & Bashan, Y. (2000). Phosphate solubilizing microorganisms associated with the rhizosphere of mangroves in a semi arid coastal lagoon. *Biology and Fertility of Soils, 30*, 460–468.

Venkateswarlu, B., Rao, A. V., Raina, P., & Ahmad, N. (1984). Evaluation of phosphorus solubilization by microorganisms isolated from arid soil. *Journal of the Indian Society of Soil Science, 32*, 273–277.

Whitelaw, M. A. (2000). Growth promotion of plants inoculated with phosphate solubilizing fungi. *Advances in Agronomy, 69*, 99–151.

Whitelaw, M. A., Harden, T. J., & Helyar, K. R. (1999). Phosphate solubilization in solution culture by the soil fungus Penicillium radicum. *Soil Biology and Biochemistry, 32*, 655–665.

Zaidi, A., & Khan, M. S. (2005). Interactive effect of rhizospheric microorganisms on growth, yield and nutrient uptake of wheat. *Journal of Plant Nutrition, 28*, 2079–2092.

Zaidi, A., Khan, M. S., & Aamil, M. (2004). Bio-associative effect of rhizospheric microorganisms on growth, yield and nutrient uptake of greengram. *Journal of Plant Nutrition, 27*, 599–610.

Chapter 23
Energy Management in Sustainable Agriculture

Abstract Energy is directly or indirectly embodied in all types of resources, physical, chemical, biological, natural or artificial. All processes, natural and human-induced, result in the transfer and redistribution of direct and embodied energy. Energy is not a sectorial issue. It undergirds all developmental issues and thus has critical linkages to food production, productivity, food security, population, poverty, environment and health, economy, foreign exchange and trade impacts, environmental degradation, including climate change, national security, gender disparity, and, access to social services. The chapter discusses, at length, the various aspects related to the above-mentioned issues with specific reference to sustainable agriculture.

Keywords Energy · Resources · Quality and content · Energy use · Benefits · Costs · Impacts · Low energy agriculture · Energy budget · Energy self reliance · Energy equivalents · Sustainable energy management · High energy system · Low energy problems · Food-Energy commonalities · Energy user · Energy supplier · Biomass energy · Agricultural sector · Livelihoods · Bioenergy · Renewable energy

Energy is directly or indirectly embodied in all types of resources, physical, chemical, biological, natural, or artificial. All processes, natural and human induced, result in the transfer and redistribution of direct and embodied energy. Energy is not a sectoral issue. It undergirds all development issues and thus has critical linkages to food production, productivity and food security, population, poverty, environment and health, economy, foreign exchange and trade impacts, environmental degradation including climate change, national security, gender disparity, and access to social services. The type, quality, and quantity of energy and the pattern and efficiency of its use are critical for the transition to sustainable development in terms of balanced and harmonious integration of ecological, economic, and social issues. This automatically, necessitates sustainable strategies for the production, utilization, and distribution of energy.

The past five decades have seen acceleration in economic growth unprecedented in the history of civilization, with the global economy having expanded seven-fold during this period (Brown 2001). This connotes a corresponding acceleration in energy consumption, because economic progress corresponds to energy consump-

K. P. Nair, *Intelligent Soil Management for Sustainable Agriculture*,
https://doi.org/10.1007/978-3-030-15530-8_23

tion. That such a steep escalation is ecologically unsustainable is clear from several environmental indicators, the most serious being global warming and the human-induced climate change brought about by the profligate buring of fossil fuels for massive industrial, infrastructure, transport, agriculture, and leisure activities.

23.1 Energy and Agriculture

Agriculture has evolved as the largest and most important human enterprise, comprising not only production, but also, processing, packaging, transport, trade and distribution of food products. Energy is needed in all stages of agriculture, from land preparation, water lifting, transport, and processing. Agriculture is an energy-conversion process which transforms solar energy into food through photosynthesis. Almost 99% of the energy for primary plant growth and productivity is derived from the sun. Sustainable management of the natural resources of land, water, air, and bio diversity is the *mantra* for sustainable agriculture. The ecological services provided by nature regulate and sustain the stability of production in natural ecosystems. However, the enhanced food productivity required to meet the demand of the current and future population growth leaves no other option, except to complement ecosystem services with external energy sources of different forms and in different measures (Saroja Raman 2006).

Modern commercial production systems have to be subsidized with large amounts of different forms of energy from outside to boost their productivity because natural means and modes of production cannot meet the current and future demand for food. Direct energy, as well as energy embodied in fertilizers and pest control chemicals, has been pivotal in coaxing increasing outputs from land and water resources. It is impossible to conceive of an effective food production system, processing, and distribution without adequate energy inputs. The quantity and quality of energy used are reflected in the quality and quantity of food produced, transformed, and consumed, all of which is vital to food security. The three-fold increase in global agricultural productivity during the last half a century has been possible because of a parallel and exponential increase in the use of chemical fertilizers, irrigation, and pest control measures. This is also the reason that the term "green revolution" itself is an euphemism (Nair 2013). It is more appropriate to call it an "industrial type agriculture," rather than green revolution, with all the concomitant adverse fallout on the environment (Nair 2013; 2014). Crop production in this phase increased from 1.1 to 2.8 billion tons globally between 1950 and 2000 (FAO 2000a).

23.1.1 Concern about the Quality and Content of Energy

The importance of energy in agriculture for global food security is incontestable. However, some deep concern surrounds the amount and form of energy inputs under different conditions. Global agriculture has reached a stock-taking stage. To what extent have energy interventions succeeded in producing their intended benefits? What have been the ecological, environmental, and social trade-offs? How far and how long can they be sustained? What improvements are possible in current practices. It is in this context that "The Nutrient Buffer Power Concept" assumes great practical importance.

Equally, if not more, important is the problem of the unacceptable asymmetry of energy availability and affordability in agriculture in different regions and segments of society. As agriculture and rural development are inseparably linked to one another in the low-income developing countries which depend predominantly on agriculture, suboptimal energy use in agriculture endangers livelihood opportunities and human development. This issue needs to figure in the debate on energy management in agriculture.

23.1.2 Pattern in Energy Use

Agricultural production by itself accounts for only a relatively small proportion of the total energy used in the industrial and developing countries. In the Organization for Economic Cooperation and Development (OECD) countries, the proportion is 3–5% of the total energy used, and in the developing countries, it is 4–8%. In contrast, in the developed countries, agricultural production and distribution, processing, transport, and trade make up a sizeable chunk of the economy and consume 30–35% of the total energy consumed.

Energy use in global agriculture follows two extremes as follows:

1. The high external input and highly mechanized factory model of agriculture of the advanced countries is 50 times as energy intensive as compared to traditional agriculture
2. The very low or nil external input agriculture which is a part of the subsistence agriculture prevalent in many of the stressed ecosystems of the tropics depends mostly on solar energy driven natural processes and on human and animal draft power Between these two highly contrasting energy use patterns, depicted above, fall the green revolution technologies especially of the developing countries, which use moderate fertilizer and pest control inputs, though the high end farmers use unbridled quantities of fertilizers and pesticides, and, farm mechanization is of low intensity, except on large farms.

23.1.3 Energy Intensification in Agriculture

23.1.3.1 Benefits

Energizing the food production chain has been an essential feature of agricultural development and a prime factor to achieve food security. A direct relationship exists between energy consumption and agricultural yield. Traditional agricultural systems depend largely on the metabolic (endosomatic) energy of humans and animals and solar energy, whereas the energy requirements of modern agriculture are almost completely met from fossil fuels, mostly petroleum and to some extent diesel (not bio diesel, though). Energy transformed outside the body (exosomatic, for example, burning gasoline in a tractor) results in much higher productivity per unit of labor. Thus, a gallon of gasoline can be transformed into the equivalent of 3 weeks of human labor (Giampietro and Pimental 1994). The efficiency of energy conversion in humans is only 25%, and in a fit person the body can use up energy up to 250 to 300 W. The capacity is approximately 500–550 W for horses and buffaloes and 25 kW for low-power tractors, which can operate rotary hoes, pumps, and threshing machines (Hicks 1997). The ratio of exosomatic and endosomatic energy is 4:1 in traditional agriculture, whereas it is almost 40:1 for modern agriculture. This results in completely lower land and labor productivity in traditional agriculture and, consequently, in lower socioeconomic standards in terms of income, health, education, services and so on. Low labor and land output was not a serious concern when population, and hence pressure on land, was small and human aspirations and needs were limited.

From early twentieth century, global population grew at an accelerating pace, which made output from low-energy input traditional agriculture inadequate to meet the growing need for food. Energy intensification of agriculture started actively in the post II World War period as a consequence of impact of industrialization and accelerated economic growth. Heavy injections of fossil energy, in the form of chemicals and machine power, in combination with high yielding crop varieties, increased the productivity of land and labor enormously in comparison to primitive agriculture. Subject to other conditions, such as soil, climate and plant variety being generally conducive, a direct relationship unquestionably exists between productivity and energy, especially from fertilizers and irrigation.

Increased productivity of labor and land through the direct and indirect contribution of fossil energy made agriculture competitive and remunerative. This resulted in increasing income and standards of living for the farmer and in more labor becoming available for other economic sectors, especially in advanced countries. But for the agricultural revolution, of post World War II phase, in the developed countries, and its spread to the developing countries, of Asia and South America, in the form of the green revolution, in the 1960–1970s, the current level of global food security would not have been attained, though the developing world, especially countries like India, paid a huge environmental price (Nair 2013; 2014).

23.1.3.2 Costs

Large scale energy infusion into modern agriculture for land preparation and irrigation and very high fertilizer inputs, often without adequate thought to the efficiency of their use, have created a trade-off between productivity on the one hand and environmental quality, resource conservation, and economic viability on the other. Excessive use of external inputs has resulted in multiple adverse impacts, including enormous degradation of soil in the form of erosion, loss of organic matter and fertility, and salinization, environmental pollution from unbridled use of chemical fertilizers and pesticides, entering surface and ground water to become a threat to human and animal health, and, damage to and extinction of aquatic species as a result of the presence of toxic chemicals in runoff. The Punjab State in India, known as the "cradle" of green revolution epitomizes all these problems, as clearly pointed out in the earlier sections of this book. The "commodity mindset" of the so-called green revolution is at the root of all these ills (Nair 2013; 2014). It is no more agriculture, but, "agri business." Excessive use of external inputs has also raised the cost of production, resulting in burgeoning subsidies becoming unavoidable to make agriculture commercially viable for farmers and to keep consumer prices low in the developed countries and in the developing countries like India, which took to green revolution with single minded loyalty, albeit due to political pressure from vested interests.

The long term productivity of agro ecosystems, depends on the conservation of the natural resources of soil, water, and biological diversity, and their quality. Intensive farming techniques impact the pattern of energy flows in ecosystems. In general, they reduce the capacity of the ecosystem to use solar energy for evapotranspiration, gross primary production, recycling of nutrients, and natural pest control, leading to a decrease in the efficiency of the use of inputs. Paradoxically, in their enthusiasm and over confidence, humans have often sought to make nature redundant in agricultural production. A very high degree of substitution of the natural processes using synthetic inputs leads to progressive degeneration of the soil and loss of resilience (Matson et al. 1997). This renders the soil a biologically dead physico-chemical medium incapable of sustaining productivity in the long run, as is happening in many intensively cultivated systems, such as rice-wheat systems of South Asia and South-East Asia, Punjab State in India, being a classic example in the former category of nations (Ladha et al. 2003). Among the most adverse impacts of high energy agricultural systems, is their high direct and indirect share of contribution to global warming, due to emission of gaseous emission, the nitrous oxide being the most important. Along with deforestation, they account for about 30% of greenhouse gas emissions, mostly through the burning of fossil fuel manufactured fertilizers and pesticides, and also through inefficient use of nitrogenous fertilizers, which enhances nitrous oxide concentrations in the atmosphere. Thus, energy use in high external input agriculture has become ecologically, economically, and environmentally, unsustainable.

In the developing countries, high external input green revolution technologies have also led to social tension among poor and marginal farmers who lack the means to invest in inputs or who achieve inadequate returns on investment. Farmers who lack sufficient knowledge have become highly indebted, sometimes forfeiting their assets and even being driven to the extreme of suicide in extreme cases. The plight of the cotton farmers in the Vidharba district of Maharashtra State in India, many of whom were led to suicide due to failure of the Bt cotton leading them to indebtedness, and also in Andhra State, is well documented. Crop failure caused by pesticide-resistant pests drove many of the cotton farmers to commit suicide. The most recent example, at the time of completing the writing of this book, is the utter failure of Bt cotton in Punjab State, which took to it whole heartedly in 2003, but, a decade later succumbed to severe white fly infestation. So many farmers have committed suicide and it has become a national disaster and an Indian shame. Such tragedies persist even in many other parts of India, like Rajasthan State and Haryana State. The Bt cotton, originally touted as a safety net against the dreaded American boll worm is now fast succumbing to the white fly and thousands of farmers in Punjab and Rajasthan who have lost their crop and out of pecuniary difficulties have committed suicide.

23.1.3.3 Impact of Low Energy Use in Agriculture

Inadequate external energy infusion, as in the low input traditional agricultural systems of the developing countries, results in low land and labor productivity and lack of food security. Agricultural practices based on animal and human power with inadequate use of fertilizers and pesticides and lack of storage, preservation, and transportation facilities cannot meet the needs of the burgeoning population. The most glaring example of this situation is in Sub-Saharan Africa, where mechanization and per capita fertilizer use are lowest globally (WRI 1994). As a result, agricultural production grew at a rate of just 1.8% per year against population growth of 2.7% in the 1990s (FAO 1995), whereas in all other countries it grew ahead of the population growth, though in India, at the time of writing this book, population growth has outstripped food grain production with a myriad of problems associated with the green revolution. The Sub-Saharan countries will not be able to attain the desired food security standards even by 2030 (FAO 2000a). Globally, inefficient production systems and a lack of means to buy food left 800 million people, approximately 18% of the population of the developing countries, hungry and undernourished. Even as profligate overuse of energy is turning out to be a problem in the developed countries, some 1.5–2 billion rural poor in the developing countries, mostly engaged in subsistence agriculture, do not have access to modern energy sources. They have to depend on traditional fuels like firewood or agricultural residues for cooking and heating, and on human or animal power for primitive agricultural operations. Adequate and appropriate energy for income generating opportunities is simply not available. Primitive energy sources are used inefficiently and without concern for health hazards, yet, have heavy costs in terms of time, money, drudgery, and poverty. Many countries in Africa continue to be among the lowest per capita energy consumers in the world, resulting in continued low agricultural productivity and impaired economic growth and environmental degradation.

Lack of energy input in agriculture has also led to enormous land degradation in the form of erosion and depletion of soil organic matter and soil fertility, which has further lowered productivity and increased environmental pollution. This situation can be improved only by upgrading agricultural operations and daily lives with the infusion of energy of appropriate quality and quantity to improve agriculture productivity and provide livelihood opportunities to counter poverty.

23.1.3.4 Advantages and Disadvantages of Low-Energy Agriculture

Low external energy input agriculture is, without doubt, environmental friendly. Natural ecological processes are certainly superior to external inputs in terms of crop production because they are regenerating, non polluting, practically free, and resource conserving. However, they are not invariably beneficial at the societal level. In the low input primitive systems of production, the ratio of exosomatic to endosomatic energy is low. When a society has a very low exosomatic/endosomatic energy ratio and uses labor-intensive techniques to save capital and fossil energy, its per capita labor and land output remains low, resulting in a lower standard of living, health, and education (Giampietro and Pimental 1994). Agricultural output is already insufficient to meet human needs for food and fuel in many parts of the world. Depending on ecological processes alone for our current needs is grossly inadequate.

Ecological processes cannot substitute completely external inputs in modern crop production because they have their own pattern and pace of mass and energy flow and distribution. Ecological processes are mediated by a diverse array of soil microorganisms, singly and in consortium, subject to the availability of appropriate substrates and other congenial biophysical conditions, especially rich organic matter, for growth and proliferation. Although excellent sources of plant nutrients are present in organic form, such as crop residues and animal and poultry excreta, their utilization may be constrained by their rate of mineralization and lack of synchrony with plant growth needs. Thus, although these nutrients may be comparable and even superior to mineral fertilizers in terms of their nonpolluting, resource-conservation characteristics, and even economic efficiency and easy availability, their substitution for mineral fertilizers, wholly or in part is constrained. They cannot support the much rapid and more rigorous growth of the high-yielding fertilizer responsive cultivars of modern agriculture (Woomer and Swift 1994). This has been both a boon and a bane, at the same time, because, it is this very property which prompted many farmers in South Asia, especially in countries like India, to go for the "get rich quick" route of unbridled use of chemical fertilizers to harvest more crop yield from the same patch of land, which led to a hazardous environmental fallout (Nair 2013, Nair 2014). The consequences are all there for one to see in Punjab State, in India, the "cradle" of India's green revolution. It is against the background of these very adverse consequences, wherein "The Nutrient Buffer Power Concept," while emphasizing the usefulness of chemical fertilizers, aims at moderation, enhancing efficiency of nutrient use, while at the same time containing environmental pollution, and, as a bonus, enhance economy of scale.

Both high external input and low energy primitive agriculture systems are unsustainable in the long run. The former route is flawed in its over dependence on fossil fuel, as an energy source, and on account of ecological, environmental, and societal stress which they create, and the latter because of their inadequate land and labor productivity to sustain food security and livelihood adequately.

23.2 Sustainable Energy Management in High Input Agriculture

Following the principles of thermodynamics, a key indicator of sustainability is the ratio of energy equivalents of all of the inputs and outputs (Andiscott 1995). The higher this ratio, the more sustainable the system, as depicted in the following figure (Fig. 23.1)

In terms of this irrefutable criterion, modern agricultural production systems, which have an input/output energy ratio approximately of an order of magnitude lower than that of traditional systems, are less sustainable. The most basic way to improve sustainability of modern agriculture is to decrease both the direct and embodied external energy component by increasing their efficiency of use, the primary aim of "The Nutrient Buffer Power Concept," in fertilizer use, and, in addition all other agricultural practices (Nair 2004). Maintaining high productivity with an acceptable impact on resource, environment, and economics is the key to sustainability. The use of non polluting alternative sources of energy is also conducive to improving sustainability.

A multipronged approach to achieving better energy efficiency along with resource conservation, environmental safety, and economic prudence has been adopted in the past decade through technologies which are holistic and make integrated use of natural ecological processes and external inputs. Conservation tillage, integrated pest management, integrated plant nutrient supply, crop rotation, micro irrigation techniques, and precision farming belong to this category. Conservation tillage practices were initially introduced to reduce the costs of mechanization, but, have turned out to offer several ecological and economic benefits, such as low soil erosion, reduced run off, more natural pest control, greater water storage and infiltration, increased cropping intensity, and economy of scale in input use. Conservation tillage has become very popular where farms are large.

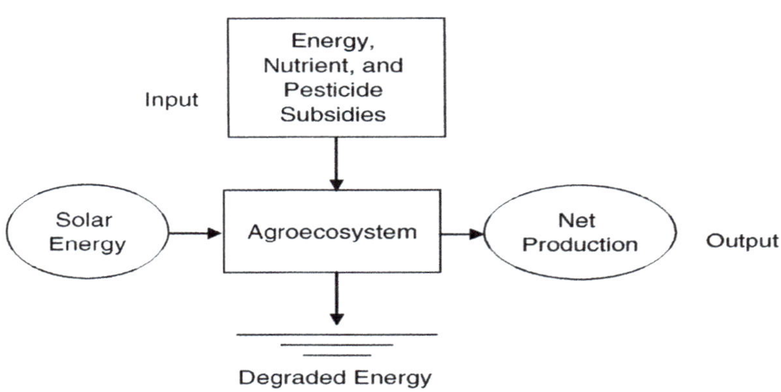

Fig. 23.1 Energy budget in crop production: sustainability = high output to input ratio

Similarly, managing water use more efficiently, using drip and sprinkler irrigation, and using mulch to improve water storage and reduce evapotranspiration, are simple ways to improve productivity of water, which is likely to be the most limiting resource for agriculture in the future. These practices not only result in energy conservation, but also attract other bonuses such as lower nutrient loss, higher efficiency of input use, and freedom from water logging and salinization. Integrated pest management combines several virtues, including cost reduction, environmental and human safety, prevention of pesticide resistance and pest resurgence, with a favorable energy balance. Biodiversity enhancement through polyculture, as opposed to the monoculture system (rice-wheat continuous cropping during the green revolution phase in Punjab State, India), crop rotation, mixed cropping, and strip cropping has been shown to have a variety of positive effects, such as conservation, efficiency of use of nutrients, and water, erosion control, biological pest and disease control, and improvement in soil quality. Integrated technologies are the best path to balance all factors related to sustainability, stability, and productivity of agroecosystems (Altieri 1995). Recently, several investigations have been conducted on energy audit of the above - mentioned technologies, and all have validated the enormous reduction in energy needs, cost of cultivation, environmental pollution, and expenditure for a "clean up," and their favorable input/output ratio with regard to energy use.

23.3 Energy Audit of High Energy Systems

The amount of energy used in modern agricultural production was reviewed in the 1970s on account of the hike in gasoline prices. According to Altieri (1995), the output/input energy equivalent ratio has actually declined by as much as ten-fold in high energy input commercial/industrial type agriculture, as exemplified by the so-called green revolution. Early assessments prompted the fear that diminishing returns to energy input had set in for intensively cultivated corn in the USA (Pimentel et al. 1973). However, subsequent calculations by several researchers have alleviated this fear for corn (Smil et al. 1983), rice (Flinn and Duff 1985), or wheat grown under widely different conditions (Bonny, 1993). According to Evans (1998), agriculture with increasing input energy has a definite yield advantage. He pointed out that the calculation of relative energy consumption in modern agriculture and traditional agriculture is flawed because the energy required for maintaining animals and the farm workers is not accounted for adequately. Recent calculations have corroborated this (Pimentel et al. 2003). The ratio of energy output to energy input is only marginally different in the developed and developing countries: 1:5 for the former and 1:4 for the latter. Many examples may be given of diminishing returns to any single factor (for example nitrogenous fertilizer) in modern agriculture, but, the conversion of all inputs and outputs to a common energy base has unambiguously shown that yields per hectare can be increased without diminishing returns to non solar energy consumption. Thus, more than the production process, it is the processing and distribution sectors which are actually large energy guzzlers, and not always with correspondingly good/positive results. "The Nutrient Buffer Power Concept" has a refreshingly new outlook on this aspect, certainly, inasmuch as fertilizer use is concerned, and, fertilizer, for sure, is a big energy guzzling product.

23.3.1 Sustainable Energy Management in Low – Input Systems

Low-input agricultural systems are the only viable option in low potential areas, which are mostly the complex, diverse, and risk prone regions of the tropics, home to the majority of the world's poor. They suffer chronic and multiple constraints of poverty, hunger, and poverty of energy, which are interconnected on account of low productivity, low quality of life, and lack of livelihood opportunities. Inadequate access to energy has meant chronically poor quality of life measures, such as hunger, poverty, drudgery, malnutrition, and disease. In addition to improving agricultural production and productivity, upgrading quality of life is a pre-requisite of functional food security and other aspects of human development. Very low/nil external input agriculture practiced on mostly low fertility and fragile lands results in not only low productivity, but also in severe land degradation through erosion and fertility mining. Degraded resources cannot provide regenerating ecosystem services such as nutrient cycling owing to low resilience and impaired soil diversity. Primitive low/nil external input agricultural systems need to be energized with at least basic additions of chemical fertilizers to replenish what has already been mined, along with adequate moisture (Buresh et al. 1997). The injection of an appropriate amount of energy inputs, direct and indirect, into primitive agriculture can lead to increased yields and production and to other cascading benefits such as improved incomes, employment opportunities, and agroindustrial growth (Hazell 1998).

23.3.2 Responding to Low Energy Problems

Access to appropriate energy can enhance productivity and food security, create livelihood opportunities, and reduce poverty (WEC/ FAO 1999). Hopes to improve the lives of the rural peasants and landless poor of the developing countries hinge largely on some level of energy infusion into their cultivation processes and into their daily lives. The realization of minimum standards of living is possible only through proper energy management (FAO 1996b). Fortunately, global forums have reacted rapidly and appropriately to come to terms with the deficiencies of the current model of energy management. The FAO's initiative in fostering appropriate energy strategies for the developing countries for agriculture and rural development dates back to the early 1980s. It has also played an active role in toning down excessive use of energy in modern systems. The Rio Earth Summit (UNCED 1992) highlighted the importance of efficiently managing fossil energy and making use of other forms of renewable energy such as solar, wind, geothermal, and biomass energy in sustainable development. The gist of the recommendations made with respect to agriculture is the following:

1. The process of transition to environmentally sound energy resources using structured and diversified alternative and renewable energy sources should be initiated in rural communities
2. Energy inputs should be increased to rural households and for agro industrial needs
3. Rural programs for energy self-reliance in the sustainable development of renewable energy sources should be implemented

Considerable attention is being focused on addressing energy accessibility, availability, and acceptability for sustainable socioeconomic development (UNDP/IEI/SEI 1997; WEC 2000). The World Summit on Sustainable Development (WSSD 2002) made energy the second most crucial issue after water.

23.4 Food-Energy Nexus and Integration

Energy is a driver for overall development. Localized, inexpensive, and easy-to-handle energy is needed in rural area, where the construction of sophisticated energy grids may not be possible. However, in contrast to other sectors, the relationship between energy and agriculture is more than just econometric. It has societal, political, and environmental implications, especially in rural areas. In the rural areas of the less industrialized countries, the scope to meet energy needs from the synergy between energy and food is enormous.

Food and energy have several commonalities. Their regular and continuing availability is a pre-requisite for human survival. Similar to food security, energy security at the household level is essential for a minimum quality of life, so much so that the entitlement approach to food advocated by Sen (1981) has been recommended for energy (FEN 1989). The need for both food and energy to attain a minimum quality of life calls for an integrated approach to the food-energy system.

23.5 Agriculture as Energy Provider

Agriculture can play the dual role of energy user and energy supplier. Biomass and by products of agriculture can be converted into energy to improve production and daily living in rural areas (Best 1997). It is, therefore, paradoxical that agricultural production and development should lag as a result of the lack of energy availability in rural areas. Agriculture can also deliver a wide range of non food goods. In addition to food security, it can also help achieve sustainability in rural development, energy, and the environmental integrity maintenance, at the local, national and global levels. Whereas the energy-consuming aspect of agriculture has been well researched, the equally important role of agriculture as an energy provider has not been adequately appreciated or acknowledged or fully exploited. This needs to be corrected, especially in the rural areas of the developing countries, where other conventional forms of energy are physically inaccessible and are not cost-effective. It has been estimated that 30–70% of the input costs of crop production in developing countries are directly or indirectly related to energy. Against this, agriculture provides 20–90% of primary energy through biomass, agricultural residues, and by products. Energy crops that can be grown on wastelands and used directly or indirectly as energy resources are also becoming popular. Worldwide, 300 million ha are estimated to be available for reforestation and agroforestry, and a large portion of this is situated in the developing countries (FAO 1999). Thus, food production systems can be integrated into energy producing systems, to form an ecological loop, where the byproducts of the food system are used as feedstock for the energy system, which can refuel the production system. Integrated food energy systems provide a synergistic solution to food and energy problems (FEN 1989). The potential for agriculture to energize the agriculture sector and rural areas has been receiving a new thrust in the past two decades in both the developing and the developed countries, and biomass is recognized as a potential source of energy.

23.6 Biomass Energy and Rural Livelihoods

Sustainable sources of energy can improve economic growth, alleviate poverty, and create income-generating livelihood opportunities, all of which are essential to achieve food security at the household level and for minimum human development needs. Bioenergy can act as a catalyst to rural development. The development of biomass energy offers significant opportunities for the creation of additional jobs. Globally, approximately 1 billion people are unemployed or underemployed and hence lead a life deprived of the basic essentials. Integrated food energy systems can operate at various scales, ranging from the household level to industry-sized operations, depending on needs and on the availability of by-products, technology, resources, and community awareness. The adoption of agricultural and industrial technologies which allow maximum utilization of by-products, diversification of raw materials, production on a small scale, recycling and economic utilization of residues, and harmonization with the food system can create immense opportunities for sustainable livelihoods in the rural areas, predominantly in countries which are primarily agricultural.

Biomass energy from agricultural by-products can be used in small-scale industries in rural areas, such as sugar, rubber, and coconut processing, rice par boiling, fish and meat drying, brick making, lime kilns, ceramics, and pottery. Indirect infusions of energy from biomass, such as straw for mushroom culture and cattle waste in pisciculture, recycling of biowastes such as banana leaves, and fodder banks are innovative means of livelihood creation, as seen in the Biovillage model in Pondicherry (Union territory, India) as detailed by MSSRF (2000).

Various other forms of biomass, including wood waste, from saw mills, vegetables, fruits, dried leaves, and cattle dung, municipal sewage, and human waste are useful as small-scale sources of energy in rural areas. Biomass has evolved as an important source of energy in South Asian countries, which have a warm and wet climate, and therefore, are endowed with plentiful biomass. In the ASEAN countries of Thailand, Indonesia, Malaysia, the Philippines and Vietnam, energy from biomass represents 40% of total energy consumption, mostly in the domestic sector and small-scale industries. Industrial activities such as tarring of road, soap making, candle making, tire retreading, paper making, and boat making are also possible using biomass. Similarly, in India, wood fuel is extensively used in rural homes and in industries such as brick kilns, metal works, and food processing (FAO 1997).

Mobilizing biomass as a source of energy has several advantages, also disadvantages. These have to be considered in evaluating the potential for the implementation of biomass energy programs in developing countries. Biomass is a stored fuel and can provide a demand-responsive supply. It is a highly flexible source of energy and can be burned or converted into other forms such as ethanol from the fermentation of sugarcane, corn, beet, etc., and methane-rich biogas through anaerobic digestion of organic wastes and agricultural residues. In China, India and Nepal, more than five million small-scale biogas digesters operating on small quantities of dung and vegetable and human wastes provide fuel for cooking and lighting. Energy crops and plantations are expected to gain in importance in the future.

Biomass is a carbon-neutral-energy source, renewable over a relatively short-time frame. All biomass utilization can be considered as part of a closed carbon cycle and should be encouraged. Fuel production, collection, and supply are labor-intensive, which are important employment factors in a rural set up. However, biomass is a diffuse source, and transport costs and the stability of supply may become constraints.

23.6.1 Biomass Energy in Commercial and Industrial Production

Biomass is stored energy that is not subject to seasonal vagaries; it can produce all forms of energy, such as heat, electricity, gas, and liquid fuel (both for motoring and cooking). These can be harnessed to create small and big industries in rural areas to provide employment and also to increase the profitability of the agricultural sector. This will also prevent the pernicious problem of migration to the cities. Popularizing the use of biomass energy will come from changing the present inefficient use. Modern combustion, gasification, and fermentation technologies to generate electricity, heat, and gaseous and liquid fuels are available for large-scale energy production from biomass, which can be used for commercial and industrial production. These technologies are cost-effective and can facilitate decentralized rural electrification and industrialization. Combined heat-power generation technologies have proved very effective in the sugarcane industry, where the byproduct bagasse is used as biomass fuel. Most recently an Australian Company "Zero Waste Ltd" has come forward with a unique technology where bagasse (by product of sugarcane crushing) is converted into an organic material, commercially marketed, now as "Geogreen" in India, with a collaboration of a company, Rallis India Ltd. (A Tata Group Company). This material has proved to be extremely successful and useful in Indian soils highly depleted of native carbon by the mindless onslaught of the so-called green revolution, where unbridled use of chemical fertilizers, especially nitrogenous like urea, applied to soil to boost cereal (rice and wheat) yields (as for example in Punjab State, the cradle of green revolution) led to severe carbon depletion in soil, leading to yield depression and degradation of soil. These technologies are cost-effective and can facilitate decentralized rural electrification and industrialization. Combined heat-power generation technologies have proved very effective in the sugarcane industry, where bagasse is used as the biomass fuel. Such technologies have the potential to produce electricity competitively in rural areas, using biomass feed stocks from agricultural waste as well as those grown on dedicated biomass energy plantations. This will attract industry to the rural areas, create jobs, prevent rural migration to cities, and thus improve economic infrastructure development (UNDP/IEI/SEI 1997, FAO 2000b). Sugarcane-derived ethanol becomes a cost effective replacement for petrol if electricity is cogenerated from cane residues.

23.6.2 Examples of Benefits to the Agricultural Sector and Livelihoods

The Spark program in China (He Kang 1994) and the Brazilian program on ethanol production from sugarcane (Dobereiner 1994) are two of the most outstanding and successful programs showcasing the impact of energy production in terms of boosting the benefits of the agricultural sector which creates vast rural livelihood opportunities. The Spark program is a massive rural enterprise development project in which rural energization centered on agribusiness and small-scale industry has been effectively used to create a large number of off-farm livelihood opportunities. This has resulted in surplus people moving away from the agricultural sector, without depeasantization, and has created economic growth, rural infrastructure development, and development based on science and technology. It has prevented large-scale rural migration to cities and supported agricultural modernization through capital infusion within.

In Brazil, alcohol production from sugarcane, sweet sorghum, and sugar beet is carried out on a medium and very large industrial scale. Sugarcane bagasse is used for electricity generation. Alcohol is used as a fuel for agricultural mechanization and for use in other machinery, including transport facilities. The Brazilian experience in producing ethanol from sugarcane has created 700, 000 jobs (UNDP/IEI/SEI 1997). The role of agriculture in providing both a source of renewable energy for rural economic development and a substitute for fossil fuels is an appropriate subject for future policy and technology development in developing countries. More training, research, and capacity building are required to deliver the energy function of agriculture in a cost-effective and market-oriented manner. Positive political and community cooperation and initiatives are needed towards this end.

23.6.3 Opportunities for Expanding Bioenergy Initiatives

The Kyoto Protocol (UNFCCC 1997) is designed to provide binding quantitative limits to future emissions of greenhouse gases (Grubb et al. 1999). It allows flexibility under the Joint Implementation and Clean Development Mechanism, which allows donor countries to fund projects which reduce emissions in a host country. The donor country receives a carbon credit which contributes to its own emission targets. The provision for carbon trading will encourage industrial countries to promote biomass conversion enterprises in developing countries. Funds under the Global Environment Facility (GEF) are available for such countries. These initiatives have been continually discussed against the background of global warming, due to excessive fossil fuel burning, as an energy source, and culminated in the Paris talks on climate control, just concluded during the finishing stage of the writing of this book.

23.6.4 Other Renewable Sources of Energy

The oil crisis of the 1970s started the debate on using renewable sources of energy, which are clean and potentially inexhaustible if managed properly. The UN Conference on Development and Utilization of New and Renewable Sources of Energy, Nairobi, 1981, recommended the extensive use of renewable sources of energy, especially in the developing countries. However, the transition to significant use of renewable sources of energy was handicapped by a lack of efficient and economical technologies to convert them to appropriate forms such as heat and electricity. The initial momentum was lost as prices stabilized when more sources of fossil fuel, coal, and natural gas were discovered.

The current revival in the use of renewable sources of energy has been triggered more by concern about unchecked carbon dioxide concentration in the atmosphere, global warming, and climate change. These have the potential to lead us into chaotic paths of climatic unpredictability, ecological changes, and uncertain effects on food production and health. Ever since the Intergovernmental Panel on Climate Change (IPCC 1996) confirmed that global warming is due to the increase in atmospheric carbon dioxide concentration, renewed efforts have been under way to replace fossil fuel by other clean or less polluting energy sources such as solar energy, hydrogen biomass, and natural gas. Solar energy, wind energy, biomass energy, and hydro and geothermal tidal energies are being evolved as potential forms of renewable energy. They are non polluting, non emission sources and are suitable for small to medium scale operations, especially in rural and remote areas. The growth of different energy types is depicted in the following table (Table 23.1).

Solar, wind, and bioenergy are in particular useful in rural areas because of their local availability, adaptability to dispersed small and medium scale energy requirements, reliability, and environmental safety. It has been estimated that these could provide up to 57% of commercial needs of Africa, 33% of Latin America, 22% in the Asia–Pacific region (Hicks 1997).

Direct solar energy is used in several areas in agriculture, for instance, drying of harvested grains, crop, and fish, for household cooking, water pumping, water heating, lighting, and refrigeration. Solar energy is the second fastest growing energy source. Photovoltaic systems have great proven potential as a decentralized form of energy in rural areas and have been integrated into large rural electrification programs in many parts of the world, including India, Argentina, South Africa, USA, Mexico and Zimbabwe (FAO 1996a, b). Millions of homes worldwide, approximately 70,000 in developing countries alone, are solar energy powered. Solar power projects of 130–300 MW capacity have been set up in the Rajasthan State in India. The cost of solar cells is decreasing; it has fallen from more than US $ 70 per watt of production capacity in the 1970s to less than US $ 35 per watt now and is expected to fall further as technology improves and capacity increases. Future advances such as photovoltaic roofing material which acts as a power plant promises further spread of solar energy. Wind energy is one of the fastest growing forms of alternative energy source in the USA and is growing in some developing countries as well. On a very small scale, it is taking root in India as well, especially in Central and Western

Table 23.1 Trends in energy use based on source

Energy source	Annual rate of growth (1990–2000 in percentage)	Growth in 2000 (in percentage)
Wind power	25	32
Solar cells	20	43
Geothermal power	4	–
Hydroelectric power	2	–
Natural gas	2	–
Oil	1	2
Coal	−1	−4.1
Nuclear power	0.8	<1

Source: Modified from Brown (2001)

India, in the arid and semi arid regions, which are given to fast blowing winds. Wind energy to generate electricity was launched in the early 1980s. The globally installed capacity rose from 3100 MWe in 1993–4800 MWe in 1995, which is an increase of 54.8%, which works out to more than 25% annually. Wind power generation capacity in India reached 650 MWe in 1996 (Brown 2001).

References

Altieri, M. A. (1995). *Agroecology: The Science of sustainable agriculture*. Boulder: Westview Press.
Andiscott, T. M. (1995). Entropy and sustainability. *European Journal of Soil Science, 46*, 161–168.
Best, G. (1997). *Food security from an energy standpoint. Occasional paper*. Rome: Food and Agriculture Organization.
Bonny, S. (1993). Is agriculture using more and more energy? A French case study. *Agricultural Systems, 43*, 51–66.
Brown, L. R. (2001). *Who will feed China? Wakeup call for a small planet*. New York: W. W. Norton.
Buresh, R. J., Sanchez, P. A., & Calhoun, F. (1997). *Replenishing soil fertility in Africa SSSA Special Publication 51*. Madison: Soil Science Society of America.
Dobereiner, J. (1994). The Brazilian biofuel programme solution to social and environmental problems. In M. S. Swaminathan (Ed.), *Ecotechnology and rural employment: A dialogue* (pp. 119–128). Chennai: Mcmillan India.
Evans, L. T. (1998). *Feeding the Ten Billion: Plants and population growth*. Cambridge: Cambridge University Press.
FAO. (1995). In N. Alexandratos (Ed.), *World Agriculture Towards 2010*. Chichester: Food and Agriculture Organizatyion/Wiley.
FAO. (1996a). *Control of water pollution from agriculture* (Irrigation and Drainage Paper 55). Rome: Food and Agriculture Organization.
FAO. (1996b). *Energy for sustainable development and food security in Africa*. Rome: Food and Agriculture Organization.
FAO. (1997). *Biomass energy in ASEAN member countries. FAO-SD Dimensions/Energy for Development*. Rome: Food and Agriculture Organization.
FAO. (1999). *The multiplicational character of agriculture and land: Bioenergy. FAO-SD Dimensions/Energy for Development*. Rome: Food and Agriculture Organization.
FAO. (2000a). *Agriculture Towards 2015/30 technical interim report, Global Perspectives Unit*. Rome: Food and Agriculture Organization.
FAO. (2000b). *The State of Food and Agriculture: Lessons from the past 50 years*. Rome: Food and Agriculture Organization.
FEN. (1989). *Strategies for sustainable development: Food-Energy Nexus programme*. New York: United Nations University.
Flinn, J. C., & Duff, B. (1985). *Energy analysis, rice production systems and rice research* (IRRI Research Paper Series 114). Manila: International Rice Research Institute.
Giampietro, M., & Pimental, D. (1994). *The tightening conflict: Population, energy and the ecology of agriculture*. N.P.G. Forum (Negative Population Growth), October 1993, 1–8
Grubb, M., Vrolijk, C., & Brack, D. (1999). *The Kyoto protocol: A guide and assessment*. London: Earthscan.
Hazell, P. (1998). *Why invest more in sustainable development of less-favored lands* (IFPRI Report 20) (Vol. 2). Washington, DC: International Food Policy Research Institute.
Hicks, A. (1997). *Power and food security. Paper presented at the Solar World Congress, August 24–30*. International Solar Energy Society (ISES), Tejon, Republic of Korea

IPCC. (1996). In J. T. Houghton, L. G. M. Filho, B. A. Gallander, N. Harris, A. Kattenberg, & K. Maskell (Eds.), *Climate Change 1995: The science of climate change. Inter-Governmental Panel on Climate Change*. Cambridge: Cambridge University Press.
Kang, H. (1994). The Spark programme and rural township enterprises of China. In M. S. Swaminathan (Ed.), *Ecotechnology and Rural Employment* (pp. 142–146). Chennai: India Macmillan India.
Ladha, J. K., Pathak, H., Padre, A. T., Dave, D., & Gupta, R. K. (2003). Productivity trends in intensive rice-wheat cropping systems in Asia. In J. K. Ladha, J. Hill, R. K. Gupta, J. Duxbury, & R. J. Buresh (Eds.), *Improving the productivity and sustainability of the Rice-Wheat systems: Issues and impact* (ASA Special Publication 65) (pp. 45–76). Madison: American Society of Agronomy.
Matson, P. A., Parton, W. J., Power, A. G., & Swift, M. J. (1997). Agriculture intensification and ecosystem properties. *Science, 277*, 504–509.
MSSRF. (2000). In K. V. Raman (Ed.), *Biovillage oriented development plan for the Union Territory of Pondicherry*. Chennai: M.S.Swaminathan Research Foundation.
Nair, K. P. P. (2004). Sustaining crop production in the developing world through "the nutrient buffer power concept" – A case study with Black pepper (Piper nigrum) growing at low pH in the laterite soils of India. In *Proceeding of 16th international symposium on plant-soil interactions at low pH* (p. 392), Sendai, Japan, August 1–5.
Nair, K. P. P. (2013). The buffer power concept and its relevance in African and Asian soils. *Advances in Agronomy, 121*, 447–529.
Nair, K. P. P. (2014, August 18). Green revolution or silent spring? *The Hindu Business Line*.
Pimentel, D., Hurd, E., Bellotti, A. C., Foster, M. J., Oke, I. N., Sholes, O. D., & And Whitman, R. J. (1973). Food production and the energy crisis. *Science, 182*, 442–449.
Pimentel, D. R., Doughty, R., Cowthers, C., Lamberson, S., Bora, N., & Lele, U. (2003). Energy impacts in crop production in developing and developed countries. In R. Lal, D. Hansen, N. Uphoff, & S. Slack (Eds.), *Food security and environmental quality in the developing world* (pp. 129–151). Boca Raton: Lewis Publishers.
Raman, S. (2006). *Agricultural sustainability: Principles, processes and prospects* (pp. 13904–11580). Birmingham: The Hawoth Press Inc.
Sen, A. (1981). *Poverty and famines: An essay on entitlement and deprivation*. Oxford: Clarendon Press.
Smil, V., Nachman, P., & Long, T. V. (1983). *Energy analysis in agriculture: An application to the US Corn production*. Boulder: Westview Press.
UNCED. (1992). Chapter 24: Global action for women towards sustainable and equitable development. In *Agenda* (Vol. 21, pp. 349–354). Rio de Janeiro: United Nations Conference on Environment and Development.
UNDP/IEI/SEI. (1997). *Energy after Rio: Prospects and challenges*. New York: United Nations Development Program in collaboration with the International Energy Institute and Stockholm Environmental Institute.
UNFCCC. (1997). *Kyoto protocol for and United Nations framework convention on climate change*. Kyoto, Japan.
WEC. (2000). *Energy for Tomorrow's world: Acting now*. London: World Energy Council.
Woomer, P. L., & Swift, M. J. (Eds.). (1994). *The Biological management of tropical soil fertility*. Chicester: John Wiley.
WRI. (1994). *World Resources 1994–95*. Washington, DC: World Resources Institute.
WSSD. (2002). *World summit for sustainable development. United Nations conference on environment and development*, August 26 – September 4, Johannesburg, South Africa

Chapter 24
Measurement of Agricultural Sustainability

Abstract Recent times have witnessed a keen urgency of sustaining agriculture on account of the pressures to improve crop productivity against the background of burgeoning global population increase, especially in developing countries like India. The aftershock of the green revolution, which, in fact, was nothing but a high-input industrial type agriculture, with huge environmental adverse impact, especially on the soil resources of South Asia, especially India, left a deep scar on the Indian psyche. The chapter highlights the importance of the measurement of agricultural sustainability with reference to the various factors, mentioned above.

Keywords Sustaining agriculture · Productivity · Biophysical indicator · Total factor productivity · Economic index · Total natural resource productivity · Environmental and socio economic parameters · Current food production patterns · Strengths and concerns · Sustainability · Future prognosis · Production targets · Pathways to sustainability · Physical constraints · Natural resource constraints · Land · Water · Biodiversity · Environmental · Ecological · Holistic and reductionistic technologies · Ecologist's value system · Productivity indicators · Gross domestic product · Technology · Natural resources · Conservation

Recent times have witnessed a keen urgency of sustaining agriculture on account of the pressures to increase crop productivity against the backdrop of burgeoning global population and severe environmental fallout, an aftershock of the green revolution, which, in fact was nothing but a high input industrial type agriculture which took a severe toll on natural resources like soil, water and biodiversity, in south Asia, especially in countries like India (Nair 2013, 2014). An added measure of urgency is also the question of providing reasonable economic returns to the farmers, as many farmers are shifting away from agriculture – a phenomenon witnessed in India. There is a general refrain that "agriculture is no more a profitable occupation," and many farmers are abandoning their lands and are looking for alternative avenues of livelihood. A national survey in India, carried out not long ago, showed that more than 40% of farmers would willingly vacate agriculture for other economic avenues of livelihood. Tangible indicators prove that production resources, especially land, are eroding and that environmental degradation has assumed global

K. P. Nair, *Intelligent Soil Management for Sustainable Agriculture*,
https://doi.org/10.1007/978-3-030-15530-8_24

dimensions and reaching unacceptable levels. Globally, the growth rates in yields of wheat and maize decreased during the period from 1961 (start of the green revolution) to 1997 (widespread yield plateauing or decline), (WRI 1998). In some intensively cultivated areas of Asia, like India, trends of yield stagnation in rice-rice and rice-wheat systems, as practiced in the State of Punjab, the "cradle" of green revolution, are worrying (Pingali et al. 1995). It is inevitable to upgrade management systems, so that any decline in sustainability indicators are identified and properly addressed.

24.1 Dimensions of Agricultural Sustainability

Agricultural sustainability denotes the ability of agricultural systems to remain efficiently and indefinitely productive; it is a complex concept involving many dimensions. Sustainability is a systems issue and can be hierarchically disaggregated in terms of space and scale. Evaluation of sustainability is still in its early stages, in particular, because sustainability is an evolving and adaptive process, rather than a fixed goal. From the point of view of farmers and scientists, the lowest level of the hierarchy (which is the field or cropping system level) is the most appropriate one at which to monitor and evaluate agricultural sustainability, because, productivity and sustainability are the main concerns at this level. Fewer, new biological, physical, and chemical influences intervene at higher levels and where economic aggregation dominates (Spencer and Swift 1992). Sustainability has temporal context too, because agricultural systems change over a period of time.

There are three components to assessing agricultural sustainability, which are the following: biophysical, economic, and social (Herdt and Steiner 1995)

1. Biophysical: It is related to the quantity of output per unit resource input
2. Economic: It is the component related to the net monetary value of the net output per net unit input
3. Social: This aspect of agricultural sustainability is the most complex to define and almost impossible to quantify. It comprises a number of tangible and intangible factors and can be broadly regarded as sustaining the overall well-being of the farming and rural community involved in agriculture and related operations

24.2 Partial Indices of Agricultural Sustainability

Ideally, sustainability should be assessed holistically with reference to all its components: productivity, ecological effects, economic and social viability. However, the mutual interaction of these dimensions and the complex tradeoffs required among them in different systems in different contexts results in a lack of clarity about the weightage to be given to the different components. This makes dealing with all aspects simultaneously formidable. Partial indices with reference to individual components are more practical and quite effective and can serve as early warning signals of a threat to sustainability.

24.3 Productivity/Biophysical Indicator

At the most basic level, the sustainability of a cropping system is assessed by the trend in output from the system. A parsimonious definition of sustainability is that a sustainable cropping system is one in which the trend of annual output is non declining (Spencer and Swift 1992). Stability of output over time is the most important prerequisite for sustainability. Fluctuations in factors such as climate, pest attack, and disease damage, will cause fluctuations in ultimate crop yield from year to year or cropping season to cropping season, but, these should be within statistically acceptable levels. This criterion can be represented mathematically as follows:

$$Y = P/A > 1 \tag{24.1}$$

Where P = Annual production and A = Area under production. This indicator is incomplete because it does not take into account the changes in inputs which may be required to maintain or increase productivity.

24.4 Total Factor Productivity (TFP)

Total Factor Productivity (TFP) measures an index of total output relative to an index of total inputs, which is a better measure of sustainability than merely the final crop yield, because, the TFP takes into account all the inputs as well, though it would be a more difficult route to arrive at the exact TFP. This is because, there are constraints to production like climatic effect (temperature variations on crop yield, both at the soil level and ambient level which are difficult to quantify precisely during crop growth), water usage covering both total quantum used and evapotranspirative losses, lamination intensity (total light and dark periods received during crop cycle) and so on. As productivity per unit area of land is the most important attribute of sustainable agriculture, TFP can be measured in terms of actual productivity per unit area with reference to the cost of all inputs in the complex production system. A quantitative measure could be mathematically described as follows:

$$\text{TFP} = \frac{P}{{}_{i=1}^{n}\{(R_i C_i)\}} \tag{24.2}$$

where P = Productivity per unit area, R = The resources used and C = Cost of a unit of resource used.

24.5 Economic Index

As economic returns are crucial in commercial agriculture, an economic index that is a performance audit, in terms of monetary returns, is an appropriate expression of sustainability. It is defined as the total monetary value of all outputs produced by the system during one cropping cycle divided by the total economic value of all inputs used by the system during the same cropping cycle (Ehui and Spencer 1990) as follows:

$$(\text{TFP})\text{M} = \frac{\mu f(\text{M})}{{}_{i=1}^{n}\{(\text{R}_i\text{C}_i)} \tag{24.3}$$

Where μ (M) is the monetary value of the output, R = The resources used and C = Cost of a unit of resource used.

Generally, the outputs and inputs are confined to those recognized as economic variables, such as purchased inputs, labor cost, and the total value of the harvest. If TFP shows a constant or upward trend over a period of time, the system is sustainable. The factors that contribute most to sustainability or the lack of it can be segregated from TFP and monitored as partial factor productivity. For instance, the response to a single crucial resource such as fertilizer N or fertilizer P is often useful as a way of monitoring sustainability. If fatigue or a negative trend is discernible in the form of constant or decreasing factor productivity, the possible reasons can be analyzed and amendments can be made to restore or improve the factor productivity.

24.6 Total Natural Resource Productivity (TNRP) or Social Factor Productivity (SFP)

The indices of sustainability described above take into account only the costs and benefits of managed inputs and outputs which can be assigned numerical or monetary values. They do not take into account the positive or negative impacts to natural resources during the cultivation process. Thus, neither the different types of degradation of croplands and environmental pollution which invariably accompany intensive cultivation practices, as in the green revolution, nor the benefits to the environment in terms of carbon sequestration, as in low input sustainable agriculture where legume inter cropping and ploughing into soil is resorted to, in the improvement in soil quality, proliferation of beneficial microorganisms, and so on are accounted for.

The social factor productivity (SFP) concept offered by Herdt and Steiner (1995), rechristened as total natural resource productivity (TNRP) by Lal (1998), takes into account both the positive and negative impacts on natural resources in addition to the benefits accrued in terms of capital and physical resources. It integrates productivity with economic and natural resource factors as depicted in the following figure and can be mathematically expressed as follows, and is figuratively shown below (Fig. 24.1):

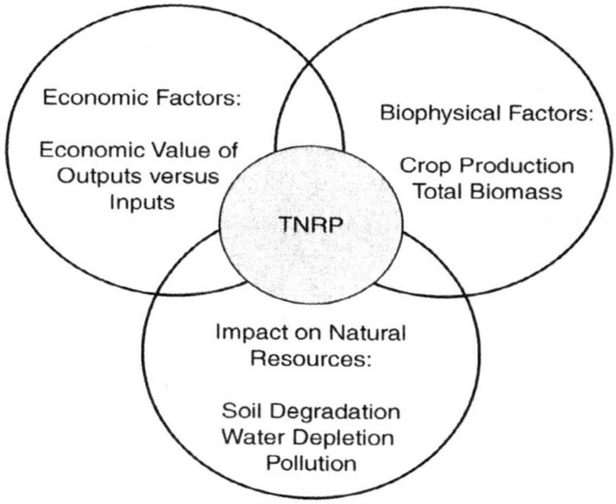

Fig. 24.1 Total natural resource productivity (TNRP): Sustainability = Nondeclining TNRP with Time. (Source: Modified from Lal 1998)

$$\frac{\text{TNRP} = \text{TFP}}{\Delta S} \tag{24.4}$$

Where ΔS = The cost of depreciation in social or natural capital.

Problems of resource depletion, degradation, and imbalances such as nutrient and water mining, deterioration of soil physical properties, and erosion need to be identified and assigned monetary values to calculate TNRP. Modern environmental and ecological economics are evolving a workable protocol for this. Attempts to reverse the depreciation or eliminate its impact can improve TNRP, and hence, sustainability.

24.7 Soil Quality as an Indicator of Sustainability

Sustainability in its totality is a multi-dimensional and complex concept encompassing economic, ecological, and social components, all of which render it a contextual relevance. Thus, although the need for the development and evaluation of sustainable management systems has been stated since the 1990s, the formidable task of resolving the weightage to be given to different components, for example, farm productivity, profitability, environmental quality, soil resource conservation, and socio-cultural factors, has remained incomplete. Evaluating the sustainability of management systems in terms of design and performance is more straight forward (Pierce and Larson 1993).

Soil quality is the most useful determinant to monitor and evaluate the sustainability of agricultural management systems because it represents a holistic measure by which to assess their ecosystem health. The physical, chemical, and biological properties of a soil are all components of soil quality. Sustainable production of healthy and nutritious crops is contingent to the maintenance of good soil quality. Deficiencies in the management of other resources such as water, biodiversity, energy, and various technologies either emanate from soil or are translated into a diminution of soil quality (Saroja Raman 2006).

The quantitative measurement of soil quality indicators and the soil quality index is practical and has evolved as a useful means to quantify the current state of a soil and to serve as a base from which to monitor changes over time with reference to different management practices. Defining, monitoring, evaluating, and then enhancing soil quality is the most credible and fundamental pathway to sustainable agriculture. It is also the most validatable way to evaluate the sustainability of farming systems. Soil is the mediator among farming practices, agricultural chemicals, and the environment. The dynamic relationship between management practices and soil quality needs to be well understood to sustain agricultural production.

The link between the sustainability of a management system and soil quality is based on the premise that a management system is sustainable only when soil quality is maintained or improved. Thus, quantitative changes in soil quality should provide a measure of the sustainability of the management system. One simple approach to the evaluation of sustainable management is a comparative assessment of the performance of a system in relation to alternative management options. The characteristics and outputs of alternative systems are compared at some time (t) with respect to biotic and abiotic soil system attributes. A decision on the relative sustainability of each system is made with reference to the differences in the magnitude of the measured soil quality parameters. For example, Reganold (1989) compared several soil quality parameters for farms using conventional and organic farming practices and concluded that the organic farming system was superior in terms of maintaining soil tilth and crop productivity and reducing soil erosion. The limitation of this method is that it does not provide information on either the processes which created the measured condition or the design and operation of the system.

24.8 The Dynamic Assessment Approach

In the dynamic assessment approach, the system is assessed in terms of its actual performance by measuring soil quality parameters over time. Although overall soil quality is important in land evaluation, knowledge of changes in certain specific indicators of soil quality that are most impacted is enough to validate sustainable management. Acceptable ranges for the indicators chosen are decided according to the principles of statistical quality control. The variation of one or more of these indicators with time dq/dt, is used to evaluate the dynamics of soil quality mathematically depicted as follows:

$$\frac{dQ}{dt} =^{ni=1} \frac{\{dq_i}{dt} \quad (24.5)$$

where

$$\frac{dq_i}{dt} = \frac{qit_2 - qit_1}{t_2 - t_1} \quad (24.6)$$

Positive and negative dQ/dt indicate improving and degrading soil quality, respectively. Changes in soil quality indicate the sustainability or otherwise of a specific management system; changes in land use are decided on the basis of this.

The dynamics of the soil quality index can be used to evaluate the impact and sustainability of different management systems and to plan improvements. It is not always necessary to ascertain the full data set of indicators. Only those that are sensitive need be monitored for future action. Details of this type of approach can be found in the literature (Harris et al. 1996).

24.9 Future Prospects

Although an obvious link exists between soil quality and agricultural sustainability, no direct cause-effect relationship exists between the two on the basis of which direct and focused remedial measures can be put in place. Often, a multitude of interacting factors, including inherent soil properties, climate, and crop and soil management, bring about soil quality deterioration. Thus, quantification of the effect of soil quality even on agricultural productivity, which is one of the simplest criteria for sustainability, is difficult. The relationship between the soil quality index or even individual soil quality indicators and the sustainability of management systems needs to be studied more extensively before benchmarks can be set for various agroecological sites for various types of land use. Geographical information system technique to optimize the factors related to soil quality may be useful in predicting soil quality variation over longer periods of time under different management and land use scenarios relating to productivity and environmental degradation. Soil quality monitoring will help in the restoration of deficient soils or declining quality by introducing specific ameliorative measures. Long term experiments are required to monitor soil quality and its effect on sustainability parameters to safeguard future productivity and intensification. The world's soil resources are finite, and intensification of agricultural activities on existing lands is unavoidable for at least another half a century. The link between soil quality and agricultural sustainability seems to be the most practical and logical way to monitor and ensure the sustainability of agriculture because the methodologies are well standardized, databases are available or can be built up, and information – analyzing technologies are available.

24.10 Framework for the Evaluation of Sustainable Land Management

Soil constraints on the intensification of agricultural production were not sufficiently or widely stressed in the initial research agenda for the green revolution by the Consultative Group on International Agricultural Research (CGIAR). The International Board for Soil Research and Management (IBSRAM) was established in 1993 with headquarters in Bangkok to strengthen soil research for the development and dissemination of sustainable land management technologies. IBSRAM, in collaboration with several national and international organizations, developed a "framework for the evaluation of sustainable land management" (Smyth and Dumanski 1993), based on the comprehensive definition of sustainable land management given by Dumanski (1993). The framework takes into account productivity, environmental, and socioeconomic parameters and is designed as a logical pathway for the analysis of the sustainability or the probability of attainment of sustainability of land management systems. It has a broader scope than soil quality, which is restricted to the field level. It can be used to assess sustainability of current land management systems in different environments and to analyze innovations in technology and policy for sustainable land management. It requires extensive data on all dimensions of agricultural sustainability.

24.11 Challenges to Global Agricultural Sustainability

In the hierarchy of human needs, in fact, for all living beings, food occupies the topmost position. This has motivated the constant upgrading of food production to match increasing needs by combining the productive potential of natural resources with human resourcefulness. Thus, the pattern of the pursuit for food changed from the hunter-gatherer mode of the pre-Neolithic period to settled agriculture around 8000 years BC. What started as shifting cultivation changed progressively to slash and burn, long fallow, short fallow, traditional agriculture, and eventually the present day high input package of fertilizer, pesticide and water, propping up "high yielding" (which no more is, despite high inputs) crop genotypes, principally wheat and rice. Human engineered alien plants, in the form of genetically modified crops, are the latest weapon, in our arsenal, though most of them are riddled with very adverse environmental fallout, as one can witness all round, in our fight to enhance crop production for survival. Concerns are growing about the sustainability of the current pattern of production because it has seriously eroded the ecological base which is the foundation of agriculture. The future has to contend with the daunting task of producing more to balance the growth in demand from a smaller set of natural resources and limiting further deterioration. This warrants a realistic analysis of present and future scenarios and a re-engineering of the current strengths and weaknesses of production systems to cope with future threats and opportunities. This chapter will discuss the strengths and weaknesses of the current system and the challenges which have to be overcome to realize the targets envisaged by global food models for the future.

24.11.1 Current Strengths and Concerns

The latter half of the twentieth century was extraordinarily eventful in terms of the 10,000-year history of settled agriculture. Unprecedented improvements in agricultural production techniques to increase productivity anchored firmly in advances in science and technology, and suited to various geographical areas, ensured that total global cereal production almost tripled from 631 million tons in 1950 to 1835 million tons in 2000 (FAO 2000). This was possible almost exclusively because of a productivity increase from 1 to 2.8 ton ha^{-1} during this period. Despite an increase in population from 2.5 billion to more than 6 billion during this period, the daily average intake has gone up from 2450 kcal to 2760 kcal per person, ironically, a mere 13% increase in half a century, which works out just about a miniscule 0.3% per annum. There are various reasons for this, like, unequal distribution of the produced food, speculative prices, market manipulations, production enhancement in countries like India confined to only some regions, where the build up of infrastructure, like irrigation facilities, substantially improved, as for instance in Punjab State, due to political vested interests, to the neglect of other regions covering vast areas. Current global supplies are adequate to meet the nutritional requirements of everyone if food is distributed according to needs (Pinstrup-Anderson 2002).

The transition from a position of despair in 1798, when Malthus prophesied a mismatch between food production and population growth, even when the global population was only 750 million, to a position of sufficiency and more for over 6 billion is a truly remarkable feat, though, paying a tremendous price at the environmental level, as one has witnessed in Punjab, in India, the "cradle" of green revolution (Nair 2013, 2014). The basis for this achievement lies in the rapid advances in science related to plant growth in the nineteenth century and the sophisticated technologies which translated ideas into action, especially during the latter half of the twentieth century. The advent of high external-input cultivation revolutionized largely traditional low-input peasant agriculture into a factory model of commercial agriculture in the post World War II decade and thereafter, in the industrially developed US and countries of western Europe, though it was labelled as "green revolution," most ironically, in the developing world, with all the attendant adverse consequences on the environmental front (Nair 2013, 2014). Massive use of land saving technologies of irrigation, fertilizers, high yielding crop varieties of global staples like rice, wheat and maize, and pest control technologies to check pests and diseases, together with labor-saving mechanization, boosted land and labor productivity several-fold. International cooperation and the commitment of the developing countries led to a similar model being replicated in the countries of Asia and Latin America, with a critical food deficit in the 1960s and 1970s to usher in the green celebration. This, while ensuring food security on a superficial scale, left many untouched, and a severe adverse environmental fallout. India is the most cited example (Nair 2013).

The increase in food production in the past five decades, in tandem with accelerated population and economic growth, is truly a vindication of the optimistic prediction of Boserup (1965) that population would act as a catalyst for technological innovations. To some extent this positive change, arising out of human innovations, is to be welcomed. But, it must be remembered that since the last two decades, when the green revolution has run into all kinds of problems, scientists are waking up to the urgent need that there is an imminent need for an alternative, sustainable path, on production technology, and, it is in this context that the "Buffer Power Concept" in soil management to sustain crop production assumes great relevance.

24.11.2 Prognosis for the Future

24.11.2.1 Predictive Models

The creation of an exhaustive database of natural resources, population, climate, policies, and other variables has facilitated the task of making realistic supply and demand projections for food in various regions of the globe and formulating the strategies to meet them. International organizations such as the United Nations Food and Agriculture Organization (FAO 2000), the International Food Policy Research Institute (IFTRI 1997), the World Bank have developed their own visions of agriculture for the next two to three decades. All these models envisage that a steadily improving global economy and intensification of the avenues for production and productivity growth currently pursued will generate a fairly comfortable food security situation with an increase in per capita cereal availability for the next two to three decades. Per capita cereal production is expected to rise continuously, assuming the United Nations median projection for population growth. According to this model, the developing countries will face deficits of 190 million tons of cereals and coarse grain in 2020, which amounts to 11% of their consumption, as opposed to 3% in 1960s. This simply demonstrates that things are getting worse as of now, and, much of this problem has been created because of population increase, but, equally so due to an unjust food distributive system, caused by speculation in grain trade, improper distribution of food grain within a country of its production, and so on. The model envisages meeting this demand from increased supply from the developed countries, at the rate of 1% growth per annum in their cereal productive capacity.

According to the FAO (2000, 2002), average global nutritional standards are expected to reach 3100 kcal per capita by 2030, with a revised lower estimate for global population of 8.112 billion by 2030 (UN 1999). To achieve this, global food grains production will have to add an increase by 1000 million tons, from the current production level of 1840 million tons, to reach 2840 million tons by 2030. This is greater than the amount of growth recorded in the past three decades. However, the actual increase required in growth rate will be lower, at 1.7% per year, compared with a growth rate of 4.1% during the past three decades. The corresponding figures for the developing countries are 1.9% and 3.6%, respectively. The significantly lower growth rates required in production are due to a reduction in the growth rate of population and the gradual saturation in terms of per capita food consumption in certain parts of the world. The rate of growth in food production will remain low for the sub-Saharan African countries, at 1.5%, even by 2030. The situation is no more rosy even for India, with an increase in population outstripping food production, by 1.5% over at 1.3%, respectively, per annum. The dependence on imports of the developing countries will continue to grow, from 107 million tons in 1995–1997 to 270 million tons by 2030. This means that the current exporting countries will have to upgrade their collective production from the current 800 million tons to 840 million tons by 2030, corresponding to an annual growth of 1% annually.

A new worrisome factor has arisen for India where the central government has enacted a Food Security Act which will need the country to not just mop up all the saleable grains within the country, but, also import, breaching the norms of the Aggregate Measure of Support (AMS), as prescribed in the Agreement on Agriculture (AOA), prescribed by the World Trade Organization WTO. At the time of writing of this book, discussions are going on in Bali, Indonesia, as to how to resolve the grave problem.

24.12 Proposed Models and Pathways for Sustainability

24.12.1 Proposed Pathways for Achieving Production Targets

Crop production in the developing countries is projected to go up by 70% by 2030 with reference to the base year 1997/1998. This will be achieved through the following three pathways:

1. Land expansion will bring about 20% increase
2. Productivity will bring about 70% increase
3. Enhancing cropping intensity will add the remaining 10% increase

Almost all of the 20% increase in land area (120 million ha) will be confined to the land-excess regions of Africa and Latin America, such a scope in South Asia, where the population increase is significant, being almost absent. The yield growth rate is pegged at 1% per annum, as opposed to 2.5% in the previous three decades, with a concomitant increase in fertilizer use of 1% per year. The share of irrigated agriculture is predicted to increase from the current 40% to 47% by 2030 through an expansion in area by 23%, resulting in a total irrigated area of 45 million ha by 2030. In terms of harvested area, the increase works out to be 34%. This will entail an additional 12% water withdrawal for agriculture, with an increase in efficiency of use from 43% to 50% in the 93 developing countries analyzed by the FAO (1998).

24.12.2 How Realistic Are the Models?

The business-as-usual, or more-of-the same, as before, models expected to steer the world towards agricultural sustainability need to be carefully assessed because they raise a number of questions to which the answers are not clear. Conway (1997) has analyzed the merits and drawbacks of such approaches and their implications for future food security has provided a lucid critique of these models. First, the models used to predict future supply and demand are econometric models based on the purchasing power of individuals. Consequently, they do not take into account ecological and sociological imperatives and constraints. As economic models, they assume that supply meets demand in a market economy. Hence, they do not include the food needs of the poor and hungry, who are priced out of the market; their needs are hidden and not accounted for in the market demand. The gap between supply and demand, which the models expect to close, is, therefore, the market gap and not the actual gap. This in reality means that the food needs of the 800 million or so, currently hungry poor people in the world, do not figure in the current demand projections for 2015, 2020, and 2030 because they are outside the market economy. In other words, the current marginalization of the hungry, the undernourished, and the poor will continue. At the end of three more decades, the numbers of the undernourished and hungry will remain sizeable, at 400 million, belying the expectations of the World Food Summit (FAO 1996a, b) and also the Millennium Project where it is expected to reduce the number of hungry by half by 2015. The year 2015 has passed with no tangible positive results on the hunger front. An approximate indicative value of the amount of food needed by the currently poor and hungry is 359 million tons, at a rate of 3000 calories per person per day. This means that the actual gap in food needs to be bridged for the developing countries will be phenomenally higher than the current assessments; however, the actual figures vary according to the different assumptions involved.

It cannot be assumed that the food-surplus-exporting countries will continue to bridge the supply-demand gap of the developing countries to the required extent, even though they may have the ecological and economic capability. Production trends for cereals in the developed countries show a continuous fall in growth rates since the 1960s (Brown and Kane 1994). The cuts can be attributed to declining prices, stagnant exports, and accumulating stocks, among other factors. The huge surplus that the US and European Union (EU) currently produce and trade in the international market has been made possible through the massive subsidies, more than US $300 billion annually, in the form of domestic subsidies for agriculture and export of agricultural products. Within the EU, France is at the forefront of these subsidies. In the recent India-EU Free Trade Agreement (FTA) discussions, the latter is poised to benefit enormously through its cheap exports of agricultural products into India, pushing out Indian products. Dairying sector will be the classic example of this newly emerging scenario. Indian branded products like AMUL, which has captured both the national and global attention, will be the loser through this agreement. These countries resist any sizeable cut in these subsidies to maintain their export advantage, while keeping domestic and international prices low and steady, unlike that of the developing countries like India, where market fluctuations are increasingly affecting food supply, and, against the currently raging (at the time of writing this book) food inflation is very adversely affecting the poor and hungry. The steeply Indian Rupee- US $ parity is the newest worry on this score, because, all imports need to be paid in hard currency. There is no easy Indian Rupee – Russian Rouble trade with Western Europe or North America, where only the dollar or euro calls the shots.

The resistance is growing to the share of agriculture in environmental pollution and degradation, especially of soil resources in many developing countries. India is a classic example. Pollution, of the soil, water and environment, is a result of unbridled use of chemical fertilizers and pesticides, which has been a bone of contention between agriculturists and environmentalists since the 1960s in the US and Western Europe and has raised widespread concern about the quality and safety of the environment and the services it provides. It is only after the failure of the green revolution in India, starting from the early 1980s that people are becoming increasingly aware of the hazardous outcome of unbridled use of chemical fertilizers and pesticides on the environment, it's quality, as a whole, and the serious question has arisen, whether it was worth all the trouble going in for a high input, industrial type of agriculture, bandied around as a "green revolution," which though helped the nation harvest more food grain initially, left a disastrous peril on the ground and in the environment (Nair 2013, 2014). Efforts are underway to restrict the use of these inputs and to adopt a more environmentally benign technology through legislative measures. A new phenomenon that has raised its head in India is the clamor for genetically modified crops, (GM crops), though, they do not pollute the soil the way the high input technology does, through excessive use of chemical fertilizers and pesticides. Nevertheless, there are far more serious hidden dangers in the use of GM crops, like genetic contamination and creation of super weeds, through pollen transfer, and, eventual, wipe out of biodiversity. The civil rights movement and concerned scientists have joined together to raise objections, and the most important outcome of their concerted efforts is the indefinite moratorium clamped on Bt brinjal, which agribusiness giant of USA, Monsanto, through its Indian arm, Maharashtra Hybrid Seed Company (Mahyco) tried to push onto Indian fields. Even legislative measures are adopted by some states to proclaim them as "GM Free States," as for example, Kerala and Odisha States in India. On the technology side, there are examples like the Farm Bill of 1995 in the US which made conservation tillage mandatory in erosion-prone regions of the country. Recently, serious attempts have been made to encourage the conversion of agricultural enterprises to more benign ones, such as, habitat development, ecotourism enterprises, and organic farming. In 2016 Kerala State will be totally organic. These changes will bring about multiple and simultaneous benefits on several fronts. They will not only help to take land away from intensive farming, and curtail environmental pollution and resource degradation, without financial loss to the farmers, but also, concurrently, cut down on the government's financial commitment to farmers for various conservation measures and production. Farmers are paid premium prices for the organic products they produce. Such measures will also help to manage the problem of overproduction and low international prices and give these countries a price and export advantage in a global food-deficit scenario. Countries such as the Netherlands have even enacted legally binding restrictions on the use of external inputs, and also, certain modes of agricultural cultivation. The Kyoto Protocol encourages such sustainability-enabling moves (McNaughton 1998). A trend is evident in the developed countries towards producing organic crops, which fetch better commercial prices and augment farmers' income. The increasing emphasis on more sustainable practices will lead to a decrease in cereal production and/or a shift to the cultivation of other crops, or to a habitat conservation and ecotourism for additional income (Patter 1998). Unfortunately, there are also many instances, as in Kerala State in India, where in the name of ecotourism, the environment is ravaged indiscriminately, in the name of "development" and income generation. As long as the "GDP (Gross Domestic Product) mindset" continues with the policy makers, and people at large, environment will, needless to add, only meet with a very bad bargain.

The assumptions regarding the supply side of food production become all the more untenable in the current context of the developments in the globalization and liberalization of trade of agricultural products, vis-à-vis the WTO stipulations inasmuch as the developing world is concerned. The actual outcome of the processes of globalization raises questions about the plight of developing countries almost two decades hence. What pressures will be placed on the developing countries under the provisions of the World Trade Organization (WTO) in terms of environmental and labor standards, and how these will impact their food security and revenue are matters that merit serious scrutiny. Already rumblings are heard about the methane emissions from rice production and crop residues being used to feed ruminants. Trade in agricultural commodities is guided not by altruistic intentions, but, by purely commercial interests. Altruism is hard to find in an economically competitive world. The ongoing rounds of the WTO have unequivocally demonstrated that the developed countries are more concerned about their own economic gains, rather than the uplift of the developing and least developed countries (LDC). The stipulations on cotton trade clearly demonstrate this in the WTO meet held in Bali, Indonesia, in 2013. Cotton is the economic mainstay for poor Central African countries like Benin, yet, there is nothing in the WTO deliberations to safeguard it's economic interests. Developments coming out of the trade "facilitation" (as the new jargon of WTO), are, as of now, totally unpredictable. The increase in the export of agricultural commodities from the developing countries anticipated to arise from decreased levels of production and export incentives in developed countries has not occurred, yet. The failure of the Doha rounds of the WTO, followed by the negotiations in Cancun, Mexico, all conducted at tremendous expense to the national exchequer of several countries, clearly highlights the conflict of interests between the developed and developing worlds, in particular, in agriculture. And, to cap it all, the latest stringent stance taken by the negotiators, especially from North America and Europe, on subsidies makes things extremely difficult for India. The country has a Food Security Act (FSA) in place since 2013, and to harness food grain to meet its requirements, the Government of India will need to enhance the minimum support price (MSP) for the grain purchased from farmers. For this, subsidy is needed from the national exchequer, and the WTO negotiators, in particular those from North America and Europe, would want to cap the subsidies. The long range objective is to put India in a difficult situation wherein, to meet the grain requirement of the FSA it will have no other option but to import grain from USA, Canada or Europe.

Finally, the World Bank prognosis for the future is based on an assumption of sufficient economic growth in the developing countries to pay for the import of grain from developed countries. These econometric models patterned on balancing supply and demand assume that demand will stimulate the technological innovations and infrastructure investments in the developing and developed world, required to ensure a sufficient supply of food. In reality, the per capita GDP growth rate in many of the developing countries, especially in the sub-Saharan Africa, shows that the fight against poverty will fall short of the goal set by the international community. It has become clear that the expected halving of poverty, especially of food deficit, by 2015 is certainly not realizable (World Bank 2000). At the time of completing this book, 2015 has passed by, and the "Millennium Development Goal" (MDG), on reducing hunger by 50%, faded into memory as a mirage. The objective of reducing the number of food-insecure people to 400 million, globally, as set out in 1996 in the World Food Summit, has been pushed back to 2030 at the earliest (FAO 2000, 2002). The plight of the sub-Saharan Africa is uncertain because they have to make path-breaking advances in economic development, in social and political arenas, and in their foreign exchange reserves to be able to pay for their imports. Their traditional export crops are not of high value, and in the absence of any real hope for impressive industrial growth, it is difficult to be very optimistic about their ability to achieve the capacity to support such large imports.

24.12.3 How Effective Are the Proposed Pathways for Sustainability?

The pathways proposed to meet the additional food demand in the next two to three decades are based on mostly further intensification of the irrigated high potential lands in the developing countries to achieve the targeted growth rate of 1% per annum. Although this is a modest goal, it will not be easy to accomplish because water is predicted to be the most limiting resource in the future, especially for agriculture. The increase in population in water-deficit countries, most of them developing nations, will be six-fold. Considering the onerous task of even maintaining the current level of appropriation of water for agriculture, especially in the developing countries, hoping for a 12% in consumption coupled with the anticipated improvement in efficiency of use is akin to walking up a downward escalator. Further, documented resource degradation and environmental pollution in intensively cultivated areas are a cause for concern. The intensively cultivated green revolution areas of the developing countries of South Asia have already suffered resource degradation, especially that of soil resources, in the form of salinization, waterlogging, and loss of fertility due to carbon depletion due to organic matter loss – a consequence of unbridled use of chemical fertilizers containing N like urea, and continuous cereal monoculture, of rice-wheat, in the "cradle" of the green revolution in Punjab state in India. A parallel situation is seen in South-East Asia as well, where soil fatigue leading to yield plateauing has been observed (Ladha et al. 2003). Unlike these conditions, the proposed increase of 1% in yield annually is a formidable task. The technologies available are more or less the same as in the past, as no path-breaking innovations have been added. However, "The Nutrient Buffer Power Concept" is full of promise in alleviating the adverse situation and shows a new direction in soil management, in these regions. There is, thus, some hope in narrowing the gap between potential yields and actual realized in South and South-East Asia, South America and Africa.

The proposed increase in land area for production by 20% is mostly concentrated in sub Saharan Africa and Latin America, even though much of the former area is of poor quality land and prone to extensive soil erosion and most of the latter is under primary forests. The clearance of such land may lead to many problems in soil fertility through early development of nutrient deficiencies and build-up of soil acidity. Little room is but available for the expansion of arable land in South Asia and North Africa. In Asia, per capita arable land will decline from 0.18 ha in 1950 to 0.09 ha by 2020, and in China the land under cereals is likely to decline to 0.05 ha per person (Gardner 1997).

Looking at these legitimate resource constraints and other global socioeconomic trends, it is difficult to be optimistic about the effectiveness of the business-as-usual path. Very real and formidable ecological, socioeconomic, and demographic challenges face the developing countries, if they are to improve their food production and productivity to the required level in the next about two decades. Assuming that they do pull off a miracle of sorts to fulfill these targets, they will still be saddled with having almost 15% of their population, mostly living in the rural areas and dependent on agriculture, hungry and undernourished. Neither science and technology nor market forces are relevant to these poor people, who are mostly confined to ecologically stressed areas such as highly drought-prone or hilly regions.

24.13 Future Challenges for Agricultural Sustainability

Past practices for achieving global food sufficiency have left serious impediments to the transition of agricultural systems to sustainability in the future. Tangible ecological, economic, and social indicators show that serious challenges lie ahead in terms of addressing future productivity growth rate requirements in a sustainable manner. Most of these are a result of the direct or indirect impact of population pressure and the adverse impact of unbridled use of high-external input technologies, which have been further exacerbated by various flawed socioeconomic policies which are not conducive to sustainability. It is becoming increasingly clear that the technological innovations of the second half of the past century, which enabled the rapid growth in food productivity, have not been an unmixed blessing. Boserup (1965) did not envisage the adverse impact of technological innovations on the ecological base for food production or the socioeconomic fabric. The challenges to future sustainable intensification can be grouped into the following categories (Saroja Raman 2006):

1. Population pressure and demographic changes
2. Natural resource constraints
3. Environmental safety concerns
4. Ecological constraints
5. Technological constraints
6. Climate change concerns
7. Socioeconomic constraints

24.13.1 Physical Constraints

24.13.1.1 Population Growth and Demographic Changes

The most important challenge to sustainable agriculture in the future comprises unrestrained population growth, which is not expected to stabilize for another century, at the minimum, and the demographic changes which are likely to occur as a consequence of future economic growth and global awareness, especially in the developing countries. These will trigger accelerated demand for more, better quality, and diversified forms of food. The world population is expected to increase to more than 8.1 billion by 2030, an increase of 30% based on nearly 80 million people being added each year (UN 1999). This represents quite a downsizing from the earlier predicted estimates, and, it is hoped that the rate of growth will become less daunting in the latter half of the century. Predictably, most of this population increase will occur in the developing countries, which are already facing food shortage. For instance, the countries in the Middle East, the countries of North Africa from Morocco right through Egypt, and sub Saharan Africa, will all experience a large increase in population, which, together with their water-short and grain-importing status, will have severe consequences for food security. Even China and India, which are more or less food sufficient at present, though in the latter case there are unbridgeable bottlenecks in the distributive pattern due to corruption at the governmental level, leading millions going hungry and starving, which has now forced the central government in New Delhi to enact a Food Security Act in 2013, may encounter food shortages by 2030 (Brown 1995).

Urban migration will accelerate, and by 2050 more than 50% of the global population will live in urban areas. Urbanization will lead to a 1% loss per annum of fertile croplands and will outstrip losses through degradation. Increasing affluence and the consequent expectations of a higher living standard, globalization, and the information and communications explosion will dramatically change the dietary habits of people, predictably towards a more sophisticated mix of grain, horticulture, poultry, meat and fish products. All this is expected to lead to a 60% increase in demand in grain production in the developing countries in the next about two decades. This will require per capita grain availability to grow at a higher rate than the rate of increase in population growth in the developing countries (FAO 2000).

24.13.2 Natural Resource Constraints

Sustainable agricultural production is dependent on the quality and quantity of the natural resources of land, water, air, and biodiversity available for cultivation. The increase in population since the beginning of the twentieth century has led to a corresponding decline in per capita entitlement to these resources. The expansion of agriculture and unsustainable practices of capital-intensive high external input crop production practices in the past five to six decades has further compounded this problem. The impact of the vast expansion of agricultural area, the extent and type of damage that agricultural systems, especially of the high input technology of the green revolution, have wreaked on the natural resources of land, water, air, and bio diversity, and the challenges for their conservation and sustainable use have been highlighted in other chapters of this book. A few additional points are included here.

24.13.2.1 Land

Availability of arable land is critical for food security. Global per capita grain area shrank from 0.24 ha in 1950 to 0.16 ha in 1985, that is a 33% decrease in 35 years, which is, undoubtedly, a huge decrease. The grain area shrinking is expected to decrease further to 0.09 ha by 2020. This shows that the per capita grain area has decreased by 62.5% in 70 years, that is almost 1% per annum, which, undoubtedly is a huge decrease. The consequences of this on world grain production needs no special mention to prove the disaster in the making on the food front. Urbanization in the developing countries will take away a huge chunk of land. By 2020, several countries in Asia will become "arable land scarce"; that is, they will have less than 600–700 m^2 per person. This is the bare minimum area capable of supplying a vegetarian diet for one person without the use of artificial chemical inputs or loss of soil or soil nutrients. This is also taken as the import dependency threshold, below which a country in this category has to import more than 20% of its grain requirements. Many countries in Asia already fall into this category, and 10–15 are expected to fall into this by 2020. Between 1960 and 1990 the number of people living in arable-land-scarce countries almost tripled, from 97 million to 253 million. The figure is currently approximately 415 million and is expected to rise to between 600 and 986 million by 2015 (PAI 2002).

24.13.2.2 Water

As pointed earlier in the chapter on water management in sustainable agriculture, shortage of good quality fresh water will be the most serious constraint to agriculture on account of its diversion to more pressing uses such as civic needs and the maintenance of ecosystem health and services. Estimates of the volume of water available show that water productivity in agriculture will have to be raised considerably in the near future to avoid severe constraints on food production. This is a formidable challenge and will be further aggravated by the erosion of irrigated land at the rate of 1.5 to two million ha per year as a result of salinization, waterlogging, and inefficient and populist policies of water pricing. More than 10% food is already produced through the mining of ground water, and the number of people in water-deficient countries will multiply five-fold within a couple of decades. Import dependency for grain will concomitantly increase, and this may lead to fragile relationships between exporting and importing countries and may require surrendering of sovereign rights or end in conflict. The Millennium Declaration sought to freeze the level of water consumption in agriculture at the current level until 2020 by increasing the efficiency of use and minimizing wastage during use merits urgent attention. The World Summit on Sustainable Development held in Johannesburg in The Republic of South Africa (2002) put water at the top of the list of constraints for sustainable development. Yet, water gets the short shrift in global policy making. Many nations simply squander, very thoughtlessly, this precious natural resource.

24.13.2.3 Biodiversity

The threat to the species biodiversity of edible plants and crops, and, more importantly, erosion of the genetic diversity of the global staples may result in grave challenges in the future in combating biotic and abiotic threats. Loss of species and genetic diversity is irreversible and poses a serious threat to the sustainability of agriculture, because with severe pressure on land and water, future gains in productivity rely primarily on genetic improvements. Recent studies show that species and genetic diversity is important for ecosystem functioning and sustainability and that we may have gone too far in sacrificing biodiversity in the pursuit of higher crop production (Altieri 1999). There is a concerned scientists global discussion going on against genetically modified crops, being promoted by some of world's leading agribusiness companies. In India, at the time of writing this book, there is much discussion on the subject of GM crops even at the top political level in the country, and the consensus seems to be firmly against them, though, vested interests are playing politics to push them on unsuspecting Indians from behind the stage, most unfortunately, without knowing the grave environmental consequences, or, perhaps, knowing it fully well, yet, willfully peddling it for selfish gains.

24.13.2.4 Environmental Constraints

Agriculture and environment have a very close, complex, dynamic, and reciprocal relationship. The requirement to increase food production to match the needs of a global burgeoning population, especially in the developing countries, which will simply continue to grow unless checked with policy decisions at governmental levels, until the end of the current century, have grave environmental and economic implications, mainly because the past five decades of agricultural intensification and industrial expansion have taken a serious toll on the natural capital resources of land, water, air and biodiversity. Setting aside reductions in the actual quantity of these stocks, their performance value in terms of the ecological goods and services and the life-support functions they provide, without which life on this planet will not be possible, is the more critical issue.

Agriculture has three major functions to play. It feeds the people, is a source of income, and provides employment for almost 50% of the world's population. It is a major user of natural resources and hence, has a considerable impact on the life-support services provided by nature. The expansion of the food system during the past five decades has placed a heavy burden on the environment, resulting in the degradation of forest, cropland, rangeland, and oceanic fishery ecosystems (Oldeman 1994), which provide all the food and raw materials for our activities, except minerals and fossil fuel. In addition, high levels of chemicalization, mechanization, and irrigation have given rise to large-scale environmental pollution and resource degradation. The intense and extended assault on natural capital during the past few decades portends a shift from an era of limiting human and technological advances (as at the time of Malthus, who predicted population growth outstripping food production) to an era when the remaining natural capital would, inevitably become a limiting factor. Environmental pollution and its consequent degradation is not exclusively the outcome of a high-input industrial type agriculture, like the so-called green revolution, which left a huge adverse fallout on the environment, in particular, on soil, water bodies and bio diversity, but, can also be caused by low-input resource-poor farming, as in the case of the Dust Bowl of the Midwestern US in the 1930s, and the resource-extractive agriculture practiced in the poor countries of Africa, in particular, and, also Asia – the forest clearing, slash and burn cultivation etc. The latter type of farming is extensively practiced even now in some poor regions of North-Eastern India, for instance in the State of Assam, and is locally called ***Jhum*** cultivation.

Although many of the adverse impacts of agriculture on the environment were fairly localized and amenable to technological amelioration, to some extent, in the past, the world is now confronted with more serious and almost irreversible environmental damage on a global scale. Some of the most worrying are:

1. A 30% share of the global increase in atmospheric green house gases (GHGs) brought about by deforestation and intensive and inefficient use of nitrogenous fertilizers
2. Loss of bio diversity as a result of over emphasis on a limited number of crop varieties and the continuous practice of monoculture, the most classic example of this being the continuous rice-wheat mono culture in Punjab State, India, the "cradle" of Indian green revolution.
3. Extensive groundwater pollution through unbridled use of chemical fertilizers and pesticides, which will take hundreds of years to clean up. The most classic example is the ground water contamination with nitrate residues in Punjab, due to excessive use of nitrogenous fertilizers to boost rice and wheat yields, as an established paradigm of the green revolution, making the ground water non potable. On a social scale, a district Gurdaspur, in Punjab, has now come to be known as the "Capital of Cancer" in India, consequent to unchecked use of carcinogenic pesticides, again, an offshoot of the green revolution
4. Mining of ground water and extraction, and over exploitation of fossil water, which represents irrevocable damages to the environment and
5. The recent rapid expansion of areas planted to genetically modified transgenic crops such as the Bt cotton (supposedly resistant to the dreaded American pink boll worm) on a very large-scale in India, Bt maize in the US, and also herbicide tolerant soybean in the US and Brazil, (Round Up Ready Soybean, which is now spreading to India), despite much concerned scientific and environmental activists' opposition. These crops pose the threat of large-scale emergence of new or resistant pests and super weeds due to pollen transfer from the GM crops. Also, the emergence of new pests, as for example the recent severe devastation of Bt cotton in Punjab State by white fly (Punjab farmers were the first in India who wholeheartedly embraced the Bt cotton in 2003, when the Genetic Engineering Approval Committee, of the Government of India, permitted its commercial cultivation). The most recent development in Punjab is the severe infestation of Bt cotton by white fly resulting in so many Punjab cotton farmers committing suicide due to total failure of the Bt cotton, leading to their bankruptcy, which has become a national disaster. The possible adverse effect of GM crops on the environment, food security, and the health of humans and animals, is an unresolved issue, and, at the time of writing this book, has become a hot topic of contention at the highest governmental level, with concerned scientists, activists and politicians, all in the fray. There is now a widespread belief that it is the commercial interests of some global agribusiness giants, involved in the production and propagation of these Bt crops, which drive the momentum on the genetically modified crops. Very recently two of the agribusiness giants, one from the US and another from Europe, were jointly awarded the 2013 World Food Prize, the decision facilitated by the intervention of the Chairman of the Selection Committee, a known protagonist of Bt crops from India. This has led to much criticism in the country from concerned scientists, farmers and anti Bt crops activists.

24.13.2.5 Ecological Limits

The most serious constraints on the path of sustainable increase in food production in future will be the ecological limits to sustainable intensification. Technological innovations are often wrongly perceived to be a route to endless increases in productivity. This is not true, in entirety, because, all factors of production are complementary and not substitutes for capital, *per se.* Recent advances in the science of ecology have shown that biological productivity is limited by ecological capability and efforts to endlessly continue to increase productivity, independent of ecological limits, will fail irrespective of the technologies employed (Nair 2014). Natural resources have threshold levels beyond which the ecosystem functions cannot operate smoothly and a steep decline in productivity can take place. This is what happened in the case of the high input technology of the green revolution in South Asia, especially in India, starting early eighties, when soil fatigue led to yield plateauing or steep fall in many cases, irrespective of the application of recommended rates of external inputs, like fertilizers, irrigation and pesticides (Nair 2013, 2014). To be sustainable, agriculture and other activities have to be confined within the boundaries of natural resources. This is what "The Nutrient Buffer Power Concept" has very brilliantly proved, focusing on the native capacity of the soil, quantified precisely by its buffering power for specific and important nutrients like P, K and Zn, which are very crucial to optimal production in many crop species, yet, became very problematic due to the mismanagement of soil resources arising out of the so-called misconceived green revolution technology.

Limits to biological productivity can occur in the form of biophysical factors as well as agronomic, physiological, and genetic factors. All seem to be heading towards saturation in the highly intensified areas of production. Global agricultural productivity, which grew at 2.3% per annum between 1950 and 1990, has considerably slowed down, slowing to less than 1% growth per annum during 1990–1996, whereas population grew at 1.6% per annum during this period as depicted in the following table (Table 24.1). Clearly, the Malthsian prophecy is unfolding.

According to the World Resources Institute (WRI 1998) annual yield growth rates for wheat declined from 2.92% during 1961–1970 to 1.78% during 1980–1997. The corresponding figures for maize are 2.88 and 1.29% during the same period. Rice yield growth remained stable at 1.95% per annum. This raises doubts about the possibility of increasing productivity by 60% in the coming three decades. It is a tall order to raise land productivity fast enough to offset future increases in population growth and future shrinkage in crop area per season. Clearly, Malthus is winning!

24.13.2.6 Genetic and Agronomic Limits

The 2.8-fold increase in world grain-land productivity since the 1950s has come from advances in plant genetics, agronomic improvements, and, synergies between the two. Genetic improvements made to redistribute a more favorable share of a

Table 24.1 Annual change in grain yields (%) from 1950 to 1995

	Total grain (world and country)			Individual grain (world)			
Year	World	USA	China	Rice	Wheat	Maize	Others
1950–1960	2.0	3.8	–	1.4	1.7	2.6	–
1960–1970	2.5	3.6	–	2.1	2.9	2.4	2.3
1970–1980	1.9	1.9	2.7 (1950–1977)	1.7	2.1	2.7	0.4
1980–1990	2.2	1.0	7.4 (1977–1984)	2.4	2.9	1.3	1.7
1990–1995	0.7	–	1.6 (1984–1995)	1.0	1.0	1.7	–0.8

Source: Compiled from Brown (1997, 1998)

plant's photosynthates into harvestable grain improved the harvest index to almost 55% in the semi dwarf varieties of wheat, rice and corn from an earlier 32%. This is close to 60%, which is considered to be the physiological limit, given that the plant has basic requirements for roots, stems, etc., to produce a sheath and sufficient leaves for photosynthesis.

On the agronomic front, irrigation, fertilizer and pesticide use, and mechanization have been the chief engines for increases in productivity. Fertilizer use increase tenfold from 1950 to 1990, from 14 million tons to 140 million tons, is tapering off in many countries owing to cost and environmental concerns. With genetic yield potential close to the physiological limits, further advances depend on the use of basic inputs such as fertilizers and irrigation to realize the plant's full genetic potential. According to Sinclair (1994), the physiological limits to crop yields may well have been reached under experimental conditions in most cases, and so farmers using the highest yielding crop varieties and the best agronomic inputs may have but few options to raise crop land productivity further. However, the limit to biological productivity of crop land has not yet been reached in large measure, despite intensification with external inputs. The same is not true of marine and grassland ecosystems. The sustainable limits to ocean fishing have already been reached, and no further productivity enhancement seems possible. It is important that growth in cropland productivity is sustained for at least another half a century, until global population stabilizes. If not, humanity will run into great disasters, and millions, especially on the African continent will succumb to hunger. There is no room for complacency because decreasing growth rates and factor productivity are being witnessed in many intensively cultivated areas. It is in this context that "The Nutrient Buffer Power Concept" assumes great importance, because, while it brings in great economy in fertilizer use, it also keeps a close check on environmental pollution. Alternatively, smaller quantities of fertilizers efficiently used can cater to larger areas of crop lands, as against the current scenario of excessive and unbridled use of fertilizers concentrated in lesser irrigated areas with diminishing returns.

24.13.2.7 Ecological Economics and Ecological Limits

Conceptually, the limits of ecology to sustainable agriculture can be explained in terms of the principles of ecological economics. Ecology views the flow of matter and energy in a pyramidal hierarchy where the soil forms the base; green plants that depend on soil for nutrients and water come next, and then herbivorous animals and insects which survive on plants, which are less abundant. Carnivores form the top segment of the pyramid, with humans on top. Soil and water form the most important components of terrestrial ecosystems, and any loss or depreciation of these will be automatically transmitted to the other segments. The area of each segment denotes its importance in the ecological pyramid.

Neoclassical economics, on the other hand, believes that supply and demand control markets. Each level in the neoclassical economists' pyramid is measured by the monetary value of the products produced and consumed at that level. This leads to an inverted pyramid, with soil at the top of the pyramid, followed by field crops, processed foods, and marketed consumer products. The higher the technological advancement, the broader the pyramid becomes, giving greater and greater weightage to monetary value. As opposed to the ecologists' pyramid, where natural resources are valued most, the economists' pyramid gives natural resources the least value. This, in fact, is a contradiction, of sorts. The global economy is sustainable only if the economists' pyramid is confined within the ecologists' pyramid. As it gets bigger, it spells the decline of the ecological system and becomes less sustainable (Davidson 2001). These concepts are illustrated through the following figures (Figs. 24.2 and 24.3) as follows:

The nexus between ecological and economic systems is nowhere more critical than in agriculture, where the output is intricately dependent upon the collective

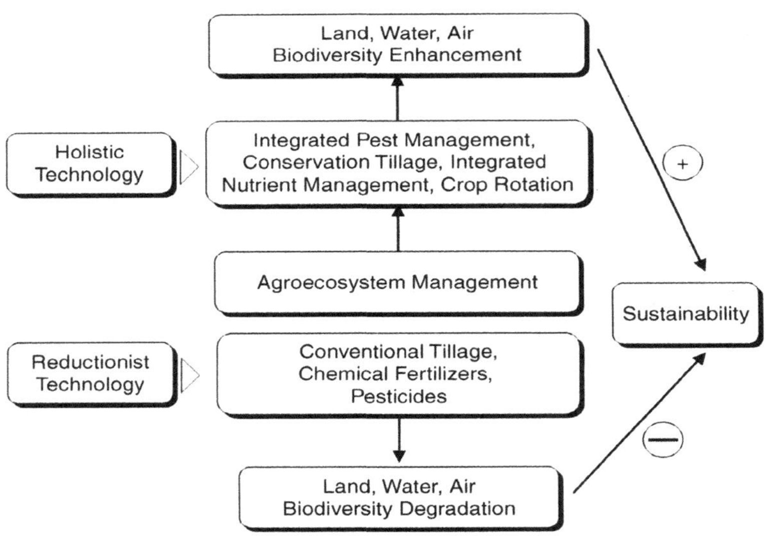

Fig. 24.2 Comparative impacts of holistic and reductionistic technologies

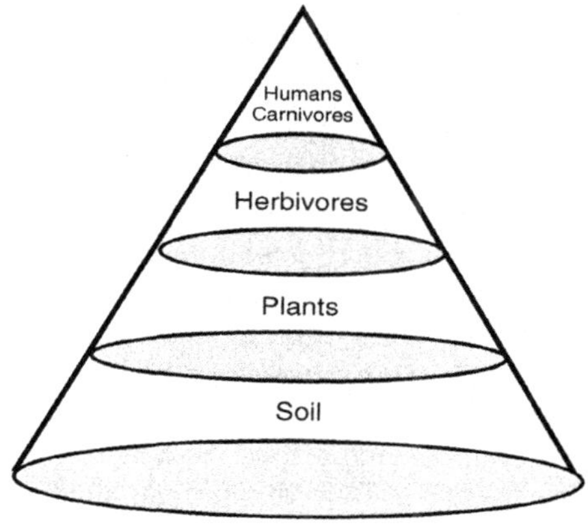

Fig. 24.3 Ecologists' value system

effect of human-made capital such as fertilizers, high yielding crop varieties, pesticides, as well as natural capital stock and its flows. It is also subject to environmental perturbations, such as climate change, and to socioeconomic pressures, such as globalization, market forces, and individual governmental policies. Limits to biological productivity will translate into limits to the economic potential of commercial agriculture (Saroja Raman 2006).

24.14 Conflicting Trends

It is clear that the impressive productivity indicators of the past and the expectations for the future increases in productivity are not in balance with environmental or socioeconomic indicators, and hence, perpetuation of the business-as-usual model is a threat to sustainability. It has already been made clear that economic growth cannot be sustained if it flouts ecological rules. Any number of agricultural production indicators can be cited to show that the current economic environment is in conflict with the earth's natural systems. Among these are collapsing fisheries, shrinking forests, degradation of both crop lands and range lands, desertification, excessive soil salinization, waterlogging, global warming and consequent melting of the polar icecaps, falling water table, drying up of rivers, vanishing of bio diversity of both plant and animal origin. The larger the growth of the economy in the agricultural subsystem, the greater will be its impact on the larger ecosystem. If growth is pushed beyond the earth's natural limits, mutual destruction cannot be halted. Whereas climate change will have a global impact on agricultural production (although this is not yet fully well understood, though there are speculations on the magnitude of yield decline with corresponding increase in ambient temperature), the economic aspects of the environmental disruption have so far fortunately been mainly local. Nevertheless, local damage can have a cascading effect, and eventually, adversely affect global economic trends. For instance, India experienced a devastating drought in the year 2003 and its consequence was felt not merely on the food prices within the country, but, also triggered speculation in the main centers of food export, mainly wheat, from countries such as the US, Canada, Australia and Russia. This is especially true in the age of globalization and trade liberalization in agricultural products, leading to a unified global food system, much under the focus of the WTO, as amply demonstrated in the just concluded (December 2013, at the time of writing this book) Bali (Indonesia) meet of 146 of the developed, developing and least developed nations.

Brown (2001) in his book Eco-Economy, has argued powerfully and convincingly, the case of controlling the global and regional economy within the limits of ecological capability. According to the author, given the scale and rate of degradation of the natural resources of land, water, air and bio diversity, current trends in environmental degradation cannot continue indefinitely without severely adversely affecting the global economy. Specifically, he argues that the food system is likely to be the most important sector through which environmental deterioration will eventually translate into economic decline. This is difficult to dispute, considering that 50% of future economic growth is expected to come from agricultural growth (Brown 1997). Shortages and disruption in food supply will lead to higher prices and overwhelming economic burden for the poorer importing countries in addition to other impacts on human health, productivity, and performance. One of the most tragic outcomes of the process of globalization and liberalization initiated in India since 1991 has been the severe toll it took on the agricultural viability of the country. At the time of writing this book, the country is suffering under a double digit food inflation, and prices of all essential agricultural products, be it grains, vegetables, fruits, fish or poultry have increased exponentially, making them out of reach for ordinary Indians. The country focused on the secondary sector, manufacturing and tertiary sector, services, like the information technology, neglecting agriculture, the primary sector, and, in the process, though it clocked a high (as high as 9%) GDP, the common man is now suffering due to food inflation. By early 2016, when this book is almost complete, things have improved, thanks to a new political dispensation in New Delhi. While the earlier regime was noted for its blatant corruption, the new one seems to have reined in corruption, and this has had a positive impact on inflation, though concerns still exist on the food front. The fluctuating climate pattern is threatening wheat production in India, and, one just wonders when the inflationary process will be set in motion, once again, on the food front. In any bargain, the ordinary Indian is at the receiving end!

24.15 Technological Constraints

Our hopes of balancing the higher demand for food in the future with the supply, hinge on the more efficient use of natural resources through improvement in existing technologies and through creating more efficient technologies. No better testimony can be given of the power and role of technology in safeguarding the sustainability of agriculture than the transition of world agriculture (which remained "subsistence agriculture" until the beginning of the twentieth century) to the current global sufficiency for a more than six-fold growth in global population. But, the warning signals are already flashing since the beginning of the 1980s, when the chemically powered green revolution began to falter, clearly demonstrating that the natural resources, primarily soil and water, have been exploited beyond sustenance. To put the role of newer technology as a tool for agricultural sustainability in the future into perspective, its limitations and direction in which it should move, should be made clear. It is against this background, that "The Nutrient Buffer Power Concept" assumes considerable relevance in food production, *vis-à-vis* intelligent soil management. Earlier discussed success stories from Asia and Africa prove the concept's relevance in modern farming.

Simple technological interventions such as the use of synthetic fertilizers, pesticides, irrigation, and tillage have been the primary drivers of the threefold increase in productivity since 1950. However, it must be acknowledged that these interventions have left very dirty ecological footprints which cannot be easily erased out (Saroja Raman 2006). India is a classic case (Nair 2013, 2014). Of the total geographical area of the country at 328.73 mha, 120.40 mha have been ruined mainly due to soil degradation by many factors, of which unbridled use of chemical fertilizers to prop up the green revolution has been the major culprit. This amounts to 36.6% of the total geographical area of the country. The early externalities of technology such as surface – water pollution and soil erosion were localized and amenable to amelioration through more technology. However, as the impacts grew larger and more widespread, as the current climate change shows, alarming rates of extinction of species, both of plant and animal origin, and genetic erosion of crop species, aggravated through the increasing cultivation of GM crops, the remedial technologies become more and more challenging or simply non existent.

To describe the limitations of technology to solve the problems, Davidson (2001) proposed two new laws termed the "Laws of Techno Dynamics" as described below:

1. Conservation of Problems: Problems do not go away; they are merely substituted one for another. The solution of one problem creates another
2. Technological Challenges always increase: As the human population increases and natural resources remain constant or degrade, technological challenges will increase in size, number, and complexity

True to these two above-mentioned laws, even if technological innovations solve many of our resource and environmental problems, they would generally create new, unanticipated ones in turn. Thus, the problem of improving productivity was solved by use of chemical fertilizers, which in turn gave rise to environmental pollution locally, which has now cascaded into global and irrevocable problems such as the increase in atmospheric green house gases (GHGs) and groundwater pollution. The same is true of excessive use of pesticides, which has led to the evolution of pesticide-resistant crop strains and the consequent emergence of newer pests. The classic example is of the Bt cotton introduced into India in 2002. It was promoted by the Government of India as a safety measure against the dreaded pink boll worm

(American Bollworm). For a time there appeared to be a respite, though the Bt cotton did completely fail in rainfed areas, as it needed assured water supply and ample fertilizer input, including infrequent sprays to control the boll worm. And, where cotton farmers could not afford more irrigation water, and provide ample fertilizer supply, the crop totally failed, as it happened in many rainfed regions of India. And, when the crop failed, many cotton farmers committed suicide due to bankruptcy, because they had raised monetary loans at high rates of interest from unscrupulous money lenders, which they could not repay and hence took the extreme step. But now, another new devastating pest, the white fly, has appeared on the scene and the Punjab cotton farmers are in total disarray, and many have committed suicide due to crop loss and bankruptcy. It has become a national crisis. The other feature of technology is that it can enhance the capacity of natural resources, but, it cannot substitute for the natural capital of land, water, or air. These resources are complementary to each other but not supplants. For example, the productivity of a piece of land depends on the efficiency of use of the technology employed (for example, the quality of fertilizer, water and land preparation). Quantitatively, the interaction of technology (T) and natural resources (NR) can be equated as follows:

$$\mathrm{NR\,X\,T} = \mathrm{Productivity}\left(\mathrm{thermodynamic}\right)$$

The maximum theoretical efficiency of technology is 100%. In practice, however, it is much lower. Hence, the limit to biological productivity, including crop productivity, is ultimately guided by ecological threshold limits and not by technology *per se*. Thus, the ultimate limitation of technology is that it is circumscribed by ecological limits. Technology can never outperform the ecological capability of any natural resource. If it could, the chemically powered green revolution would not have fallen on its face after a decade and a half of "cascading success," as some of its die hard enthusiasts believed (Nair 2013, 2014). It is also equally important to note that it is against the background of this miserable failure that "The Nutrient Buffer Power Concept" came up as an ameliorative and shifting paradigm. But, it would be well worth remembering and accepting the inevitable, that is, this revolutionary concept will also come face to face against a wall, when it would have run its course. Mercifully, it is not trapped in an assumption that the nature's most precious resource, soil, can be exploited, thoughtlessly, *ad nauseam* (Nair 2015a, b, c).

The current signs of a lower rate of productivity growth or yield plateauing in the intensively cultivated systems of agriculture indicate lower total factor productivity and are a reflection of the saturation of current technologies, which "The Nutrient Buffer Power Concept" has succeeded in correcting, inasmuch as the principal input in crop production, plant nutrients, are concerned. Pesticide resistance, pest resurgence, and new pests emerging with prolonged pesticide use are less tangible signs of technology limits. The most serious flaw in agricultural technology development in the past was that technology became geared to agricultural growth, rather than development and sustainability. However, one should not discount future breakthroughs, in terms of new and more effective technologies. Many cutting-edge technologies such as genetic engineering, functional genomics, and precision agriculture are emerging and may be able to fill the feared void. That "The Nutrient Buffer Power Concept" has met with several success stories on the African and Asian continents is proof that this is a possibility. To this extent, we need not be overly pessimistic about the future advances in technology for sustainable agriculture.

References

Altieri, M. A. (1999). The ecological role of agriculture in ecosystems. *Agriculture, Ecosystems and Environment, 74*, 19–31.

Boserup, E. (1965). *The conditions of agricultural growth: The economics of Agrarian change under population pressure*. New York: Aldine.

Brown, L. R. (1995). *Who will feed China? Wakeup call for a small planet*. New York: W. W. Norton.

Brown, L. R. (1997). *The agricultural link: How environmental deterioration could disrupt economic progress* (World Watch Paper 136). Washington, DC: World Watch Institute.

Brown, L. R. (1998). Struggling to raise cropland productivity. In L. R. Brown (Ed.), *State of the World*. Washington, DC/New York: World Watch Institute/W.W. Norton.

Brown, L. R. (2001). *Who will feed China? Wakeup call for a small planet*. New York: W. W. Norton.

Brown, L. R., & Kane, H. (1994). *Full House: Reassessing the earth's population carrying capacity*. Washington, DC/New York: World Watch Institute and W.W. Norton.

Conway, G. R. (1997). *The Doubly Green Revolution: Food for all in the 21st century*. London: Penguin Books.

Davidson, E. A. (2001). *You can't eat the GNP: Economics as if ecology mattered*. Cambridge, MA: Perseus Publishing.

Dumanski, J. (1993). *Sustainable land management for the 21st century* (Vol. 1). Bangkok/Ottawa: International Board for Soil Research and Management and Agriculture Canada.

Ehui, S. K., & Spencer, D. S. C. (1990). *Indices for measuring the sustainability and economic viability of farming systems* (RCMP Research Monograph 3). Ibadan: International Institute of Tropical Agriculture.

FAO. (1996a). *Control of water pollution from agriculture* (Irrigation and Drainage Paper 55). Rome: Food and Agriculture Organization.

FAO. (1996b). *Energy for sustainable development and food security in Africa*. Rome: Food and Agriculture Organization.

FAO. (1998). *Crops and drops: Making the best use of water for agriculture*. Rome: Food and Agriculture Organization.

FAO. (2000). *Agriculture Towards 2015/30 technical interim report, Global Perspectives Unit*. Rome: Food and Agriculture Organization.

FAO. (2002). *World Agriculture: Towards 2015/2030*. Rome: Food and Agriculture Organization.

Gardner, G. (1997). Preserving global cropland. In *The State of the World 1997* (World watch institute report) (pp. 42–59). New York: W.W. Norton.

Harris, R. F., Karlen, D. L., & Mulla, D. J. (1996). A conceptual framework for assessment and management of soil quality and health. In J. W. Doran & A. J. Jones (Eds.), *Methods for assessing soil quality* (SSSA special publication 49) (pp. 61–82). Madison: Soil Science Society of America.

Herdt, R. W., & Steiner, R. A. (1995). Agricultural sustainability: Concepts and conundrums. In V. Barnett, R. Payne, & R. Steiner (Eds.), *Agricultural sustainability: Economic, environmental and statistical considerations* (pp. 1–17). New York: Wiley.

IFPRI. (1997). *The world food situation: Recent developments, emerging issues and long term prospects* (Vision 2020 Food Policy Report). Washington, DC: International Food Policy Research Institute.
Ladha, J. K., Pathak, H., Padre, A. T., Dave, D., & Gupta, R. K. (2003). Productivity trends in intensive rice-wheat cropping systems in Asia. In J. K. Ladha, J. Hill, R. K. Gupta, J. Duxbury, & R. J. Buresh (Eds.), *Improving the productivity and sustainability of the Rice-Wheat systems: Issues and impact* (ASA Special Publication 65) (pp. 45–76). Madison: American Society of Agronomy.
Lal, R. (1998). Soil quality and agricultural sustainability. In R. Lal (Ed.), *Soil quality and agricultural sustainability* (pp. 3–12). Chelsea: Ann Arbor Press.
McNaughton, N. (1998). Earning carbon credits: A future market for farmers. *Holistic Management Quarterly* 59 (April).
Nair, K. P. P. (2013). The buffer power concept and its relevance in African and Asian soils. *Advances in Agronomy, 121*, 447–529.
Nair, K. P. P. (2014, August 18). Green revolution or silent spring? *The Hindu Business Line.*
Nair, K. P. P. (2015a, February 13). The debt story less told. *The Hindu Business Line.*
Nair, K. P. P. (2015b, April 7). When there is no soil left in this world. *The Hindu Business Line.*
Nair, K.P.P. (2015c). The nutrient buffer power concept – A revolutionary soil management technique for sustainable agriculture. In *Keynote address at the National Seminar on "issues, challenges and strategies in sustaining soil health"*. Kerala Forest Research Institute, Peechi, Kerala State, India, December 10–11, 2015.
Oldeman, L. R. (1994). The global extent of soil degradation. In D. J. Greenland & I. Szabolcs (Eds.), *Soil resilience and sustainable land use* (pp. 99–118). Wallingford: CAB International.
PAI. (2002). *People in the balance: Population and natural resources at the turn of the Millennium.* New York: Population Action International.
Patter, C. (1998). *Against the grain: Agricultural environment reform in the United States and European Union.* Wallingford: CAB International.
Pierce, F. J., & Larson, W. E. (1993). Developing criteria to evaluate sustainable land management. In J. M. Kimble (Ed.), *Proceedings of the 8th international soil management workshop: Utilization of soil survey information for sustainable agriculture land use* (pp. 7–14). Lincoln: USDA-SCS National Soil Survey.
Pingali, P., Hossein, M., & Gerpacio, R. V. (1995). *Asian rice bowls: The returning crisis.* Wallingford: CAB International.
Pinstrup, A. P. (2002). Towards a sustainable global food system: What will it take? *Keynote presentation for the Annual John Pesek colloquium in sustainable agriculture, March 26–27*, Iowa State University
Raman, S. (2006). *Agricultural sustainability: Principles, processes and prospects* (pp. 13904–11580). Birmingham: The Hawoth Press Inc.
Reganold, J. P. (1989). Comparison of soil properties as influenced by organic and conventional farming systems. *American Journal of Alternative Agriculture, 3*, 144–154.
Sinclair, T. R. (1994). Limits to crop yield. In *Physiology and determination of crop yields.* Madison: American Society of Agronomy, Crop Science Society of America, and Soil Science Society of America.
Spencer, D. S. C., & Swift, M. J. (1992). Sustainable agriculture: Definition and measurement. In K. Mulongoy, M. Gueye, & D. S. C. Spencer (Eds.), *Biological nitrogen fixation and sustainability of tropical agriculture* (pp. 15–24). Ibadan: International Institute of Tropical Agriculture.
UN. (1999). *World population prospects: The 1998 revision.* New York: United Nations Publications Division.
World Bank. (2000). *World development indicators.* Washington DC: World Bank.
WRI. (1998). *World Resources 1998–99. Food production: Have yields stopped rising?* Washington, DC: World Resources Institute.

Chapter 25
Climate Change and Agricultural Production

Abstract A spate of high-energy industrial, agricultural, infrastructural, trade, and other economic activities, based on the use of mindboggling amounts of fossil fuel, have resulted in a rate of global warming and climate change in the past few decades, not seen in the previous 1000 years. The unprecedented increase in the atmospheric concentration of green house gases, as a result of intensified anthropogenic activities has been confirmed as the cause of climate change. The chapter discusses the consequences of these changes on sustainable agriculture, especially with regard to the high emissions of Nitrous Oxide, during urea hydrolysis, owing to unbridled use of urea, to prop up the green revolution.

Keywords International Panel on Climate Change · Green House Gases · Nitrous Oxide · Agriculture · Climate change · Deforestation · High intensity farming

A spate of high energy industrial, agricultural, infrastructural, trade, and other economic activities, based on the use of mind-boggling amounts of fossil fuel, have resulted in a rate of global warming and climate change in the past few decades not seen in the previous 10,000 years (Saroja Raman 2006).

The unprecedented increase in the atmospheric concentration of greenhouse gases as a result of intensified anthropogenic activities has been confirmed as the cause of climate change. Greenhouse gases (GHGs) consist predominantly of carbon dioxide and to a lesser extent methane, oxides of nitrogen, ozone in the troposphere, and a family, particularly of, insidious synthetic gases, such as, Sulfur hexafluoride SF_6 and chlorofluorocarbons (CFCs), used in refrigeration engineering and aerosols industries. These gases inhibit the escape of heat radiated by the earth and trap it in the atmosphere, leading to warming of the earth's surface.

The International Panel on Climate Change (IPCC 2001) in its third assessment report concluded that the evidence for humanity's influence on global climate is much stronger than ever before. Assuming that no new policies to reduce emissions are put in place, carbon dioxide concentrations will increase from the current 350 ppm to 490 ppm to 1260 ppm by 2100. However, the most recent report of the IPCC (at the time of writing this book in 2014–2015) indicates that, in fact, the rate of warming over the past decade and a half from 1998 to 2012 at 0.05 °C per decade,

K. P. Nair, *Intelligent Soil Management for Sustainable Agriculture*,
https://doi.org/10.1007/978-3-030-15530-8_25

is comparably less than that during the six decades between 1951–2012 at 0.12 °C per decade. In absolute terms, between 1951 and 2012, that is six decades, the increase was 140%, that is nearly 24% per decade, which is a substantial increase given the global scale. This simply shows that the rate of increase of late is much lower than what pessimists try to make out. This, in no way, is room for solace. We have many other contributory factors which are quite worrying.

Of all human activities, high input agriculture contributes substantially to global warming, which many are unaware of. Most environmentalists pin down carbon dioxide as the biggest culprit in global warming. But, we have one factor, nitrous oxide N_2O, a by-product of chemical changes from nitrogenous fertilizer application to soil, in particular, urea, a far bigger culprit than carbon dioxide, contributing to global warming. N_2O naturally occurs in atmosphere as part of the earth's cycle, and has a variety of natural sources, chemically driven agriculture being the main one. N_2O molecules stay in the atmosphere for an average of 120 years before being removed or destroyed through chemical reactions. A pound of N_2O causes over 300 times more to global warming than a pound of carbon dioxide, emanating from industrial activities or automobile emission. In highly industrialized US agriculture, N_2O emissions from soils have increased 80% between 1990 and 2011, accounting to about 5% of all green house gases (GHGs) combined, contributing to global warming. Agricultural soil management is the largest source of N_2O emissions in the US, accounting for 69% of total emissions in 2011.

Climate change is a major imponderable for agricultural production and food security in future on account of uncertainties concerning its magnitude and the way it will impact crop production. Changes in temperature, precipitation, evapotranspiration, and carbon di oxide concentration have the potential to affect crop growth both directly and indirectly through their effects on other biophysical resources, including deforestation, desertification, inundation of cropland, and changes in soil properties and pest patterns. Almost all the secondary effects except carbon dioxide enhancement are expected to be adverse and more virulent in the tropics. The absence of firm data regarding various individual weather parameters on a regional scale and the uncertainty about their effect on crop growth make things all the more serious in relation to the sustainability of agriculture.

In addition to the above-mentioned discussion on N_2O, it must be noted that methane CH_4 also contributes to global warming. Agriculture is the largest anthropogenic source of methane and accounts for approximately 40% of emissions. The concentration of methane in the atmosphere is much lower than that of carbon di oxide, at only 1.72 ppm (IPCC 1996), but, methane is approximately 20 times more deleterious than carbon di oxide. The major agricultural sources of methane are rice paddies, ruminant animals (from flatulence), and bio mass burning. Methane emission is projected to rise to approximately 130–145 million tons per annum by 2020.

Even small-scale crop responses to these changes in the initial stages of destabilization at local levels have the potential to cascade into profound changes in agricultural production and agricultural economy of entire regions and eventually the world, as a whole. This is the last outcome anyone wants given the necessity of higher production from diminishing per capita natural resources of lower potential in the future.

25.1 Challenges to Agriculture from Climate Change Parameters

Climate change is expected to increase the frequency and severity of extremes such as heat waves, intensified rainfall, greater flooding, and tropical cyclones. Such events are bound to affect rural agricultural societies in an adverse manner. Coping with the normal variability of the climate is itself becoming more and more challenging because of the growing numbers of people living in marginal areas under stressful economic conditions. Crafting strategies to adapt to climate change is complicated by the uncertainties in the predictions and projections at the regional level, and, further, by a plethora of interacting socioeconomic, political, and ecological stresses. The landless poor, who are already burdened with multiple dimensions, will be most vulnerable.

Any increase in temperature can be disastrous in such regions as the tropics where the temperatures are already high and resources for adaptive measures are low. Precipitation is the most important factor for productivity of crops. Global warming leading to an increase in mean global precipitation and changed hydrological regimes can affect productivity in a complex location – and – crop specific manner. Extreme events such as brief spells of high temperature, torrential downpour, and droughts can affect productivity very adversely. A sea level rise will pose a threat to agriculture in low-lying coastal areas, where impeded drainage of surface water and ground water, sea water intrusion into estuaries and aquifers, and land inundation will pose problems for agriculture. It is estimated that a rise of 1 m in sea level would affect six million people in Egypt, with 12–15% loss of agricultural land; 13 million in Bangladesh, with 16% of national rice production lost; and 72 million in China, with a loss of tens of thousands of hectares of agricultural land. Other adverse effects, including erosion and damage to coastal infrastructure, salinization of wells, and loss of biotic resources such as mangroves, will be serious. Deltas will be particularly hit by global warming as they are the most vulnerable fertile areas, and, are often heavily populated. The most affected deltas will be the Mekong Delta in Vietnam and the Ganges and Brahmaputra deltas in India and Bangladesh (Rosenzweig and Hillel 1998).

In general, ecosystems that are already stressed, such as those found in the arid and semiarid topics, and low-lying coastal areas, will be at much greater risk of harsh changes. Their socioeconomic vulnerability will be an additional handicap. Unfortunately, these regions are very vulnerable to food insecurity, even now.

Agriculture contributes in the following ways to global warming:

1. Clearing of forests for crop production -Amazon is the classic example
2. Inefficient use of heavy doses of nitrogenous fertilizers
3. Use of heavy machinery for tillage, harvest, and processing of harvested produce
4. Cultivation of rice and paddy on a large – scale, especially under flooded conditions

5. Very large increases in the rearing of ruminant animals and burning of crop residues
6. Production of N, P and K fertilizers (Jackson 1993)

References

IPCC. (1996). In J. T. Houghton, L. G. M. Filho, B. A. Gallander, N. Harris, A. Kattenberg, & K. Maskell (Eds.), *Climate Change 1995: The science of climate change. Inter-Governmental Panel on Climate Change*. Cambridge: Cambridge University Press.

IPCC. (2001). In R. T. Watson (Ed.), *Climate change 2001: Synthesis report*. Cambridge: Cambridge University Press.

Jackson, R. B., IV. (1993). Greenhouse gases and agriculture. In R. A. Greyer (Ed.), *A global warming forum: Scientific, economic and legal overview* (pp. 417–444). Boca Raton: CRC Press.

Raman, S. (2006). *Agricultural sustainability: Principles, processes and prospects* (pp. 13904–11580). Birmingham: The Hawoth Press Inc.

Rosenzweig, C., & Hillel, D. (1998). *Climate change and the global harvest: Potential impacts of the greenhouse effect on agriculture*. Oxford: Oxford University Press.

Chapter 26
Achieving Agricultural Sustainability – The Future Challenge

Abstract The road to sustainable development, globally, is so much rooted in sustainable agriculture. Although the Rio Earth Summit in 1992, in Rio de Janeiro, Brazil, kindled hope for a realistic approach to sustainable development, the World Summit For Sustainable Development (WSSD 2002) in Johannesburg, The Republic of South Africa, a decade later, found the results to have fallen far too short of the intended goals and the efforts to be lacking in commitment and focus. The chapter discusses, at length, the various aspects where serious attention needs to be focused on to attain sustainable development *vis-à-vis* sustainable agriculture.

Keywords Sustainable development · Socio economic problems · Food security · Global economic, and social challenges · Food security · Sustainable intensification · High potential lands · Low potential lands · Natural ecosystems · Agricultural ecosystems · Agroecosystems · Ecology · Productivity · Stability · Sustainability · Equitability

The road to sustainable development, globally, is so much rooted in sustainable agriculture. Although the Rio Earth Summit in 1992 held in Rio de Janeiro in Brazil, kindled hope for a realistic approach to sustainable development, the World Summit for Sustainable Development (WSSD 2002), held in Johannesburg, Republic of South Africa, a decade later, found the results to have fallen far short of the intended goal and the efforts to be lacking in commitment and focus. Like the periodic World Food Summits held in Rome, under the auspices of the UN Food and Agriculture Organization, where very well-fed delegates from all over the world discuss hunger over lavish banquets, and go home after letting out much hot air, and nothing happening, as a followup, while the world's millions of poor in Africa and Asia, go to be bed each day with empty or half-filled stomachs, these "Summits" have earned the notoriety of draining national exchequers' to pamper the rich and mighty. The much cherished hope of reducing poverty and food insecurity by half by 2015, has already been pushed back by at least a decade and a half. These facts confirm the urgency of the need to intensify efforts at the micro and macro levels to inch forward to a better common future for all, at least to the extent envisaged in the Millennium Declaration (UNMD 2000). Towards this end, it is important to overcome the

K. P. Nair, *Intelligent Soil Management for Sustainable Agriculture*,
https://doi.org/10.1007/978-3-030-15530-8_26

multifaceted challenges to sustaining agriculture, as described in the earlier chapter. Ensuring greater agricultural productivity without depreciating resources and environmental quality is a complex exercise, because, productivity and conservation are interlinked and mutually compromising, and, achieving both requires a creative and innovative, yet pragmatic, approach to resource management. Inasmuch as the complete conservation of resources is a Utopian thought, the recommendation of the Keystone Center conference in 1997 in USA, to aim for the absolute minimum acceptable depreciation of natural resources for all economic activities, including agriculture, appears more practical. The current state of the environment necessitates that natural resources should not, under any circumstances, be allowed to deteriorate beyond their ecological or economic regenerative or rehabilitative capacities.

Socioeconomic problems of food security are even more complex, contextual and contentious and are a result of the skewed creation and distribution of wealth and food over a long period and of certain socio cultural biases. The causes are region specific and manifold and range from genuine economic constraints, such as population pressure, low industrialization, and harsh ecological conditions to political ideology, inefficiency, populism, and corruption in governments, to coercion and manipulation by powerful countries and international institutions, quite often to serve their own hidden agendas. Socio cultural prejudices are more difficult to explain. The rapid economic and technological changes which are occurring in the world pose tremendous challenges to social institutions to adjust. The much-needed international cooperation among various countries, even in science-based and validated issues such as climate change and unsustainable agricultural practices, has not been easy to achieve, even though the risks from non adherence and apathy are well understood and remedial measures available.

26.1 Focus on Differential Capabilities of Resources

The pressure to increase food production globally, and in food-deficit developing and least developed countries, in particular, in the coming one decade is tremendous. The required increase is possible only with sustainable intensification of production on all available croplands, both with high and low potential. It must be emphasized in this context the relevance of newer technologies. The relevance of "The Nutrient Buffer Power Concept" is not just in economizing and enhancing fertilizer use efficiency, but also, in the possibility of extending the concept to larger areas. For instance if one comes out with the finding that in a homogenous region, some fields show positive response to a specific lower fertilizer recommendation, it follows that this finding can be extended to the neighboring region without any additional expenditure on experimentation. In other words, using X units of a specific fertilizer carrying a plant nutrient like P, K or Zn, which are very important in crop production, and in whose cases the concept was found to be very effective, in region A with Y area, the same recommendation could be extended to region B with Z area, both Y and Z being of equal size.

The global environmental, economic, and social challenges facing us today will profoundly and detrimentally impact the future of agriculture and its sustainability unless a paradigm shift is effected in the manner food is produced and distributed. Natural resource conservation and the goal of food security for all need people to look beyond the business-as-usual approach and to explore alternative complementary and supplementary measures to address current problems. The aspiration toward food security for all in the near future is an especially tall order. It is imperative that the narrow, one – size-fits-all approach, be replaced by customized models to suit different contexts in terms of natural resources, human aspirations and capabilities, and regional strengths and weaknesses. A reorientation and revitalization of priorities in the agricultural sector, with a focus on the differential abilities of resources and people to function, seems to be the key to sustainable agricultural development.

26.2 Sustainable Intensification of High Potential Lands

As in the past, intensification of ecologically well endowed fertile regions and irrigated agriculture will doubtless remain the mainstay for meeting future increases in food demand. However, extra care needs to be exercised to make production processes more productive and yet more resource conserving and less polluting than the prevalent exclusively chemically powered green revolution pathway, which advocates a "fertility = fertilizer," "pest control = pesticide," and "productivity = high yielding crop variety" reductionist paradigm. It is high time the world adopted a holistic ecosystem approach to manage agricultural systems by increasing the use of integrated technologies such as conservation tillage, integrated pest management, integrated nutrient management, and crop rotations, in place of the reductionist management of individual resources. In addition, a continuous effort to explore and apply cutting-edge technologies, based on recent advances in science and technology, as for example information technology, should also be made to upgrade productivity and environmental integrity and safety, without compromising equity. Human ingenuity still has much scope to complement natural resources, because, the supply of knowledge relevant to agricultural production is high relative to the elasticity of the supply of natural resources.

26.2.1 Differential Intensification of Low-Potential Lands and the People Who Inhabit Them

Though the chemically powered green revolution generally benefited the highly endowed irrigated lands of South Asia, like the Punjab State in India, for a time, for the future, one has to look beyond this spread of benefit, and explore the possibilities for benefit to reach the poorly endowed and risk prone areas. For instance, in

India, central India is primarily rainfed, and the green revolution has not benefited this vast region at all. This is critical if we are to bring food security to people of these regions. An alternative development paradigm based on the contextual and differential capabilities of land and its occupants, combined with the goals of environmental safety and economic equity, at least to the extent of providing all with minimum standards of human development and dignity, needs to be strategized. To achieve this, it is essential to empower the rural poor by providing them with physical and economic access to sustainable food security. In addition to the biophysical and socioeconomic deterrents, primitive rural societies have problems of gender discrimination that prevent women's access to resources and hinder their realizing their full potential and being adequately rewarded for their work. All these factors show that the agenda for attaining universal food security and minimum standards of human development in an ecologically, economically, and socially equitable manner, which is the quintessence of sustainable agricultural development, should focus simultaneously on a variety of measures specific to the conditions prevailing at different places and at different times. The kinds of interventions which require attention now can be clustered into the following categories:

1. Sustainable intensification of high-potential lands using an ecosystem approach
2. Development and adoption of cutting-edge technologies
3. Sustainable intensification of stressed ecosystems
4. Ensuring food security for all
5. Enabling minimum development for the rural poor through agricultural growth as the vehicle to create sustainable livelihood opportunities
6. Striving for gender equality in peasant agriculture

The above-mentioned interventions together will help bridge the current gap in environmental care as well as the deep divide in individual entitlements to resources and human dignity. The concept of the Doubly Green Revolution (Conway 1997) flags up some of these issues. In addition, being strategically prepared to deal with anticipated, but, hitherto unencountered threats or situations is essential to safeguarding the smooth transition to sustainable agriculture. Global warming and its impact on agriculture and the globalization and liberalization of agricultural trade are two current concerns which need strategic and anticipatory planning to attenuate their adverse impacts on global and individual food security. Thus, production processes, need to be prepared to adapt to or cope with the adverse impacts of global warming and local climate changes, and an enabling climate needs to be created to ensure just conditions for the developing countries and poor farmers to survive and compete in the new regime. These topics are discussed in the following sections.

26.3 Expanding the Concept of Sustainable Intensification of High Potential Lands

26.3.1 Natural Ecosystems as Models for Agricultural Systems

Natural ecosystems are the best models of sustainable biophysical systems and the closest to agricultural systems. Understanding their characteristics and functioning is an ideal and logical route to understand and design sustainable agricultural systems. An ecosystem consists of plants, animals, and microorganisms which interact with each other, with the physical and chemical environment, with adjacent ecosystems, and with the atmosphere. Thus, ecosystems are communities of integrating organisms and the physical environment in which they live. They are not simple assemblages of species, but are holistic systems with close and symbiotic interactions between their organic, inorganic, and living forms. Synergistic feedbacks between the organisms and their environment sustain the structure and functioning of ecosystems. Natural ecosystems have lasted over long periods of time essentially because they are self-regulating and not goal- seeking, which makes productivity, stability, sustainability, and equity in-built characteristics.

The basic requirement to sustain any activity or system is the availability of adequate resources. Solar energy is the driving force which facilitates the cyclic use of the materials and chemicals needed for natural ecosystem organization and maintenance, and the atmosphere acts as the sink for the waste which is created. Ecosystems are small factories which produce a number of goods, both biological and physical, and perform a number of functions or services which help in terms of productivity, maintenance of integrity, and conservation of resources. These functions have been described in an earlier chapter.

Species and genetic diversity provide sustainability to ecosystems. They play major roles in the self- organization and maintenance of large-scale ecosystems. Species diversity provides the units through which energy and materials flow, giving the system its functional properties and contributing the resilience to respond to and recover from unpredictable stresses and shocks (Holling et al. 1995). Genetic diversity is important for the evolution of all species, (something which the GM crops attempt to nullify, as it will be the precursor to the destruction of genetic diversity), including humans. Genes preserve information, which helps the self-organization process and allows the generation of options which have better chances of survival. Equitability in the case of natural ecosystems refers to the balance, cohabitation, and coevolution of different species without any bias.

The productivity of ecosystems is sustainable because the natural capital stock (soil, water, air, biodiversity of species, and soil fertility) is renewable, certainly, only up to a limit. Soil microorganisms in conjunction with their physical environment enact the biogeochemical carbon, nitrogen, phosphorus, and sulfur cycles which play a major role in maintaining the fertility of soil, decomposition of waste, conservation of soil structure, storage of water, and provide aeration in all terrestrial ecosystems (Saroja Raman 1992). These basic cycles are aided by diverse species of above ground plants which regulate sunshine, help in nutrient mining from different soil depths, and harbor insect species required for pollination and pest control and the animals and birds required for seed dispersal and the addition of nutrients. Natural ecosystems have closed nutrient cycles; they recycle all the materials produced. In a closed ecological loop, the wastes or by-products of one process become the input for another process. This is critical for the sustainability of ecosystems.

Ecological systems play a pivotal role in supporting all levels of life on earth. Ecosystems produce renewable resources and ecological services which support and protect human activities or affect human well-being. However, the global ecosystem, which is the source of all the resources needed for the economic system, is finite and limited in its regenerative and assimilative capacities, and thus, all human activities have to be accommodated in the global system to be sustainable.

26.4 Agroecosystems

Agricultural systems are essentially ecological systems transformed by human beings to produce food and fiber, often with the goal of increasing their productivity. This leads to a high degree of simplification and modification of the original ecosystem. The great diversity of life forms in the original ecological system is reduced to a restricted assemblage of crops, pests, weeds and so on. Thus, the biodiversity of an agricultural system becomes quite simplified and tailored to the requirements of human design in terms of the crops chosen. Along with management practices, this design dictates the presence or absence of other species. Similarly, extensive land manipulation through tillage for seed bed preparation and irrigation to complement natural precipitation brings about changes in soil structural and biological characteristics.

Although the ecology of the agroecosystem is modified, basic ecological processes continue to operate. Thus, processes such as nutrient cycling, pest-predator relationships, and hydrological cycles carry on, but, they are regulated by agricultural processes and external management. Nutrient cycling in natural ecosystems is conserving because most nutrients are returned to the soil and recycled. In contrast, agricultural products are harvested and taken out of the farm, resulting in the removal of nutrients from the soil. Similarly, the choice of a limited number of crops, as in monoculture, reduces the biodiversity of pest control agents, which leads to a deficiency in the natural pest control services available (Conway 1987).

Humans seek to impose on agroecosystems the products they need and some constancy and dependability of their supply. Thus, the stability, sustainability, and equity inherent in natural ecosystems are not built into agricultural systems that have to be imposed or ensured through human management strategies. Hence, within a managed agroecosystem, trade-offs are made among productivity, stability, sustainability, and equitability depending on the goals set by humans and the management practices followed. Traditional farming systems tend to mimic natural ecosystems as much as possible by making extensive use of internal resources such as locally adapted species and land races and by depending largely on soil nutrient cycling for plant growth and on species diversity and cultural practices for pest and disease management. Such systems lead to greater ecological balance, stability, sustainability, and a high output/input ratio for energy, but, they may not match the very high productivity of intensive farming with select high yielding varieties and high external inputs.

Commercial agricultural systems operate with the goal of maximizing productivity through the addition of external inputs, which tends to discount the natural ecosystem services that undergird sustainability and trade them off against productivity. It is important that these trade- offs remain within threshold limits because the impairment of some crucial processes can irreversibly damage the system. Devoid of the natural ecosystem processes, the soil may become dead, a mere physicochemical matrix, and lose its potential to sustain production beyond a certain limit. Once it has been lost, the resilience of the soil may be difficult to recover. To ensure sustainability, agricultural systems need to be managed in as similar a manner as possible to natural ecosystems. Holistic technologies meet this need in the sustainable intensification of high-potential land, whereas the agroecological pathway is best suited to low-potential lands.

References

Conway, G. R. (1987). The properties of agroecosystems. *Agricultural Systems, 24*, 95–107.

Conway, G. R. (1997). *The Doubly Green Revolution: Food for all in the 21st century*. London: Penguin Books.

Holling, C. S., Schindler, D. W., Walker, B. H., & Roughgarden, J. (1995). Bio diversity in the functioning of ecosystems: An ecological synthesis. In C. A. Perrings, K. G. Maler, C. Folke, C. S. Holling, & B. O. Jansson (Eds.), *Biodiversity loss: Ecological and economic issues* (pp. 44–83). Cambridge: Cambridge University Press.

Raman, S. (1992). Mineral-microbe interface: A veritable biochemical factory. In *Invited lecture, annual convention of the clay mineral society of India, November 14–16, 1992, Hyderabad, India*

UNMD. (2000). United National millennium declaration. UN General Assembly Resolution 55/2, document A/RES/55/2, September 8, New York: United Nations

WSSD. (2002). *World summit for sustainable development. United Nations conference on environment and development*, August 26 – September 4, Johannesburg, South Africa

Chapter 27
Holistic Technologies

Abstract The chapter discusses different aspects of holistic technologies, such as, conservative tillage, against the background of high-intensity agricultural systems like the green revolution.

Keywords Green revolution · Resource conservation · Pollution curtail · Conservation tillage

A critical review of the so-called green revolution technology shows that a radical change needs to be introduced to blunt its very adverse effects on natural resources, in particular, soil and water, leading to a great deterioration of environmental quality (Nair 2013). The green revolution, has, without long-range vision, followed a reductionist, productivity – and profit-oriented (the agri" business" model, not agri "culture" model) path to production by treating soil, water, biodiversity, and energy as more or less independent, unconnected resources. This approach is incompatible with the very strong inter-connectivities and synergies between the physical environment and the biological communities that inhabit it. Agricultural production systems are structurally similar to natural ecosystems, which are embodiments of a harmonious inter-relationship among their individual components. Sustainability is a systems issue and has to be negotiated through a holistic strategy involving all the components of the system. Ideally, the management of agricultural production systems should follow an ecosystems approach.

Fortunately, several types of resource-conserving, pollution – curtailing, and energy-saving integrated technologies such as conservation tillage, integrated nutrient and pest management, and poly culture (e.g., crop rotations, intercropping, cover cropping etc) have been developed and to a limited extent put in place. Although most were designed to respond to specific constraints in different contexts, they have been shown to have several integrative benefits across resources not anticipated at the time of their innovation. Many have also been improved and upgraded to include multiple benefits. Thus, the practice of no-tillage, which was introduced in the 1970s in the wake of the oil crisis, following which the oil prices steeply escalated, has metamorphosed into conservation tillage, which includes several additional benefits.

K. P. Nair, *Intelligent Soil Management for Sustainable Agriculture*,
https://doi.org/10.1007/978-3-030-15530-8_27

These integrated technologies incorporate a large degree of dependence on the regenerative resources of natural ecosystems such as nutrient cycling, water storage, biological pest control methods, enhancement of soil microorganisms, fertility, resilience and soil structure. They not only are more benign and cost- effective, but, also have mutual synergy and reciprocal advantages. They are effectively hybrid technologies combining human-made technologies and natural renewable resources to facilitate intensification without overuse of fertilizer input, pesticides, and fossil fuel. Such practices are being increasingly recommended and implemented in various parts of the world and constitute a major pathway to attenuating, if not mitigating, the ecological and environmental degradation created by high-external-input intensive farming. The following sections provide a brief account of these technologies and their contribution to sustainability of agriculture.

27.1 Conservation Tillage

Conservation tillage in general refers to any form of reduced tillage system with at least 30% of the soil surface covered by crop residue after planting. The term encompasses many different soil management practices, including "no tillage," ridge tillage, mulch tillage and contour planting. Compared with conventional tillage, which involves moldboard ploughing and inversion of the soil, conservation tillage is a non inversion tillage that retains protective amounts of residue mulch on the surface and reduces soil and water loss (Mannering and Fenster 1983).

Conservation tillage became popular in the US in the 1970s (when oil prices were shooting up) with the discovery of new pre-emergence and postemergence herbicides and the development of machinery to efficiently carry on planting, with little or no tillage and with minimal disturbance of residue remaining at the soil surface. The 1985, 1990, and 1995 Farm Bills in the US made it mandatory to adopt conservation practices to mitigate soil erosion. Apart from savings on energy consumption, several other benefits can be realized through this form of soil management, and the practice is spreading fast into other countries. The need for sustainable intensification of prime lands to meet the expanding demand for food has led to the widespread adoption of conservation tillage practices in different parts of the world.

27.2 The Comparative Merits/Demerits of Conservation Tillage

Conservation tillage has a number of advantages, but, a few disadvantages, as well. These have been reviewed comprehensively by Blevins and Frye (1993). Conservation tillage is very suitable for slopy and well-drained lands, while it is unsuitable for wet soils. It reduces labor requirements and offers greater timeliness in terms of land preparation and other operations and more flexibility in planting and harvesting. In addition to these operational advantages, conservation tillage offers several ecological advantages, such as, improvement in surface water quality and soil physical, chemical, and biological properties.

The most important advantage of conservation tillage is the prevention of soil loss, which mainly stems from the fact that it requires little or no soil disturbance. Abundant evidence supports the conclusion that conservation tillage systems with substantial residue on the soil surface reduces sediment loss and loss of agricultural chemicals (especially nitrates) from the field by way of run off. Conservation tillage improves soil organic matter content, which leads to other benefits such as improvement of soil aggregation and leads to good soil structure (Douglas and Goss 1982). Much of the improvement in soil structure is attributable to high biotic activity, especially that of the earthworms, which flourish in a no-tillage residue-management system. Reduced tillage increases water infiltration through improved soil aggregate stability, which offers resistance to erosion and compaction and facilitates seedling emergence. Tillage systems with 50% more residue cover are effective in conserving moisture through increased infiltration, which is advantageous in low-rainfall areas and in soils with low water-holding capacity. The surface residues slow down surface run off and lower the rate of water evaporation.

1. Energy Benefits Energy savings from conservation tillage can accrue from many factors: less of fuel consumption, lesser time and labor consumption, the possibility of double cropping, and so on. However, some of these savings may be offset by other factors, including higher rates of herbicide use and special seeding rates and equipment for the same (Frye 1984). Although a better understanding and use of crop rotation, leading to allelopathic activity against certain weeds, may help reduce herbicide consumption, it is not generally practical.
2. *Effect on Soil fertility*

There appears to be no consistent pattern as far as the effect of conservation tillage on mineralization and the availability of N and P is concerned. As fertilizers remain at the soil surface in low tillage, waste is quite possible, especially of N, owing to run off and volatilization. In addition, since notillage soils have better structure and are undisturbed, nitrate loss through infiltration of surface water carrying oxidized nitrate ions into ground water can, firstly, deny plant roots (especially of those cereals like rice and wheat, whose roots do not traverse to lower depths) of N, but, equally importantly, overload ground water with nitrates making it non potable. This has been a major environmental hazard in Punjab, the "cradle" of green revolution, where farmers were goaded to apply huge quantities of nitrogenous fertilizers like urea, in the hope of harvesting ever more yields of rice and wheat, until it turned out to be an environmental nightmare. Of course, this did not result from conservation tillage, rather from unbridled use of nitrogenous fertilizers like urea. In temperate climates where rainfall is high, prolonged no-tillage leads to the acidification of surface layers of soil, which can be attributed to the decomposition of the layer of organic residues. Subsequent leaching of organic acids, by-products of organic matter decomposition, into soil slows down mineralization and nitrification of organic N. The undisturbed soil in a no-tillage system accumulates relatively large amounts of immobile ions, like phosphorus, zinc etc., at the soil surface (Dick 1983).

References

Blevins, R. L., & Frye, W. W. (1993). Conservation tillage in soil management. *Advances in Agronomy, 51*, 34–77.

Dick, W. A. (1983). Organic carbon, nitrogen and phosphorus concentrations and pH in soil profiles as affected by tillage intensity. *Soil Science Society of America Journal, 47*, 102–107.

Douglas, J. T., & Goss, M. J. (1982). Stability of organic matter content of surface soil aggregates under different methods of cultivation and grassland. *Soil and Tillage Research, 2*, 155–175.

Frye, W. W. (1984). Energy requirements in no tillage. In R. E. Phillips and S. H. Phillips (Eds.), *Non-Tillage agriculture: Principles and practices* (pp. 127–151). New York: van Nostrand Reinhold.

Mannering, J. L., & Fenster, C. R. (1983). What is conservation tillage? *Journal of Soil and Water Conservation, 38*, 141–143.

Nair, K. P. P. (2013). The buffer power concept and its relevance in African and Asian soils. *Advances in Agronomy, 121*, 447–529.

Chapter 28
Integrated Plant Nutrient Management

Abstract Integrated plant nutrient system involves the balanced supply of chemical, organic, and biological forms of plant nutrient sources with the aim of improving the productivity, sustainability, and stability of production in an environmentally safe and economically and socially viable manner. The chapter elaborately discusses the merits of organic fertilizers, crop rotation in sustainable agriculture, and also frontier technologies, for sustainable agriculture, like precision farming.

Keywords Plant nutrients · Organic fertilizers · Integrated plant nutrient systems · Traditional farming systems · Crop yield · Soil quality · Biological diversity · Constraints on implementing crop rotations · Information technology · Precision farming · **"The Nutrient Buffer Power Concept"** · Sustainable land management · Low potential areas · Multifaceted dependence of agricultural sustainability · Biodiversity · Agroecosystems · Agricultural biodiversity Agricultural intensification · Off-farm loss of biodiversity · Conservation and sustainable use of agricultural biodiversity

Integrated plant nutrient systems involve the balanced supply of chemical, organic, and biological forms of plant nutrient sources with the aim of improving the productivity, sustainability, and stability of production in an environmentally safe and economically and socially viable manner. The organic sources of nutrients comprise a wide variety of resources, including plant residues, farm yard manure (FYM), livestock waste, green manure from leguminous crops, biogas slurry, municipal sewage sludge, and biological N fixers. Plants build up their biomass using water, atmospheric carbon dioxide, energy from sunlight, and nutrients from soil and water. For optimum plant growth, nutrients have to be available in the proper mineralized form, in adequate and balanced amounts, when required by the crop, and in a form accessible to the root system. Management of nutrient availability is the most fundamental step in sustainable crop production.

K. P. Nair, *Intelligent Soil Management for Sustainable Agriculture*,
https://doi.org/10.1007/978-3-030-15530-8_28

28.1 Sources of Plant Nutrients and Their Impact

Plants acquire their nutrients from a variety of sources, such as, soil reserves, mineral fertilizers, organic sources, biological N fixation through leguminous crops, irrigation and rainwater. Of these, soil reserves are the most crucial because they determine the sustainable productive capacity of the soil. It is for this very reason that "The Nutrient Buffer Power Concept" assumes considerable relevance in sustainable soil management for optimum crop production, because, the concept directly addresses the crucial question how a soil maintains an optimum nutrient concentration on the plant root's surface, based on its "Buffer Power." This, in turn, directly affects the productive capacity of a soil. Soil reserves depend on the nature of different nutrients and also on the type of management practices adopted. These reserves are often in forms not immediately available to plants and are released slowly through mineralization mediated by soil microorganisms and through chemical processes. The rate of release is often too slow to match the removal of nutrients through agricultural production. For sustained intensification, nutrients have to be supplied from external sources such as chemical fertilizers and from organic sources (FAO 1998).

The total stock of nutrients available to a crop depends on the supply of nutrients from internal and external resources, plant uptake, and loss of nutrients to the environment. An optimal stock should be maintained for sustained productivity. The capacity of a soil to store easily available plant nutrients is important for efficient plant nutrition management. Agriculture inevitably removes plant nutrients from soil, and they need to be continually replenished to sustain the productivity of land over long periods. Continual nutrient mining through crop removal without adequate replacement and without balanced plant nutrition practices leads to loss of soil fertility, and, an inability to sustain adequate levels of soil productivity. This happens quite early in low-input agricultural systems and often leads to irrevocable soil degradation, especially when the soil is inherently low in fertility. It has become a major threat to future food security in areas such as, Sub Saharan Africa and, parts of South America, where, the soils are of an inherently low fertility level and are not conducive to the use of adequate quantities of fertilizers, due to various problems.

Excessive and exclusive use of mineral nutrients, as in the case of intensive high-external-input agriculture, like the so-called green revolution, poses a different type of problem. It leads to depletion of organic matter, leaving the soil a dead physico-chemical medium, and to problems of soil erosion, pollution as a result of run off, wastage, lower efficiency of nutrient use, and lower profitability. Nitrate pollution of surface water and its intrusion into the ground water have become common outcomes of high-input agriculture. Emissions of ammonia from the hydrolysis of urea and of nitrous oxide into the atmosphere from denitrification have led to acid deposition and global warming, respectively. Nitrous oxide has five times more heat trapping power compared to carbon di oxide and a molecule of it stays put in the atmosphere for 120 years, which is difficult to get rid of. Use of only chemical fertilizers has been shown to affect sustainable production in intensified commercial agricultural systems. Several long-term intensification programs are showing signs of fatigue and yield stagnation, one of the reasons being a reduction in the amount of organic matter. Punjab state, the "cradle" of Indian green revolution is the classic example of this extreme environmental fallout, where thousands of acres have degraded soil.

28.2 Merits of Organic Fertilizers

Organic matter is central to soil fertility. The greatest environmental havoc the so-called green revolution has created on Indian soils is the total depletion of native soil organic matter due to continuous cereal (rice-wheat) mono cropping, coupled with unbridled use of chemical fertilizers, especially nitrogenous, like urea. Punjab State, in India, the "cradle" of Indian green revolution is the classic example of this situation. Organic matter acts both as sink and source of plant nutrients. The resilience of the soil nutrient balance is to a large extent based on this dual role. Organic matter acts as a substrate for a variety of soil microorganisms that immobilize excess nutrients, mineralize them, and make them available to plants as they are needed. Organic matter thus prevents nutrients' loss and environmental pollution, and, above all, helps maintain soil resilience, which is the basis of sustainability (Swift and Woomer 1992). In fact, soil organic matter plays a crucial role in soil buffering. Integrated plant nutrient systems improve the efficiency of use of plant nutrients, limit their loss to the environment, and reduce unproductive uptake of nutrients. This enhances profitability.

The usefulness of organic matter sources goes beyond their nutrient content; they help improve not only soil fertility and environmental quality, but also, sustainability and stability of production. Integrated nutrition management is, in particular, useful in improving and maintaining plant nutrient stocks in soils. Both the immobilized capital stock of plant nutrients ("Quantity") and the working capital of nutrients from the mineral and organic sources ("Intensity") are important pools for optimum crop production and sustainability. It is through a mechanistic-mathematical model that Nair (1996) established a quantitative thermo dynamic relationship between these two entities, which helped quantify "Buffer Power" of important plant nutrients like Phosphorus, Potassium and Zinc, with great precision. This helped predict uptake of these plant nutrients by test plants very precisely and was used as the basic parameter to compute precise fertilizer recommendations for a number of test crops in soils of three continents, Europe, Africa and Asia. The nutrients stored in soil should not be depleted beyond a critical level because they cannot be transferred rapidly from one pool to another and from one place to another.

28.3 Additional Advantages of Integrated Plant Nutrient Systems

Integrated plant nutrient systems offer other synergistic benefits, including efficient rooting systems, better water infiltration and water holding capacity, and, amelioration of soil acidity and soil structural defects, such as, soil compaction. They make efficient use of crop residues and promote crop rotations to optimize N fixation, indirectly reducing the problems of pests of monoculture and exploiting nutrients which are released slowly from organic mineral sources. They also maximize economic returns, especially in resource-poor and marginal environments, by using locally available and low – cost materials and reducing the cost of production. The addition of optimal amounts of mineral fertilizers to the locally generated organic sources helps to prevent insidious depletion of the stock of plant nutrients in the soil, which is a hidden form of resource degradation and a threat to sustainability. The nutrient capital stock in the soil is essential for its regenerative processes and resilience.

28.4 Crop Rotation in Sustainable Agriculture

Crop rotation is a system in which different crops are grown in cyclical succession and in a planned sequence on the same land. The practice of crop rotation predates even recorded history. The history of crop rotation and all of its modern developments have been extensively reviewed (Karlen et al. 1994).

Traditional farming systems attach much importance to crop rotations, even without understanding the scientific rationale. Crop rotation was initially related to a perceived improvement in soil productivity. Legumes featured in early crop rotation practices. The beneficial effect of the residual N fixed by legumes became known during the nineteenth century through extended experiments on crop rotation at Rothamstead Experiment Station in England. Rotation with legumes such as alfalfa was common in wheat, cotton and corn in the US until 1950s to augment N availability and enhance soil fertility. Post- World War II developments in commercial agriculture – plenty of inexpensive nitrogenous fertilizers, pesticides (especially in the US and Europe, though not in Asia), large farm size, and mechanization of crop production – ushered in the culture of continuous monoculture. This model was replicated in the fields of South Asia, especially in countries like India, Pakistan and Bangladesh, which led to an exponential increase in grain yields of important staples like rice and wheat, during some years, but, after a while, became unsustainable, due to very adverse environmental fallout.

Increasing concern about the sustainability of high intensity, all-chemical agriculture, has once again brought into the fore the importance of crop rotations. Traditional ecological knowledge, coupled with current advances in agroecology, has placed crop rotations on a strong scientific platform. Evidence indicates that in addition to productivity increases through including legumes in the rotations, several other ecological benefits can accrue, such as biological, pest and disease resistance, weed control and improvements in soil quality and sustainability. Several long-term experiments lasting more than 100 years, at Rothamstead, and elsewhere around the world, have validated the multiple benefits of crop rotations.

28.4.1 Improvement in Crop Yield

Improvements in crop yield is among the strongest practical justifications for the re-introduction of crop rotations. Improvement in yield in grain-legume rotations such as corn or sorghum with soybean has been reported in a number of experiments on crop rotation in the US. Rice or wheat with legumes is a common rotation in India to augment N and to capture the residual soil moisture. Effective scavenging of residual N added to the soil from the previous crop and biological N fixation by the legume crop are conducive to increased efficiency of N use and higher yield. Several other factors, including better utilization of available water, increased efficiency of use of other nutrients (P, K and other micronutrients like Zn), the synergistic effect of the rooting pattern, improvements in soil structure, and increase in soil organic matter content, are also considered to be responsible for better yields. The increase in plant and microbial diversity associated with crop rotations is now seen as a facilitator of several other beneficial effects (Karlen et al. 1994).

28.4.2 Soil Quality

Improvement in soil quality has been recognized as the most important beneficial impact of crop rotations. This automatically leads to several improvements in the production environment, such as increased soil water storage, soil nutrient storage and availability to plant roots, improvements in soil structure, enhanced soil microbial activity and weed control, decreased insect pressure, lower nematode populations and disease incidence, and presence of growth-promoting substrates (Karlen et al. 1994). Improving soil quality is the most direct and effective way of sustaining soil productivity, conserving soil resources, and maintaining the environmental integrity. Different crop rotations may improve soil quality by mimicking natural ecosystems more effectively than current farming systems (Karlen et al. 1992). An increase in spatial, and temporal diversity across landscape is likely to positively impact several soil quality parameters such as soil structure, soil organic carbon, aggregation, bulk density, water infiltration, and retention. However, it is not so easy to demonstrate direct linkages between crop rotation and individual soil quality parameters and to quantify them (Bullock 1992).

The positive effect of crop rotation on soil quality far outweighs all other benefits in terms of sustainability enhancement. Crop rotations which include legumes or grasses are generally beneficial to aggregate stability and the formation of favorable soil structure, possibly through increases in the formation of bio pores by plant roots and soil fauna and the addition of organic matter. The net effect depends on the characteristics of the plant species being grown, the sequence of different species, and the frequency of harvest. Grasses, in particular, can improve soil quality by creating bio pores in compacted soils at depths that are not economically viable to till.

Soil bulk density is generally lowered by crop rotations in comparison to monoculture; indeed, those cropping systems that return the most residues to the soil usually result in lowered soil bulk density. The main reasons for this are the incorporation of crop residues and the formation of extensive root system leading to the formation of macrospores. Soil organic matter is the most important soil quality component because it enhances most other soil quality parameters, including water infiltration, aggregate formation and stability, and soil aeration, and reduces soil erosion and bulk density. This has a direct bearing on sustainable agriculture.

Certain useful guidelines have been laid down to exploit specific benefits from crop rotations. Strategies using crop rotations can be tailored to specific requirements. Balanced fertility build-up and exploitative cropping can be achieved by including a legume and crops with different rooting depths. Complementary root systems in a rotation sequence may be able to fully utilize available water and nutrients in the soil (Karlen and Sharpley 1990) and may become a factor in opportunistic cropping for sustainable crop production and water conservation. Combining crops with different pest, disease, and weed susceptibility and possible allelopathic interactions can introduce control mechanisms to prevent attacks by pests – insect, disease, and weed. Including a green manure crop, like Sesbania, Sunhemp or Glyricidia, sod cover, and winter cover can improve organic matter content and prevent soil erosion and run off. Crop rotations, that involve long periods of sod, pasture or hay crops generally increase soil organic matter content, which benefit the subsequent crops. Short-term rotation does not increase soil organic matter (Bullock 1992). A deeper understanding of the biological and natural environment and the complex interactions among the components of a cropping sequence can help in harnessing a greater share of internal resources (Altieri 1995).

28.4.3 The Role of Crop Rotations in Pest and Disease Control

Crop rotation is a basic tool of integrated pest management. Cultural control of insects depends on increasing the biodiversity on and surrounding farms through crop rotation, cover crops, mulching, strip cropping, and so on. This makes the environment unfavorable for specialized pests and promotes varied types of pests, together with predators and parasites, which reduce initial pest colonization, reproduction, survival and dispersal (Dent 1991). Increased temporal and spatial biodiversity in crop rotation helps to break the life cycle of pests and decreases pest pressure. Rotations are designed to avoid damaging levels of pests and diseases by excluding plants of the same species from the rotation. Greater botanical differences among the crops in a rotation yields better cultural and biological control generally. For example, the rotation of legumes with cereals, or of summer annuals with winter annuals, is more beneficial. Strategic cultural controls such as varying planting time, location, and the timing of harvests can also create a synchrony between crop and insect pest phenology and retard late colonization. However, crop rotation does not control all pests and is ineffective against highly mobile pests that can invade from other fields. Furthermore, the host range of the pest must be fairly narrow for crop rotation to be effective, and the pests must be incapable of surviving long periods without a living host.

In addition to insect control, crop rotation also helps to control specific weed species. Temporal diversity achieved through crop rotation and spatial diversity achieved through inter cropping can reduce weed population density and enhance biomass production of the principal and accompanying crop. Allelopathic plants that release chemical compounds either directly or indirectly through microbial decomposition of residues also suppress weeds (Liebman and Dyck 1993). The effectiveness of crop rotation in the prevention of diseases is not well understood. The build- up of competitive and predatory microorganisms to combat existing pathogens may be the most plausible explanation (Crookston et al. 1991).

28.4.4 Biological Diversity

Crop rotations result in more biological diversity than monoculture practices. The spatial diversity which results from a greater number of crops being grown on the landscape at any given time is useful in preventing erosion and lowering of pest concentration. The temporal diversity resulting from a sequence of crops being grown in a given field facilitates the breaking up of pest cycles and preventing the long-term build-up of pests. The increase in biodiversity also leads to other beneficial effects, including closed nutrient cycles, and generally leads to cumulative yield increases.

28.4.5 Crop Rotation Effect on Reduction of Greenhouse Gases

Of late, crop rotations have been gaining attention for the reductions they allow in energy use and the emission of green house gases (GHGs) in agricultural production by reducing the use of nitrogenous fertilizers, a very important source of GHGs. Crop rotations are among the accepted measures under the Kyoto Protocol (UNFCCC 1997) for reducing GHGs' concentration in the ambient space. Active research is under way in the area of energy audits and the beneficial impact of crop rotation on the environment.

28.4.6 Constraints on Implementing Crop Rotations

Although the benefits of crop rotations over monoculture are well documented, the reasons are not well understood and cannot be critically validated. Analysis and validation require extensive experimental knowledge because of multiple and unquantifiable effects. Crop rotations are also affected by other factors such as climate, soil type, cropping sequence, and tillage, and the benefits are site-specific. Hence, generalizations are not possible. Factors such as labor constraints and associated risks may act as deterrents to practicing crop rotations. Crop rotation may not always be beneficial, either. For example, growing sorghum reduces the yield of any other crop subsequently in the same field; the decomposition of the very high carbohydrate content of its roots gives rise to a proliferation of microorganisms which immobilize soil nutrients and make them less available to succeeding crops.

28.5 Frontier Technologies for Sustainable Agriculture

The fear of reaching ecological limits to our current technological capability has made clear the need for a continuous search for more efficient and environmentally safe practices to achieve future improvements in crop productivity. Recent advances in cutting-edge space, information, and communications technology and in biotechnology have opened up new frontiers for sustainable agriculture in the form of precision agriculture, transgenic plants, and functional genomics. These technologies hold promise for resource conservation coupled with efficient and environmentally benign production. Precision farming is based on remote sensing, satellite technology, and geographical information systems (GIS). Agricultural biotechnology based on the gene revolution is another area of promise that is still in its infancy, although valid scientific and sociological concerns have been voiced about its merits, especially its bio safety. These technologies are most appropriate for the sustainable intensification of high-potential lands of fairly large areas.

As the topic of this book relates more to soil management, the following discussion would be confined only to those topics of direct relevance to soil management, such as remote sensing, GIS, and precision farming, and, exclude discussions on gene revolution, transgenic plants etc.

28.5.1 Information Technology and Precision Farming

28.5.1.1 Remote Sensing

Remote sensing involves acquiring information about an object without coming into contact with it. It has become a powerful technique to gather onetime or continuous information about the various types of natural and artificial features on the surface of the earth. Combined with other tools to integrate information, such as geographical information systems (GIS), it has proved very valuable in the management of natural resources.

The interaction of the electromagnetic radiation with a surface causes the incident electromagnetic radiation to be reflected, absorbed, and transmitted to different extents depending on its wavelength and the molecular structure and condition of the material. Remote sensing is based on the specific and characteristic spectral signatures of various surfaces or objects. Different types of sensors are used depending on the type of radiation and the required resolution. Multispectral scanners capable of detecting and recording both reflected and emitted radiation simultaneously across several spectral bands are available. The bases on which remote sensors are mounted (remote sensing platforms) differ in type and sophistication from simple balloons to aircrafts and satellites. Satellite platforms enable large-area coverage at frequent intervals in uniform solar illumination conditions and are suitable for applications based on synoptic measurements over large areas and where periodic observations over the same area are required.

Two types of earth satellite orbits have been used: a geosynchronous or stationary orbit and a sun-synchronous orbit. A geosynchronous satellite has an orbital period equal to the earth's period of rotation; its altitude is 35900 km. Such a satellite is stationary with respect to the earth and allows continuous viewing of the part of the earth within the line of sight of sensors. Its distance from the earth and the need for multiple satellites to obtain global coverage are the disadvantages. Sun-synchronous satellites pass over the same ground track at the same local time each day, and the data collected are standardized with respect to sun angle; they can provide global coverage. The data can be in the form of photographic images of different types and resolutions and can be stored in digital form for later processing or interpretation. Specific terrestrial features can be analyzed using multi-land and multi-stage photography. Landsat imagery has many advantages, including large area coverage, repetitive coverage, computer-compatible data, low-cost, ground reduction of 20 m, and stereoscopic coverage.

Remote sensing has been very effective in gathering multi-layer information of various kinds for natural resource management or to establish an administrative framework. Reliable and comprehensive appraisal of features such as land forms, soil, vegetation, land uses, water resources, pest coverage, and crop yield is possible through visual interpretation of aerial photographs and satellite imagery and through digital analysis of computer-compatible data in conjunction with on-the ground-observation. The launch of the U.S. Landsat 1 satellite in 1972 ushered in a revolution in the surveying and mapping of forests in real time. Remote sensing from space has become the vital tool for monitoring, inventorying, and managing all natural resources. Among the major areas of use of remote sensing in agriculture are the following:

1. Monitoring forest resources, including biomass, for the planning and management of forests
2. Monitoring land use and identifying pasture land, farm land, agricultural land, cropping areas, irrigated and dry areas, salt-affected areas, wasteland and so on

3. Monitoring soil type, classification and soil moisture
4. Predicting area under agricultural crops
5. Monitoring pollution of surface and ground water resources
6. Monitoring coastal land environment
7. Identifying biodiversity changes and animal habitats
8. Predicting, monitoring, and managing of natural disasters such as drought, floods, cyclones, and earthquakes
9. Monitoring weather and precipitation

A land-use survey is useful to identify major land-use units such as cultivated lands, sandy wastes, saline lands, and rocky, gravelly wasteland within a reasonable time frame. Planning can be initiated and implemented according to population and resource constraints. Similarly, water resources can be mapped to plan for agricultural, municipal, and industrial uses, to develop cheaper resources, and to identify surface-waterbodies and ground water zones (aquifers). Watershed planning and development, in particular, has benefited from remote sensing. Combined with geographical information systems (GIS), a continuous survey is helpful in the detection of temporal changes in land quality, forest cover, water bodies, grazing land, and so on. Remote sensing has been used extensively in India in the classification and mapping of land forms and soil types, in soil classification, and in evaluating forest cover. The application of remote sensing to wasteland mapping has been a major achievement in India to estimate the employment and livelihood opportunities which could be derived from its rehabilitation.

28.5.1.2 Geographical Information Systems (GIS) and Sustainable Agriculture

Geographical Information Systems (GIS) are computerized information storage, processing, and retrieval systems, with hardware and software specifically designed to handle geographically referenced spatial data and corresponding attribute information. Remotely sensed data for various earth-surface attributes from satellite imaging or aerial photography are digitized, analyzed, and integrated with other geographical databases. The spatial data usually take the form of layers, each of which may depict a different attribute such as topography, soil types, water availability, climate, forests, population, and infrastructure. Conceptually, they can be thought of as a stacked set of map layers where each layer is aligned or registered to all other layers. Different map layers can be combined by overlaying them and integrating the desired information.

Advances in computer digitization and remote sensing techniques have facilitated the linking of different kinds of spatial data. Geographical information systems are database management systems specifically designed for the processing of spatial and related attribute data simultaneously. They have many other uses, including automated map making and computer-assisted cartography and graphics. Any necessary modification to the spatial data can be automatically incorporated into the statistical databases.

GIS has been particularly useful in agriculture and also in the management of natural resources for agricultural operations. The delineation of global and regional agroecological zones to select sites with the best potential for different crops by integrating data on soils, topography, rainfall, and so on, is a good example. The information on the location of biologically sustainable areas can be overlaid with patterns of land ownership, infrastructure facilities, labor availability, and market possibilities, for example, to determine the most economically suitable sites for these crops. The effects of future changes in these parameters, such as pricing, natural calamities (e.g., drought, excess rainfall), and changes in infrastructure, can also be predicted in making strategic plans.

In the same manner, as explained above, in aquaculture, GIS has been used to forecast development prospects by combining parameters such as soil type to assess site suitability for pond construction with climate parameters for the type of fish and its growth. Forest management can be optimized based on soil type, species suitability, growth, yield, and other data and the impact of deforestation can be monitored.

Sustainable agriculture requires an integration of natural resources that often have conflicting effects. Similarly, the socioeconomic goals of different regions and different societies can be contradictory. The contextual optimization of all these parameters is facilitated by GIS technology, which provides the tools to establish cross-sectoral or dimensional communication. It has become an important tool for the sustainable reconciliation of expanding human needs and economic enterprises.

28.5.1.3 Precision Farming and Agricultural Sustainability

Precision farming, also known as precision agriculture or site-specific agriculture, is a strategy which employs detailed local information to manage production inputs for agriculture in the optimal manner. Its aim is to map out spatial variations in crop response to external and controllable inputs such as fertilizers, seeding rate, herbicide and pesticide usage doses, and available water, with the objective of customizing soil and crop management practices to fit the variable conditions present in each field.

Large commercial farms are often mechanized and managed homogenously without reference to their inherent spatial viability. Precision farming is designed to correct this: advances in information technology and computerized machinery have enabled large modern farms to be mapped into distinct site units which are managed according to their specific requirements

The precision farming approach to crop production is a four-step interactive process of data collection (information), management planning, precise field operations, and finally monitoring and evaluation, which guides the data collection and subsequent processes during the next cycle as depicted in the following figure (Fig. 28.1).

Data collection: Crop yield monitoring is the starting point and also the most mature component of precision farming technology. The pattern of yield variation

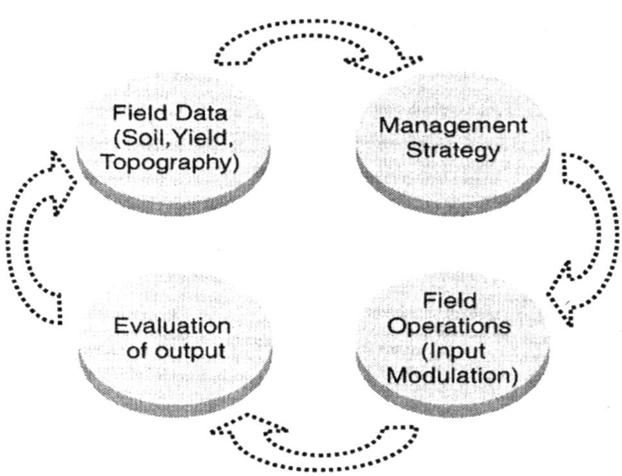

Fig. 28.1 Iterative nature of precision farming. (Source: Modified from Sudduth (1998))

within a field is indicative of the effectiveness or otherwise of the current management practices and drives the collection of information on various yield determinants. Such information can be generated using sophisticated engineering and information technology presently available for spatial referencing.

Gathering information about site-specific variations, both spatial and temporal, with respect to soil-related properties (such as organic matter content, soil pH, nutrient content, and soil moisture) and crop characteristics (such as yield, quality, weed and insect populations) is important to realize the full potential of precision farming. Mapping is especially suitable for collecting data on variables, such as soil organic matter, texture, and yield potential, that do not fluctuate from season to season, but, change slowly over the years. Accurate knowledge of the position at which samples are taken for analysis is achieved using differential global positioning systems, which provide the spatial coordinates required to generate mapped information to an accuracy of within 1 m.

A map of the desired characteristics is generated after grid-sampling the field and analyzing the soil samples. With the exception of grain yield, information is collected through laboratory analyses of the samples. The map is used to decide and to control the rate of application of the inputs: the degree of spatial variability found in the field will indicate whether differential treatment is warranted or not in certain areas.

Management Planning: Unique agronomically suitable treatment strategies for individual spatial units are decided through correlation analysis between variation in crop yield and the measured factors. A treatment map is generated to guide rate-control mechanisms in the field. GIS can be used to evaluate current and alternative management strategies by combining and manipulating data layers to account for spatial differences and interactions between adjacent areas of the field. A computerized decision support system is thus possible.

Field Operations: Variable rate application. The basic idea of precision farming is to measure field variability and supply inputs such as seed, fertilizers, and pesticides at varying rates according to almost "real-time" needs: at the desired location in the field, in the right amounts at the right time, and for the right reasons. The variable rate application machinery receives information from sensors to maintain a database on the actual application rate as a function of field position. A soil map of specific field on a computer fitted inside a tractor, along with a global positioning satellite system, indicates the tractor's position in the field and allows seeding, fertilizer and insecticidal application rates and so on, to be adjusted across the field. Farmers in the advanced countries have access to services in which satellites collect data from a variety of sources, which are then analyzed. In the near future, satellites will be able to identify the location of farm equipment to within a few square centimeters, which will facilitate the following:

1. The matching of the soil map to the location of the soil sample and laboratory results
2. The adjustment of fertilizer and pesticide application rates to fit soil properties such as clay and organic matter content and conditions such as contour drainage
3. The adjustment of tillage to suit various conditions across the field, which will help farmers to prepare efficient seedbeds, control weeds, and reduce fuel consumption and total cost
4. The monitoring and recording of yield data by simply moving across the field
5. The optimization of seeding rate to suit soil conditions and sunlight availability

Farm level controllers for variable-rates of seed, fertilizer, pesticide, herbicide, and irrigation system, and also the application of animal manures are now available (Sudduth 1998). The current systems are generally map-based, but, more of on-line systems are likely to become available in the future, as real-time sensing technologies become widespread in use. In the sensor-based approach, control decisions for variable-rate application are implemented using on-board sensors. No global positioning satellite is required, and the technique is ideal for rapidly changing variables such as nitrate concentration. The sensor-based method utilizes real-time sensors and feedback controls to measure the desired properties, usually soil properties or crop characteristics, and immediately uses this signal to control the variable-rate applications. Sensors are currently available for several soil variables, including soil organic matter content (important for herbicide and fertilizer application), moisture content (for seeding depth), top soil depth (for seeding and fertilizer application rates), weed pressure and identification (for herbicide application), and crop condition (for within-season fertilizer application).

28.5.1.4 Precision Farming and Sustainable Agriculture

Precision farming is a holistic management technique for increasing the efficiency of management of large agricultural farms by targeting inputs according to spatial variability. This in turn leads to a multitude of benefits such as the following:

1. Increase in agricultural productivity, profitability, and product quality
2. Improvement in the efficiency of use of external inputs and of the land
3. Prevention of land degradation as a result of overuse, underuse, or misuse of external inputs
4. Reduction of input costs and minimization of potential non-point sources of pollution resulting from excessive use of inputs
5. Conservation in energy used for direct operation as well as energy embodied in chemicals

Thus, precision farming leads to multiple benefits, such as environmental safety, economic gains, energy conservation, and sustainable use of resources, all of which are important in operationalizing sustainable agriculture (Blackmore 1994). Precision farming is not simply the ability to apply treatments that are varied at the local level; it must be considered as the ability to precisely monitor and assess an agricultural enterprise at the local and farm levels. It can help farmers to develop sufficient understanding of the processes involved to be able to apply the inputs in such a manner as to achieve a desired goal dictated by the context. Precision farming integrates land management with all other inputs. It subsumes the concept of stewardship of the land for future generations, preserving its potential for multiple uses in the future for perpetuity. It facilitates total resource management for agriculture for the conservation and sustainable use of natural resources and external inputs and is vital not only for sustaining agriculture, but also, for the efficient performance of our entire ecosystem.

28.5.1.5 "The Nutrient Buffer Power Concept" vs Precision Farming

The principal and most important difference between precision farming and "The Nutrient Buffer Power Concept" rests on a solid rationale that while it is a certain impossibility to put in place precision farming structures in millions of situations across the developing world, notably on the African, Asian and Latin American continents, "The Nutrient Buffer Power Concept" can be put in place in many farms, starting from small, to marginal, medium and large, because, it primarily addresses the most important single component of farming, that is, the intelligent management of soil to provide optimal nutrition to growing plants, based on the logic that nutrition contributes more than 50% of the plant's field performance. And to be sure, the nutrient factor is the least resilient to management during plant growth. One must ensure that while the productive potential of a field crop should be tapped to the fullest, one must not overlook the very important need to check both soil degradation and environmental pollution. Enhancing nutrient efficiency use, while checking environmental pollution, is the primary focus of "The Nutrient Buffer Power Concept.' The success stories, detailed in the first part of this book, obtained from Africa and Asia, validates the fundamental soundness of the concept.

28.6 Sustainable Intensification of Low-Potential Areas

Although the generation and adoption of technology have been impressive in the agricultural sector in the past five decades, many small and poor farmers in the developing countries have been left out of this development. The lack of availability, affordability, and timely supply of inputs and lack of credit facilities, markets, and infrastructure have been the constraints. In addition, capital-intensive external-input technologies are unsuitable for the complex, diverse, and risk-prone stressed areas that are home to more than 50% of the global population. These are among the reasons for the widespread low productivity, food insecurity, and poverty in vast areas of the developing world.

Strategies for the sustainable intensification of low-potential lands need to be different from the capital-intensive and productivity-oriented green revolution technologies. Low-potential ecologies require simple, inexpensive, and socially and culturally relevant technologies based on agroecological principles and human ingenuity. It is important that these practices improve farm and household income, including returns to labor and physical capital inputs. The conceptual differences between sustainable intensification of high – and low- potential lands are depicted in the following figure (Fig. 28.2).

For universal food security and sustainability, it is essential that the productivity of all types of crop land is increased in a sustainable manner.

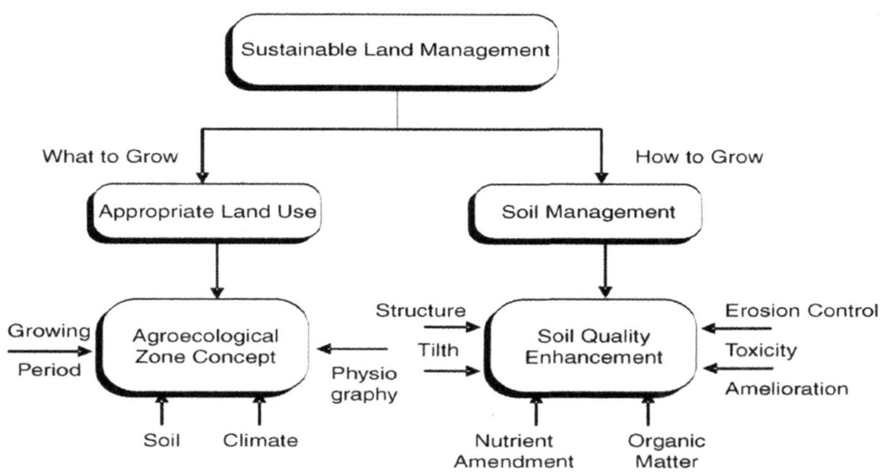

Fig. 28.2 Components of sustainable land management

In many parts of the world, the present reward to agricultural labor, whether for small farmers or for agricultural workers, is very negligible, even to the point of starvation in many cases. This is so very typical of Indian peasantry, the so-called green revolution, notwithstanding, which occurred only in hugely-landed farmers' fields. This very tragic situation is also an outcome of the omnipotent corruption that is rampant in the country, ably aided and abetted by the unscrupulous politicians, swindling the country, with the promise of giving a fair deal to the poor and faceless majority. This has been a continuing political slogan of a dynastic family wielding power during the past several decades. A political wave is sweeping the country, at the time of writing this book. This tragic situation needs to be revised to provide these helpless millions dependent on these areas with food security and acceptable minimum standards of development. Low productive potential, fragile nature, and modest socioeconomic conditions are better served by strategies based on sound ecological principles, which emphasize natural ecological processes and services. Increasing plant, animal, and soil biodiversity, human resource development, and stakeholder participation are crucial under these conditions.

28.7 Agroecological Principles and Protocols

Agroecology evolved as a science in the West in the 1970s, as a fusion of agronomy and ecology. Agroecology views agricultural fields holistically as agroecosystems to be managed along the lines of natural ecosystems. Heavy reliance on natural internal resources and renewable goods and services, such as the biogeochemial cycling of nutrients for fertility and waste disposal, soil structure maintenance, and natural biological control of pests and diseases, is typical of agroecology. In addition to cutting costs, agroecology prevents pollution and land degradation. The traditional indigenous agricultural systems of the tropics, which have thrived on knowledge gained over hundreds of thousands of years of experience, have routinely used such principles and mechanisms as the key to stability and sustainability of production and to protect crops from pests and diseases. A conceptual multifaceted dependence of agricultural sustainability on many inter-linked factors is pictorially depicted below (Fig. 28.3).

28.8 Biodiversity and Sustainable Agriculture

28.8.1 Biodiversity and Its Importance

The recognition of ecological viability as an inalienable component of sustainable development heralded new strategies for the management and conservation of biological diversity. Biodiversity has the overarching role of maintaining the stability,

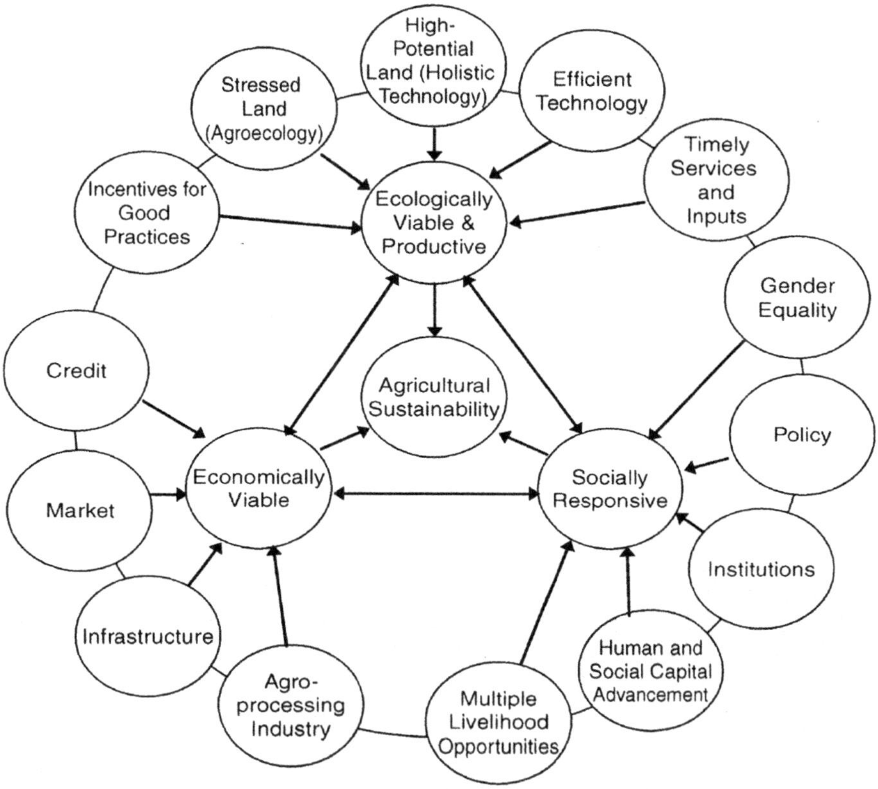

Fig. 28.3 Multifaceted dependence of agricultural sustainability

productivity, sustainability and equity of all kinds of natural ecosystems. Agenda 21 of the Earth Summit, where the Convention on Biological Diversity (CBD 1992) was adopted, provided a blueprint for the management of natural resources with unprecedented emphasis on biodiversity. The objectives of conservation of bio diversity, its sustainable use, and ensuring equitable harnessing of its benefits formed the foundation of the convention.

The convention gave a broader connotation to agricultural bio diversity by transcending the interests of commercial agriculture and including the vast pool of genetic resources and the socio-cultural diversity associated with traditional agriculture. Agricultural biodiversity has since been acknowledged as a distinct branch of biodiversity and as an overarching and essential component of sustainable agriculture and food security. The conservation and sustainable use of plant genetic resources for agriculture and equitable sharing of their benefits are among the most debated issues in the liberalized global trade in agricultural products.

28.8.2 What Is Biodiversity and What Is Its Importance?

Biodiversity refers to the rich and diverse array of living organisms of all species, the genes they contain, and the ecosystems they constitute. It is the product of hundreds of millions of years of evolution involving speciation, migration, extinction, and, more recently, human influence. The Global Biodiversity Assessment adopted the following extended definition of biodiversity (Heywood 1995):

"Biological diversity means the variability among living organisms from all sources, including, inter alia, terrestrial, marine and other aquatic ecosystems and the ecological complexes of which they are part; this includes diversity within species, between species and of ecosystems."

Thus, biodiversity is considered at the following three levels:

1. Genetic diversity Species diversity
2. Ecosystem diversity

Genetic diversity refers to the variation of genes between different populations of a species as well as between species. Species or organismal diversity refers to the variety of living species. Ecosystems are communities of organisms and their environment that function as integrated units (for example, forests, rangelands, mountain ranges, wetlands, and croplands). Ecosystem biodiversity relates to the variety of habitats, biotic communities, and ecological processes and to enormous diversity present within ecosystems in terms of habitat differences and the variety of ecological processes.

Biodiversity is recognized as the most important natural resource. It directly and indirectly influences and regulates the functioning of the other natural resources of soil, water, and air in the ecosystem. The three levels of bio diversity contribute individually and interactively to produce ecological goods and services. When ecosystems are diverse, a range of pathways are available for primary production and ecological processes, such as, nutrient cycling and waste decomposition, so that, if one is damaged, or destroyed, an alternative pathway can be used and the ecosystem can continue to function at its normal level. If biological diversity is greatly reduced, the functioning of an ecosystem is put to risk (Naeem et al. 1994). This is a serious concern for modern croplands, where, large-scale erosion of species and genetic diversity has been allowed. The "high input" technology of the so-called green revolution, is the classic example, where emphasis on monoculture, as for example in Punjab State, the "cradle" of green revolution in India, is witnessing ruinous environmental issues.

Fig. 28.4 Ecosystem abuse and cyclical destruction of natural resource. (Source: Adapted from Heywood 1995)

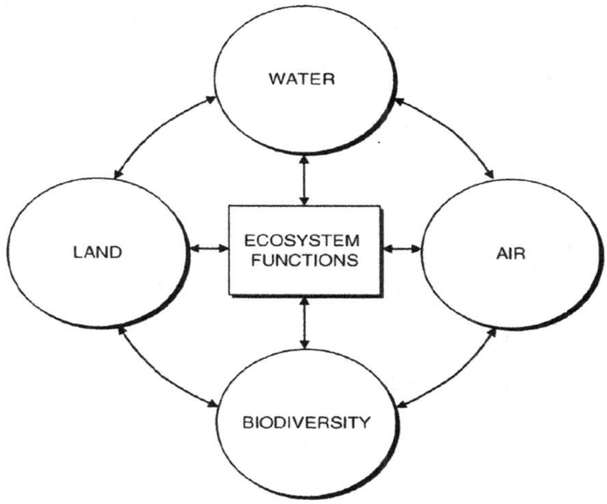

28.8.3 Agroecosystems and Agricultural Biodiversity

The progressive transition and simplification of natural ecosystems into specialized agroecosystems originated with the domestication and cultivation of plants and animals for food at the beginning of the Neolithic Revolution, around 8000 BC. The coevolution and coexistence of plants, animals, and humans, that followed, and the manipulation of plants and animals by humans to suit their own convenience, benefits, and interests, resulted in the creation of unique agroecosystems. The monoculture of the so-called green revolution, and the latter day introduction of the genetically modified organisms (importantly of crop plants, and other organisms) is the most heinous aberration of the basic beauty and harmony of biodiversity on this planet. Conway (1987) defined an ecosystem as follows:

> An ecological and socioeconomic system, comprising of domesticated plants and animals and the people who husband them, intended for the purpose of producing food, fiber or other agricultural products.

A pictorial representation of ecosystem abuse and cyclical destruction of natural resources is shown in the following figure (Fig. 28.4)

28.8.4 Importance of Agricultural Biodiversity

Although productivity is often the most highlighted function of agricultural systems, other productivity-enabling functions are all largely biological, and their persistence depends on maintaining biological diversity. Some of the following are examples:

1. Recycling of plant nutrients
2. Control of microclimate
3. Regulation of local hydrological processes
4. Detoxification of waste and toxic chemicals
5. Regulation of the abundance of desirable and undesirable organisms
6. Soil structure
7. Infiltration and runoff soil erosion
8. Natural pest and disease control
9. Pollination
10. Genetic introgression and hybridization

These are renewable, natural, and free ecological services whose value for agricultural sustainability is effectively infinite, economically and ecologically. The extended definition of agricultural biodiversity is inclusive of all these components. The CBD (CBD- Convention on Biological Diversity 1992)-FAO (Food and Agriculture Organization) workshop included ecological and Social services provided by agroecosystems, such as, soil protection and soil quality, water and air quality, carbon dioxide sequestration, and even landscape and wildlife protection.

28.8.5 Agricultural Intensification and Biodiversity

Agriculture is dependent on the vast pool of species and genetic resources within and outside the narrowly defined realm of agroecosystems. The growing demand for food in the past five decades has changed the way agriculture is practiced and has placed excessive strain on and outside the production environment. Humans have manipulated 70% of the world's temperate and tropical ecosystems to produce 98% of their food requirement and all of their wood products, and only 5% of the temperate and tropical land area is totally uninhabited or unmanaged. Land use for food production stands at 11% of the world's total land area for cropland and 26% for permanent pasture (WRI 1994). This has led to the simplification of vast areas of natural ecosystems rich in bio diversity, such as, forest lands, grasslands, and wetlands.

Agricultural practices have three types of impacts on bio diversity such as:

1. They alter the characteristics of natural ecosystems and their constituent species
2. They impact the species and genetic variability of the chosen cultivated species themselves and also their non food components
3. They affect on-and off-farm non-food biodiversity through such adverse physico-chemical effects on soil nutrient availability, soil and water erosion, salinization and pesticide pollution. The monoculture of the so-called green revolution is the most classic example of this aberration. The magnitude of these impacts varies with the intensity of the intervention. The above impacts are pictorially depicted in the following figure (Fig. 28.5).

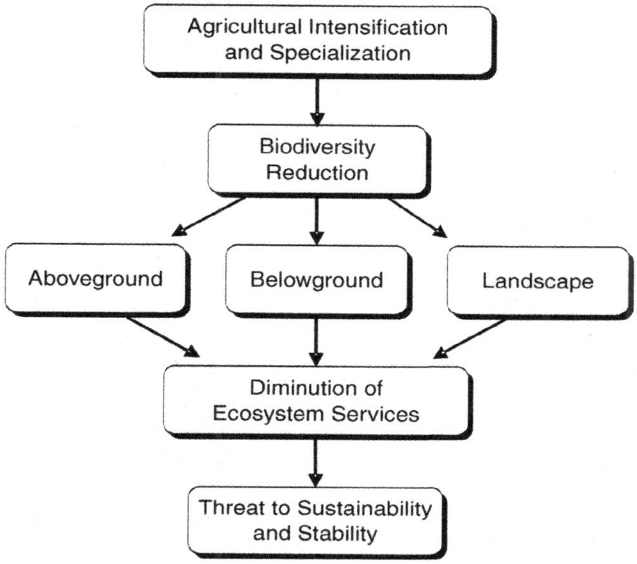

Fig. 28.5 Impact of agricultural intensification on biodiversity

Traditional agricultural systems are soft on biodiversity. Species and genetic diversity are quite high, and the impact of these traditional systems on ecosystems consequently is not as severe as that of the modern commercial systems, as for example, the chemically driven so-called green revolution, focused on monoculture, such as, wheat-rice in Punjab State.

28.8.6 Agricultural Pollution and Off-Farm Loss of Biodiversity

Agriculture is a major polluter of modern times, and, no other endeavor exemplifies it as much as the so-called green revolution does. Soil erosion, water and air pollution, pesticide poisoning, and habitat destruction are well known examples of this environmental pollution (Raman 1987; Biswas 1994; Nair 2013, 2014). The runoff from high fertilizer input agricultural operations, as in the case of the green revolution, which contains large quantities of nitrates and phosphates, gives rise to eutrophication in water bodies, resulting in the death of large fish swarms, invertebrates, plants, and microorganisms, finally leading to health risk in humans living in such regions. The use of large quantities of fossil fuel in fertilizer manufacture has given rise to acid deposition in the developed countries and has caused large-scale destruction of biodiversity (Lean et al. 1990), in addition to global warming through excessive use of nitrogenous fertilizers like urea that leads to the deposition of nitrous oxide N_2O in the atmosphere, which takes more than a century to get rid of, as discussed in earlier chapters. The abuse of pesticide use has a tremendous environmental cost. Almost 90% of pest control chemicals do not reach the target pests and lead to the destruction of natural predators of agricultural pests, soil microbial biodiversity, wildlife, insect pollinators, and so forth (Pimental and Levitian 1986). Improper use of pesticides in the US alone is estimated to cause the death of millions of wild birds, fish, animals, and insects annually. Pesticide runoff brings about the demise of many aquatic life forms through direct poisoning. The sediments from erosion that reach surface waterways such as rivers, and estuaries, which largely settle in nearby coastal marine environments, interfere seriously with the biodiversity of mangroves and corals and affect the breeding of fish and other marine creatures.

28.8.7 Conservation and Sustainable Use of Agricultural Biodiversity

The current projection that the global population is not likely to stabilize before the end of the twenty first century requires that a continuous increase in food production must balance population growth. Given that the scope for the expansion and exploitation of physical resources for food production, such as land, water, and energy, is limited, conserving the current pool of biodiversity and using it to create innovative biological resources to increase productivity and combat biotic and abiotic stresses offers hope for bridging the gap.

28.8.8 Current Lacunae

Modern agricultural production practices have created tensions related to different aspects of agricultural biodiversity, such as the following:

1. Severe reduction in the number of species of food plants cultivated
2. Loss of genetic diversity of the presently cultivated staple food crops, vegetables and fruits
3. Destabilization of pest species in the agroecosystems
4. Deficiency in soil biodiversity
5. Erosion of cultural diversity, loss of the indigenous knowledge of traditional farming systems, and environmental degradation

A pictorial representation of biodiversity management for food security is given in the following figure (Fig. 28.6).

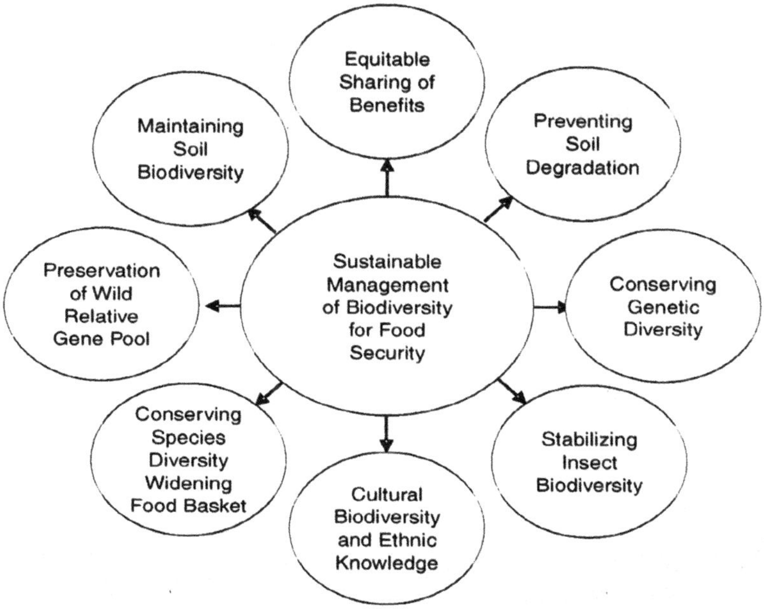

Fig. 28.6 Biodiversity management for food security. (After Saroja Raman 2006)

References

Altieri, M. A. (1995). *Agroecology: The Science of sustainable agriculture*. Boulder: Westview Press.

Biswas, M. R. (1994). Agriculture and environment: A review, 1972–1992. *Ambio, 23*, 192–197.

Blackmore, S. (1994). Precision farming: An introduction. *Outlook on Agriculture, 23*(4), 275–280.

Bullock, D. G. (1992). Crop rotation. *Critical Review of Plant Sciences, II*, 309–326.

CBD. (1992). *Convention on biological diversity*. Montreal, Canada.

Conway, G. R. (1987). The properties of agroecosystems. *Agricultural Systems, 24*, 95–107.

Crookston, R. K., Kurle, J. E., Copland, P. J., Ford, J. H., & Lueschen, W. E. (1991). Rotational cropping sequence effects yield of corn and soybean. *Agronomy Journal, 83*, 108–113.

Dent, D. (1991). *Insect pest management*. Wallingford: CAB International.

FAO. (1998). *Crops and drops: Making the best use of water for agriculture*. Rome: Food and Agriculture Organization.

Heywood, V. H. (1995). Magnitude and distribution of biodiversity. In V. H. Heywood (Ed.), *Global biodiversity assessment* (pp. 107–192). New York: Cambridge University Press.

Karlen, D. L., & Sharpley, A. N. (1990). Management strategies for sustainable soil fertility. In J. L. Hartfield & D. L. Karlen (Eds.), *Sustainable agricultural systems* (pp. 47–108). Boca Raton: Lewis Publications/CRC Press.

Karlen, D. L., Eash, N. S., & Unger, P. W. (1992). Soil and crop management effects on soil quality indicators. *American Journal of Alternative Agriculture, 7*, 48–55.

Karlen, D. L., Varvel, G. E., Bullock, D. G., & Cruse, R. M. (1994). Crop rotations for the 21st century. In D. L. Sparks (Ed.), *Advances in agronomy 53* (pp. 1–45). New York: Academic.

Lean, G., Hinrichsen, D., & Markham, A. (1990). *Atlas of the environment*. London: Arrow Books.

Liebman, M., & Dyck, E. (1993). Crop rotation and intercropping strategies for weed management. *Ecological Applications, 3*, 92–122.

Naeem, S., Thompson, L. J., Lawler, S. P., Lawton, J. H., & Woodfin, R. M. (1994). Declining biodiversity can alter the performance of ecosystems. *Nature, 368*, 734–737.

Nair, K. P. P. (1996). The buffering power of plant nutrients and effects on availability. *Advances in Agronomy, 57*, 237–287.

Nair, K. P. P. (2013). The buffer power concept and its relevance in African and Asian soils. *Advances in Agronomy, 121*, 447–529.

Nair, K. P. P. (2014, August 18). Green revolution or silent spring? *The Hindu Business Line*.

Pimental, D., & Levitan, L. (1986). Pesticides: Amounts applied and amount reaching pests. *Bioscience, 36*, 86–91.

Raman, S. (1987). Mineralogy in relation to pesticide use and soil pollution. *Clay Research, 6*(2), 50–58.

Raman, S. (2006). *Agricultural sustainability: Principles, processes and prospects* (pp. 13904–11580). Birmingham: The Hawoth Press Inc.

Sudduth, K. A. (1998). *Engineering Technologies for Precision Farming*. Columbia: United States Department of Agriculture.

Swift, M. J., & Woomer, P. L. (1992). Organic matter and sustainability of agricultural systems. In R. Merkx & K. Mulongoy (Eds.), *Dynamics of organic matter in relation to sustainability of agricultural systems* (pp. 3–37). New York: John Wiley.

UNFCCC. (1997). *Kyoto protocol for and United Nations framework convention on climate change*. Kyoto, Japan.

WRI. (1994). *World Resources 1994–95*. Washington, DC: World Resources Institute.

Chapter 29
The Salient Conclusions

Abstract The central theme of the book is a revolutionary soil management technique, which is now globally known as "The Nutrient Buffer Power Concept". The discussion on the experimental procedures, crop response both in the phytotron and fields, in three continents, namely, Europe, Africa and Asia, with the focus on economizing fertilizer use form the first part of the book. This second part, which is complementary to the first, relates to different facets of sustainable agriculture, *vis-à-vis* "The Nutrient Buffer Power Concept".

Keywords Green revolution · Text book knowledge · Revolutionary soil management technique · The nutrient buffer power concept · Buffer power · Soil fertility management · Sustainable agriculture · Water and energy management · Holistic and frontier technologies

The central theme of the book is a revolutionary soil management technique, which is now known, globally, as "The Nutrient Buffer Power Concept." The book has two parts, the first, relates to the thermodynamic aspects of soil nutrient bio availability, and the second, to sustainable agriculture and its different facets. Both complement each other. The author uses the term "Buffer Power." which is central to the understanding of the entire dynamics of soil nutrient bio availability. The approach is a significant departure from the classical and routine "text book knowledge." which attempts to quantify nutrient bio availability, with totally unreliable consequences, more often than not, when applied to field situations. Examples abound in scientific literature with reference to important plant nutrients. Discussion in the book primarily relates to the three most important plant nutrients, phosphorus, potassium and zinc, which are principally absorbed by the plant root by diffusive process. Experimental data on laboratory, controlled plant culture and complementary field data are discussed related to a number of test crops, such as, summer rye, maize, alfalfa, wheat, all of which are annual crops, and, also, perennial crops, such as black pepper and cardamom, from three continents, Europe, Africa and Asia. These discussions relate to field success stories, as well.

The second part of the book, which is complementary to the first, relates to the different facets of sustainable agriculture. The discussion starts with the historical

K. P. Nair, *Intelligent Soil Management for Sustainable Agriculture*,
https://doi.org/10.1007/978-3-030-15530-8_29

perspectives, such as the Malthusian prophecy, Faustian Bargian etc. The reader is introduced to these terms, because, without a clear understanding of these terms, discussion on sustainable agriculture would be incomplete. Having introduced the reader to these terms, the discussion moves on to key aspects of the elements of sustainable agriculture, sustainability conundrums, operationalisation of sustainable agriculture etc. Some of the key aspects discussed are soil degradation, sustaining land productivity, challenges to sustainable management of natural resources, primarily soil. There is an exclusive chapter on soil fertility and nutrient management, which primarily discusses various aspects of two most important plant nutrients, nitrogen and phosphorus, in addition to a discussion on integrated nutrient management. Under phosphate nutrition, an exclusive discussion relates to the role of phosphate solubilizing microorganisms (PSM) of vesicular arbuscular mycorrhizae (VAM) and their role in augmenting phosphate nutrition in problematic soils.

Further on, there is extensive discussion on water and energy management in sustainable agriculture. Intelligent and cost-effective water management will be the focal point, as much as intelligent management of soil fertility, *vis-à-vis* effective nutrient management, in sustainable agriculture. Water must be treated as a "public good," but, not abused in its usage, as is happening currently in many parts of the world. Under the energy management section, several aspects like energy audit of high energy systems, food-energy nexus and integration, biomass energy and rural livelihoods, bio energy initiatives, renewable sources of energy etc., are discussed. The industrial type agriculture, euphemistically known as the "green revolution," is a high energy user, principally, fossil-fuel based. With a gradual depletion of fossil fuel resources, alternative sources of energy will need to be focused on. The chapter on energy management discusses all these factors.

A thorough discussion on sustainable agriculture would be incomplete without a discussion on biodiversity. One of the principal reasons why the industrial type agriculture, euphemistically known as the "green revolution," failed, is because of the disappearance of crop diversity. Nowhere else in the world is this so illustratively seen as in the State of Punjab in India, known as the "cradle of the green revolution," where the commodity mindset ushered continuous rice-wheat monoculture, with disastrous consequences on soil fertility, depletion of soil carbon, the bed rock of soil fertility, vanishing of bio diversity, drying of aquifers, excessive salinization and loading of ground water with excessive nitrates, rendering it non potable, a consequence of indiscriminate use of chemical fertilizers, principally nitrogenous, such as urea.

The book also discusses various other aspects of agricultural sustainability, touching upon differential capabilities of resources, sustainable intensification of both high-potential and low-potential lands, agroecosystems, measurement of agricultural sustainability, challenges to global agricultural sustainability etc. There is also a chapter on climate change and agricultural production.

The concluding part of the book discusses both holistic and frontier technologies, the latter which includes information technology and precision farming, remote sensing, geographical information systems (GIS) etc. "The Nutrient Buffer Power Concept" comes under the latter category.

Photos

Photo 1 Extreme Zn deficiency and the after effects of Zn application based on "Buffer power concept" in Turkish soil

Large wheat fields in Anatolia, the main wheat growing region of Turkey, showing extreme deficiency of the most important micronutrient Zinc, and how it can be corrected employing Prof Nair's new concept. Note the extremely zinc deficient fields (yellow patches) and the adjacent fields with very green and luxuriant growth of the crop where zinc has been applied using Prof Nair's new concept. He has demonstrated a saving up to 75% in zinc fertilizer application in these wheat growing regions through his revolutionary soil management concept.

K. P. Nair, *Intelligent Soil Management for Sustainable Agriculture*,
https://doi.org/10.1007/978-3-030-15530-8

Photo 2 Luxuriant growth of Black pepper (***Piper nigrum*** L.) grown in a typical laterite soil of Kerala State, India, employing the revolutionary soil testing procedure of Prof Nair "***The Nutrient Buffer Power Concept***"

Controlled trials in demonstration plots have shown the effectiveness of Kodoth Padinhare Nair's soil testing procedure, the "nutrient buffer power concept." In the control plot in the foreground, fertilizer was applied in the conventional way. In the plot with the luxuriant growth just behind it, fertilizer was applied using the "nutrient buffer power concept."

Photo 3 Rice grown in The Republic of Cameroon employing Prof Nair's new soil management concept

Photo 4 Prof Nair and his colleagues in a very large rice field in Kerala State, India, the "rice bowl" of the State, where the crop has been grown using Prof Nair's new soil management concept

Photo 5 A cardamom (*Elettaria cardamomum* Maton.) – The "Queen" of spices field in India, where the crop has been grown as per Prof Nair's new soil management concept. Note the luxuriant growth of cardamom berries

Photo 6 A ginger crop (a very valuable spice crop of the developing world) grown in a field in India, as per Prof Nair's new concept

Photo 7 Large cardamom field in Karnataka State, in India, where the crop is grown as per Prof Nair's new concept

Photo 8 & 9 Prof Nair discussing the merits of his new concept in growing rice with a group of scientists and a group of small farmers

Photo 10 A farmer harvesting Black pepper berries, in Kerala State, India, where the crop has been grown as per Prof Nair's new concept African farmers cultivating their fields with an animal drawn plough. The land has been extensively damaged by poor soil management techniques and incorrect use of fertilizers. Kodoth Padinhare Nair is trying to overcome these problems in developing countries by promoting his soil testing procedure, the "nutrient buffer power concept"

Photo 11 Very poor farmers in Central Africa, ploughing their fields to sow the crops. Prof Nair is helping these poor farmers of the African continent with his new concept

Photo 12 Mass production of two fungus which will help contain the most dreaded disease of Black pepper **"Quick Wilt"** caused by ***Phytophthora*** sp. Prof Nair is actively involved in saving the Black pepper farmers of Kerala, India, from this dreaded disease, which is triggered by the extreme zinc deficiency in soil. His "The Nutrient Buffer Power Concept" has helped thousands of Black pepper farmers in the State of Kerala, where the crop is the economic mainstay and a very big dollar earner through export to US, Europe, Japan etc. His field research has come as a "saviour" of Black pepper farmers of the State

References

Aarts, H. F. M., Biewinga, E. E., & van Keulen, H. (1992). Dairy farming systems based on efficient nutrient management. *Netherlands Journal of Agricultural Science, 40*, 285–299.

Adams, F. (1971). Ionic concentrations and activities in soil solutions. *Soil Science Society of America Proceedings, 35*, 420–426. Anderson.

Adams, F. (1974). Soil solution. In E. W. Carson (Ed.), *The plant root and its environment* (pp. 441–481). Charlottesville: University Press of Virginia.

Adams, C. R., & Eswaran, H. (2000). Global land resources in the context of food and environmental security. In *Advances in land resources management for the 21st century. Proceeding of International conference on land resource management for food, employment and environmental security (ICLRM), soil conservation society of India, November 9–13, 2000* (pp. 35–50). New Delhi: New Delhi Angkor Publishers.

Aggarwal, P. K., Knopff, M. J., Cassman, K. G., & ten Berge, H. F. M. (1997). Simulating genotypic strategies for increasing rice yield potential in irrigated, tropical environments. *Field Crops Research, 51*, 5–17.

Ahmad, S. (1995). *Agriculture-fertilizer interface in Asia issues of growth and sustainability*. New Delhi: Oxford and IBH Publishing Co.

Algawadi, A. R., & Gaur, A. C. (1988). Associative effect of Rhizobium and phosphate solubilizing bacteria on the yield and nutrient uptake of chickpea. *Plant and Soil, 105*, 241–246.

Altieri, M. A. (1995). *Agroecology: The Science of sustainable agriculture*. Boulder: Westview Press.

Altieri, M. A. (1999). The ecological role of agriculture in ecosystems. *Agriculture, Ecosystems and Environment, 74*, 19–31.

Altieri, M. A. (2002). Agroecological principles for sustainable agriculture. In N. Uphoff (Ed.), *Agroecological innovations: Increasing food production with Participatory Development* (pp. 40–46). London: Earthscan.

Aman, R. I. (1995). Fluorescently labeled rRNA targeted nucleotide probes in the study of microbial ecology. *Microbial Ecology, 4*, 543–554.

Anderson, G. D. (1998). Grow the soils to grow the crops in Africa. In R. Lal (Ed.), *Soil quality and agricultural sustainability* (pp. 237–250). Chelsea: Ann Arbor Press.

Andiscott, T. M. (1995). Entropy and sustainability. *European Journal of Soil Science, 46*, 161–168.

Argersinger, W. J., Jr., Davidson, A. W., & Bonner, O. D. (1950). Thermodynamics and ion exchange phenomena. *Transactions of the Kansas Academy of Science, 53*, 404–410.

Asea, P. E. A., Kucey, R. M. N., & Stewart, J. W. B. (1988). Inorganic phosphate solubilization by two Penicillium species in solution culture and soil. *Soil Biology and Biochemistry, 20*, 459–464.

K. P. Nair, *Intelligent Soil Management for Sustainable Agriculture*,
https://doi.org/10.1007/978-3-030-15530-8

Azcon-Aguilar, C., Diaz-Rodriguez, R., & Barea, J. M. (1986). Effect of soil microorganisms on spore germination and growth on the vesicular arbuscular mycorrhizal fungus (Glomus moseae). *Transactions of the British Mycological Society, 86*, 337–340.
Babu-Khan, S., Yeo, T. C., Martin, W. I., Duron, M. R., Rogers, R. D., & Goldstein, A. H. (1995). Cloning of a mineral phosphate solubilizing gene from Peudomonas cepacia. *Applied and Environmental Microbiology, 61*, 972–978.
Bagyaraj, D. J. (1984). Biological interaction with VA mycorrhizal fungi. In C. L. Powell & D. J. Bagyaraj (Eds.), *VA Mycorrhiza* (pp. 131–153). Boca Raton: CRC.
Banik, S., & Dey, B. K. (1982). Available phosphate content of an alluvial soil as influenced by inoculation of some isolated phosphate solubilizing microorganisms. *Plant and Soil, 69*, 353–364.
Banik, S., & Dey, B. K. (1983). Phosphate solubilizing potentiality of the microorganisms capable of utilizing aluminium phosphate as a sole phosphate source. *Zentralblatt fur Mikrobiologie, 138*, 17–23.
Barber, S. A. (1974). Influence of the plant root on ion movement in soils. In E. W. Carson (Ed.), *The plant root and its environment* (pp. 525–564). Charlottesville: University Press of Virginia.
Barber, S. A. (1984). *Soil nutrient bioavailability: A mechanistic approach*. New York: Wiley.
Barber, S. A. (1995). *Soil nutrient bioavailability: A mechanistic approach*. New York: Wiley.
Barber, S. A., Halstead, E. H., & Follett, R. E. (1967). *Significant mechanisms controlling the movement of manganese and molybdenum to plant roots growing in soil* (pp. 299–304). Aberdeen: Trans. Joint Meeting Commissions II and IV. International Soil Science Society.
Barea, J. M., Azcon, R., & Azcon-Aguilar, C. (1983). *Interaction between phosphate solubilizing bacteria and VA mycorrhiza to improve the utilization of rock phosphate by plants in non acidic soils* (pp. 127–152). Brussels: Third International Congress on Phosphorus Compounds.
Barekzai, A. (1984). *Alterung von wasserlöslichem Phosphat-untersucht in Giefäß und Modellversuchen*. Ph.D. thesis, FB 1g, Justus Liebig University, Giessen.
Barrios, E. (2007). Soil biota, ecosystem services and land productivity. *Ecological Economics, 64*, 269–285.
Barrow, N. J. (1970). Comparison of the adsorption of molybdate, sulfate and phosphate by soils. *Soil Science, 109*, 282–288.
Barrow, N. J. (1983). On the reversibility of phosphate sorption by soils. *Journal of Soil Science, 34*, 751–758.
Bar-Yosef, B., Rogers, R. D., Wolfram, J. H., & Richman, E. (1999). Pseudomonas cepacia mediated rock phosphate solubilization in kaolinite and montmorillonite suspensions. *Soil Science Society of America Journal, 63*, 1703–1708.
Bashan, Y., Puente, M. E., Rodriquea, M. N., Toledo, G., Holguin, G., Ferrera-Cerrato, R., & Pedrin, R. (1995). Survival of Azorhizobium brasilense in the bulk soil and rhizosphere of 23 soil types. *Applied and Environmental Microbiology, 61*, 1938–1945.
Becker, M., Asch, F., Maskey, S. L., Pande, K. R., Shah, S. C., & Shrestha, S. (2007). Effects of transition season management on soil N dynamics and systems N balances in rice-wheat rotations of Nepal. *Field Crops Research, 103*, 98–108.
Beckett, P. H. T. (1971). Potassium potentials—A review. In *Potash Review* (pp. 1–41). Bern: Int. Potash Inst.
Beckett, P. H. T. (1972). Critical cation activity ratios. *Advances in Agronomy, 24*, 376–412.
Beckwith, R. S. (1965). Sorbed phosphate at standard supernatant concentration as an estimate of the phosphate needs of soils. *Australian Journal of Experimental Agriculture and Animal Husbandry, 5*, 52–58.
Belder, P., Bouman, B. A. M., Spiertz, J. H. J., Peng, S., Castaneda, A. R., & Visperas, R. M. (2005). Crop performance, nitrogen and water use in flooded and aerobic rice. *Plant and Soil, 273*, 167–182.
Bennett, J. (2003). Opportunities for increasing water productivity of CGIAR for tolerance of water defects for using crops through plant breeding and molecular techniques. In J. W. Kijne,

R. Barker, & D. Molden (Eds.), *Water Productivity in Agriculture: Limits and Opportunities for Improvement* (pp. 103–126). Wallingford: CAB International.

Berger, G. W. (2004). Photonic dating of prehistoric irrigation canals of Phoenix, Arizona, U. S. A. *Geoarchaeology, 19*, 381.

Bertsch, P. M., & Thomas, G. W. (1985). Potassium status of temperate region soils. In R. D. Munson (Ed.), *Potassium in Agriculture* (pp. 131–162). Madison: Am. Soc. Agron.

Best, G. (1997). *Food security from an energy standpoint. Occasional paper*. Rome: Food and Agriculture Organization.

Bethlenfalvay, G. J. (1994). Sustainability and rhizoorganisms in an ecosystem. *Sociedad Mexicana de la Ciencia del Suelo, 4*, 9–10.

Biemond, H., & Vos, J. (1992). Effects of nitrogen and the development and growth of the potato plant 2. The partitioning of dry matter, nitrogen and nitrate. *Annals of Botany, 70*, 37–45.

Blackmore, S. (1994). Precision farming: An introduction. *Outlook on Agriculture, 23*(4), 275–280.

Blevins, R. L., & Frye, W. W. (1993). Conservation tillage in soil management. *Advances in Agronomy, 51*, 34–77.

Boguslawski, E. V., & Lach, G. (1971). Die K-Nachlieferung des Bodens im Pflanzen experiment im Vergleich mit dem austauschbaren Kalium. *Zeitschrift für Acker- und Pflanzenbau, 134*, 135–164.

Bonny, S. (1993). Is agriculture using more and more energy? A French case study. *Agricultural Systems, 43*, 51–66.

Borlaug, N. E. (1990). The challenge of feeding 8 billion people. *Farm Chemical International*, summer issue, 1990.

Borlaug, N. E., & Dowswell, C. R. (1993). Fertilizer: To nourish infertile soil that feeds a fertile population that crowds a fragile world. *Better Crops With Plant Food, 77*(3), 6–7.

Boserup, E. (1965). *The conditions of agricultural growth: The economics of Agrarian change under population pressure*. New York: Aldine.

Boul, S. W. (1994). Soil, society's yoke to the earth. In *Proceeding XV of the ISSS symposium* (Vol. 1, pp. 89–104), Acapulco, Mexico, July 10–16, 1994.

Bouman, B. A. M. (2007). A conceptual framework for the improvement of crop water productivity at different spatial scales. *Agricultural Systems, 93*, 43–60.

Bouman, B. A. M., Van Keulen, H., Van Laar, H. H., & Rabbinge, R. (1996). The "School De Wit" crop growth simulation models: A pedigree and historical overview. *Agricultural Systems, 52*, 171–198.

Bowman, R. A., & Olsen, S. R. (1985a). Assessment of phosphate buffering capacity. 1. Laboratory methods. *Soil Science, 140*, 287–291.

Bowman, R. A., & Olsen, S. R. (1985b). Assessment of phosphate buffering capacity. 2. Greenhouse methods. *Soil Science, 140*, 387–392.

Bray, R. H., & Kurtz, L. T. (1945). Determination of total organic and available form of phosphorus in soils. *Soil Science, 59*, 39–45.

Brouder, S. M., & Cassman, K. G. (1994). Evaluation of a mechanistic model of potassium uptake by cotton in vermiculitic soil. *Soil Science Society of America Journal, 58*, 1174–1183.

Brown, L. R. (1995). *Who will feed China? Wakeup call for a small planet*. New York: W. W. Norton.

Brown, L. R. (1997). *The agricultural link: How environmental deterioration could disrupt economic progress* (World Watch Paper 136). Washington, DC: World Watch Institute.

Brown, L. R. (1998). Struggling to raise cropland productivity. In L. R. Brown (Ed.), *State of the World*. Washington, DC/New York: World Watch Institute/W.W. Norton.

Brown, L. R. (2001). *Who will feed China? Wakeup call for a small planet*. New York: W. W. Norton.

Brown, L. R., & Kane, H. (1994). *Full House: Reassessing the earth's population carrying capacity*. Washington, DC/New York: World Watch Institute and W.W. Norton.

Brussard, L., de Ruiter, P. C., & Brown, G. G. (2007). Soil biodiversity for agricultural sustainability. *Agriculture, Ecosystems & Environment, 121*, 233–244.

Bucks, D. A. (1995). Historical developments in micro irrigation. In F. R. Lamm (Ed.), *Micro irrigation for a changing world* (Proc. Fifth International Micro Irrigation Congress, April 2–6, 1996, Orlando, FL) (pp. 1–5). St Joseph: American Society of Agricultural Engineers.
Bullock, D. G. (1992). Crop rotation. *Critical Review of Plant Sciences, II*, 309–326.
Buresh, R. J., Sanchez, P. A., & Calhoun, F. (1997). *Replenishing soil fertility in Africa SSSA Special Publication 51*. Madison: Soil Science Society of America.
Burgstaller, W., Straser, H., & Shinner, F. (1992). Solubilization of zinc oxide from filter dust with Penicillium simplicissimum: Bioreactor, leaching and stoichiometry. *Environmental Science & Technology, 26*, 340–346.
Burt, C. M., Howes, D. J., & Mutziger, A. (2001). Evaporation estimates for irrigated agriculture in California. In *Paper presented at the irrigation association conference*, November 4–6, 2001, San Antonio, TX
Buyer, J. S., & Sikora, L. J. (1990). Rhizosphere interactions and siderophores. *Plant and Soil, 129*, 101–107.
Cabrera-Bosquet, L., Molero, G., Bort, J., Nogues, S., & Araus, J. L. (2007). The combined effect of constant water deficit and nitrogen supply on WUE, NUE and Δ^{13} C in durum wheat potted plants. *The Annals of Applied Biology, 151*, 277–289.
Campbell, A. (1994). Participatory inquiry: Beyond research and extension in the sustainability era. In *Paper presented at the international symposium on systems-orineted research in agriculture and rural development*. November 21–25, 1994, Montpellier, France
Campbell, C. A., Myers, R. J. K., & Curtin, D. (1995). Managing nitrogen for sustainable crop production. *Fertility Research, 42*, 277–296.
Carpenter, R. C. (1998). Can sustainability be measured? In H. S. Sharma & S. K. Chattopadhyay (Eds.), *Sustainable development: Issues and case studies* (pp. 46–55). New Delhi: Concept.
Carson, R. (1962). *Silent Spring*. Boston: M.A. Houghton Mifflin.
Cassman, K. G. (1990). The role of soil fertility research in developing sustainable food production systems. *Better Crops With Plant Food, 74*(4), 16–19.
Cassman, K. G., & Harwood, R. R. (1995). The nature of agricultural systems: Food security and environmental balance. *Food Policy, 20*(5), 439–454.
Cassman, K. G., Olk, D. C., & Dobermann, A. (1997). Scientific evidence of yield and productivity decline in irrigated rice systems of tropical Asia. *International Rice Commission Newsletter, 46*, 7–16.
Cavdar, A. O., Arcasoy, A., Cin, S., Babacan, E., & Gozdasoglu, S. (1983). Geophagia in Turkey: Iron and zinc deficiency, iron and zinc absorption studies and response to treatment with zinc in geophagia cases. In *Zinc deficiency in human subjects* (pp. 71–79). New York: Alan R. Liss. Inc.
CGIAR. (2002). *Challenge program for water and food*. Washington, DC: Consultative Group for International Agricultural Research.
Chabot, R., Anton, H., & Cescas, M. P. (1996). Growth promotion of maize and lettuce by phosphate solubilizing Rhizobium leguminosarum biovar phaseoli. *Plant and Soil, 184*, 311–321.
Chapman, A. L., & Muchow, R. C. (1985). Nitrogen accumulated and partitioned at maturity by grain legumes under different water regimes in a semi-arid tropical environment. *Field Crops Research, 11*, 69–79.
Claassen, N. (1990). *Die Aufnahme von Nährstoffen aus dem Boden durch die höhere Pflanze als Ergebnis von Verfügbarkeit und Aneignungsvermögen*. Gottingen: Severin-Verlag.
Claassen, N., & Barber, S. A. (1976). Simulation model for nutrient uptake from soil by growing plant root system. *Agronomy Journal, 68*, 961–964.
Claassen, N., Hendriks, K., & Jungk, A. (1981). Rubidium-Verarmung des wurzelnahen Bodens durch Maispflanzen. *Zeitschrift für Pflanzenernährung und Bodenkunde, 144*, 533–545.
Claassen, N., Syring, K. M., & Jungk, A. (1986). Verification of a mathematical model by simulating potassium uptake from soil. *Plant and Soil, 95*, 209–220.
Clement, R. M., & Horn, S. P. (2001). Pre-Columbian land-use history in Costa Rica: A 3000 – year record of forest clearance, agriculture and fires from Laguna Zoncho. *Holocene, 11*, 419–426.

Cobb, C., & Cobb, J. (1994). *The Green National product: A proposed index of sustainable economic welfare*. New York: University Press of America.
Colbert, S. F., Hendson, M., Ferri, M., & Schroth, M. N. (1993). Enhanced growth and activity of a biocontrol bacterium genetically engineered to utilize salicylate. *Applied Microbiology, 59*, 2071–2076.
Coleman, J. S. (1990). *Foundation of social theory*. Cambridge, MA: Harvard University Press.
Conway, G. R. (1987). The properties of agroecosystems. *Agricultural Systems, 24*, 95–107.
Conway, G. R. (1997). *The Doubly Green Revolution: Food for all in the 21st century*. London: Penguin Books.
Conway, G. R., & Barbier, E. B. (1990). *After the Green Revolution: Sustainable agriculture for development*. London: Earthscan.
Cope, J. T., & Evans, C. E. (1985). Soil testing. *Advances in Soil Science, 1*, 201–228.
Cordero, A., & Espinosa, J. (1994). Maximum economic rice yield in Costa Rica. Research for maximum yield in harmony with nature. In *XV world congress of soil science* (pp. 29–39), Acapulco, Mexico, July 10–16.
Costanza, R., & Daly, H. E. (1992). Natural capital and sustainable development. *Conservation Biology, 6*, 37–46.
Costanza, R., d'Arge, R., de Groot, S., Farber, M., Grasso, B., Hannon, K., Limburg, S., Naeem, R. V., O'Neill, J., Paruelo, J., Raskin, R. G., Sutton, P., & van den Belt, M. (1997). The value of world ecosystem services and natural capital. *Science, 387*, 253–260.
Crookston, R. K., Kurle, J. E., Copland, P. J., Ford, J. H., & Leuschen, W. E. (1991). Rotational cropping sequence effects yield of corn and soybean. *Agronomy Journal, 83*, 108–113.
Crosson, P. (1994). Conditions for achieving a sustainable global agricultural system. In *Environmental agriculture: Rethinking development issues for the 21st century. Proceeding Symposium in honor of R. D. Havener, May 5–6, 1993*. Morriltown: Winrock International Institute for Agricultural Development.
Crosson, P. (1995). Soil erosion costs and estimates. *Science, 269*, 461–464.
Cunningham, J. E., & Kuiack, C. (1992). Production of citric and oxalic acids and solubilization of calcium phosphate by Penicillium bilaji. *Applied and Environmental Microbiology, 58*, 1451–1458.
Curtin, D., & Smillie, G. W. (1983). Soil solution composition as affected by timing and incubation. *Soil Science Society of America Journal, 47*, 701–707.
Daily, G. C. (1997a). *Nature's services: Societal dependence on natural ecosystem*. Washington, DC: Island Press.
Daily, G. C. (1997b). What are ecosystem services? In G. C. Daily (Ed.), *Nature's services: Societal dependence on natural ecosystem* (pp. 1–7). Washington, DC: Island Press.
Davidson, E. A. (2001). *You can't eat the GNP: Economics as if ecology mattered*. Cambridge, MA: Perseus Publishing.
De Willigen, P., & van Noordwijk, M. (1987). Roots for plant production and nutrient use efficiency. Ph.D thesis Agricultural University, Wageningen, The Netherlands 282 p.
Delgado, J. A., Shaffer, M., Hu, C., Lavado, R., Cueto-Wong, J., Joosse, P., Sotomayer, D., Colon, W., Follett, R., DelGrosso, S., Li, X., & Rimski-Korsakov, H. (2008). An index approach to assess nitrogen losses to the environment. *Ecological Engineering, 32*, 108–120.
Dent, D. (1991). *Insect pest management*. Wallingford: CAB International.
Dent, D., & Young, A. (1981). *Soil survey and land evaluation*. Wallingford: CAB International.
DeWit, C. T. (1992). Resource use efficiency in agriculture. *Agricultural Systems, 40*, 125–151.
Dick, W. A. (1983). Organic carbon, nitrogen and phosphorus concentrations and pH in soil profiles as affected by tillage intensity. *Soil Science Society of America Journal, 47*, 102–107.
Dimitrakopoulos, P. G., & Schmid, B. (2004). Biodiversity effects increase linearly with biotope space. *Ecology Letters, 7*, 574–583.
Dobereiner, J. (1994). The Brazilian biofuel programme solution to social and environmental problems. In M. S. Swaminathan (Ed.), *Ecotechnology and rural employment: A dialogue* (pp. 119–128). Chennai: Mcmillan India.

Dobermann, A., & Cassman, K. G. (2002). Plant nutrient management for enhanced productivity in intensive grain production systems of the United States and Asia. *Plant and Soil, 247*, 153–175.

Doran, J. W., & Parkin, T. B. (1994). Defining and assessing soil quality. In J. W. Doran, D. C. Coleman, D. F. Bezdicek, & B. A. Stewart (Eds.), *Defining soil quality for a sustainable environment* (SSSA Special Publication 35) (pp. 3–21). Madison: American Society of Agronomy and Soil Science Society of America.

Doran, J. W., Sarrantonio, M., & Leibig, M. A. (1996). Soil health and sustainability. In D. L. Sparks (Ed.), *Advances in Agronomy* (Vol. 56, pp. 1–54). New York: Academic.

Douglas, G. K. (1984). *Agricultural sustainability in a changing world order*. Boulder: Westview Press.

Douglas, J. T., & Goss, M. J. (1982). Stability of organic matter content of surface soil aggregates under different methods of cultivation and grassland. *Soil and Tillage Research, 2*, 155–175.

Downey, J., & Van Kessel, C. (1990). Dual inoculation of Pisum sativum with Rhizobium leguminosarum and Penicillium bilaji. *Biology and Fertility of Soils, 10*, 194–196.

Dregne, H. E. (1990). Erosion and soil productivity in Africa. *Journal of Soil and Water Conservation, 45*, 431–436.

Drew, M. C., Nye, P. H., & Vaidyanathan, L. V. (1969). The supply of nutrient ions by diffusion to plant roots in soil. I. Absorption of potassium by cylindrical roots of onion and leek. *Plant and Soil, 30*, 252–270.

Dubey, S. K. (1996). Response of soybean to rock phosphate applied with Pseudomonas striata in a typic chromustert. *Journal of the Indian Society of Soil Science, 44*, 252–255.

Dubey, S. K. (2001). Associative effect of nitrogen fixing and phosphate solubilizing bacteria in rainfed soybean (Glycine max) grown in vertisols. *Indian Journal of Agricultural Sciences, 71*, 476–479.

Dumanski, J. (1993). *Sustainable land management for the 21st century* (Vol. 1). Bangkok/Ottawa: International Board for Soil Research and Management and Agriculture Canada.

Duponnois, R., Kisa, M., & Plenchette, C. (2006). Phosphate solubilizing potential of the nematofungus Arthrobotrys oligospora. *Journal of Plant Nutrition and Soil Science, 169*, 280–282.

During, C., & Duganzich, D. M. (1979). Simple empirical intensity and buffering capacity measurements to predict potassium uptake by white clover. *Plant and Soil, 51*, 167–176.

Edmeades, G. O., Cooper, M., Lafitte, R., Zinselmeir, C., Ribaut, J. M., Habben, J. E., Loffler, C., & Banziger, M. (2001). Abiotic stresses and staple crops. In J. Nosberger, H. H. Geiger, & P. C. Struik (Eds.), *Crop science: Progress and prospects* (pp. 137–154). Wallingford: CAB International.

Ehui, S. K., & Spencer, D. S. C. (1990). *Indices for measuring the sustainability and economic viability of farming systems* (RCMP Research Monograph 3). Ibadan: International Institute of Tropical Agriculture.

Eickhout, B., Bouwman, A. F., & van Zeijts, H. (2006). The role of nitrogen in world food production and environmental sustainability. *Agriculture, Ecosystems & Environment, 116*, 4–14.

Elgala, H. M., Ishac, Y. Z., Abdel-Monem, M., El-Ghandour, I. A. I., Hang, P. M., Berthelin, J., Bollag, J. M., Mc Gill, W. B., & Page, A. I. (1995). Effect of single and combined inoculation with Azotobacter and VA mycorrhizal fungi on growth and nutrient content of maize and wheat plants. *Environmental Impacts of Soil Component Interactions, 2*, 109–116.

Elgawhary, S. M., Lindsay, W. L., & Kemper, W. D. (1970). Effect of EDTA on the self-diffusion of zinc in aqueous solution and in soil. *Soil Science Society of America Proceedings, 34*, 66–70.

Erickson, C. L. (1992). Prehistoric landscape management in the African Highlands – Raised field agriculture and its environmental impact. *Population and Environment, 13*, 285–300.

Erisman, J. W., Bleeker, A., Galloway, J., & Sutton, M. S. (2007). Reduced nitrogen in ecology and the environment. *Environmental Pollution, 150*, 140–149.

Eswaran, H., Beinroth, R., & Reich, P. (1999). Global land resources and population supporting capacity. *American Journal of Alternative Agriculture, 14*, 129–135.

Evangelou, V. P., Wang, J., & Phillipis, R. E. (1994). New developments and perspectives on soil potassium quantity/intensity relationships. *Advances in Agronomy, 52*, 173–227.

Evans, L. T. (1998). *Feeding the Ten Billion: Plants and population growth*. Cambridge: Cambridge University Press.

FAO. (1976). *A framework for land evaluation, FAO Soils Bulletin* (Vol. 32). Rome: Food and Agriculture Organization.

FAO (1978–1991) *Agro-ecological zone project*. Rome: Food and Agriculture Organization.

FAO. (1993). *Prevention of water pollution by agriculture and related activities*. Rome: Food and Agriculture Organization.

FAO. (1994). *Land degradation in South Asia: Its severity, causes and effects on People*. Rome: Food and Agriculture Organization.

FAO. (1995). In N. Alexandratos (Ed.), *World Agriculture Towards 2010*. Chichester: Food and Agriculture Organizatyion/Wiley.

FAO. (1996a). *Control of water pollution from agriculture* (Irrigation and Drainage Paper 55). Rome: Food and Agriculture Organization.

FAO. (1996b). *Energy for sustainable development and food security in Africa*. Rome: Food and Agriculture Organization.

FAO. (1997a). *Biomass energy in ASEAN member countries. FAO-SD Dimensions/Energy for Development*. Rome: Food and Agriculture Organization.

FAO. (1997b). *Management of agricultural drainage water quality* (Water Reports 13). Rome: Food and Agriculture Organization.

FAO. (1998). *Crops and drops: Making the best use of water for agriculture*. Rome: Food and Agriculture Organization.

FAO. (1999). *The multiplicational character of agriculture and land: Bioenergy. FAO-SD Dimensions/Energy for Development*. Rome: Food and Agriculture Organization.

FAO. (2000a). *Agriculture Towards 2015/30 technical interim report, Global Perspectives Unit*. Rome: Food and Agriculture Organization.

FAO. (2000b). *Resources Working Paper 4, Rome: Food and Agriculture Organization. The State of Food and Agriculture: Lessons from the past 50 years*. Rome: Food and Agriculture Organization.

FAO. (2000c). *The State of Food and Agriculture: Lessons from the past 50 years*. Rome: Food and Agriculture Organization.

FAO. (2002). *World Agriculture: Towards 2015/2030*. Rome: Food and Agriculture Organization.

FAOSTAT. (1999). *FAOSTAT Database*. Rome: Food and Agriculture Organization.

FEN. (1989). *Strategies for sustainable development: Food-Energy Nexus programme*. New York: United Nations University.

Flinn, J. C., & Duff, B. (1985). *Energy analysis, rice production systems and rice research* (IRRI Research Paper Series 114). Manila: International Rice Research Institute.

Föhse, D., Claassen, N., & Jungk, A. (1991). Phosphorus efficiency of plants. II. Significance of root radius, root hairs, and cation–anion balance for phosphorus influx in seven plant species. *Plant and Soil, 132*, 261–272.

Fornara, D. A., & Tilman, D. (2008). Plant functional composition influences rates of soil carbon and nitrogen accumulation. *Journal of Ecology, 96*, 314–322.

Fox, R. L. (1981). External phosphorus requirements of crops. In R. H. Dowdy et al. (Eds.), *Chemistry in the soil environment* (ASA Special Publication 40) (pp. 223–239). Madison: ASA and SSSA.

Fox, R. L., & Kamprath, E. J. (1970). Phosphate sorption isotherms for evaluating the phosphate requirement of soils. *Soil Science Society of America Proceedings, 34*, 902–907.

Frederic, B. G., Estefania, A., Jordi, B. F., Charles, A. A., Dolors, M., & Manel, P. (2000). Assessment of microbial community structure changes by amplified ribosomal DNA restriction analysis (ARDRA). *International Microbiology, 3*, 103–106.

Fresco, L. O., & Kroonenberg, S. B. (1992, July). Time and spatial scales in ecological sustainability. *Land use policy, 9*, 155–167.

Frissel, M. J. (Ed.). (1978). *Cycling of mineral nutrients in agricultural ecosystems*. Amsterdam: Elsevier.
Frye, W. W. (1984). Energy requirements in no tillage. In R. E. Phillips & S. H. Phillips (Eds.), *Non-tillage agriculture: Principles and practices* (pp. 127–151). New York: van Nostrand Reinhold.
Gapon, E. N. (1933). Theory of exchangeable adsorption in soil. *Journal of General Chemistry (USSR), 3*, 144–163. [Abstract in Chem Abs. 28, 41–49, 1934.].
Gardner, G. (1997). Preserving global cropland. In *The State of the World 1997* (World watch institute report) (pp. 42–59). New York: W.W. Norton.
Gaume, A. (2000). Low P tolerance of various maize cultivars: The contribution of the root exudation. Ph.D dissertation, Swiss Federal Institute of Technology, Zurich, Switzerland.
Gaur, A. C. (1990). *Phosphate solubilizing microorganisms as Biofertilizers* (p. 176). New Delhi: Omega Scientific.
Ghassemi, F., Jackerman, A. J., & Nex, H. A. (1995). *Salinization of land and water resources: Human causes, extent, management and case studies*. Sydney: University of New South Wales Press.
Giampietro, M., & Pimental, D. (1994). *The tightening conflict: Population, energy and the ecology of agriculture*. N.P.G. Forum (Negative Population Growth), October 1993, 1–8
Glick, B. R., & Bashan, Y. (1997). Genetic manipulation of plant growth promoting bacteria to enhance biocontrol of phytopathogens. *Biotechnology Advances, 15*, 353–378.
Goldstein, A. H., Rogers, R. D., & Mead, G. (1993). Mining by microbe. *Bio Technology, 11*, 1250–1254.
Goulding, K., Jarvis, S., & Whitmore, S. (2008). Optimizing nutrient management for farm systems. *Philosophical Transactions of the Royal Society B, 363*, 667–680.
Graham, R. D., & Rengel, Z. (1993). Genotypic variation in zinc uptake and utilization by plants. In A. D. Robson (Ed.), *Zinc in Soils and Plants* (pp. 107–118). Dordrecht: Kluwer Academic Publishers.
Graham, R. D., & Welch, R. M. (1994). Breeding for staple – Food crops with high micronutrient density: Long-term sustainable agricultural solutions to hidden hunger in developing countries. In *IFPRI workshop on food policy and agricultural technology to improve diet quality and nutrition*. Annapolis, Maryland, January 10–12
Greenland, D. J. (1997). *The sustainability of Rice farming*. Wallingford: CAB Inernational.
Grimme, H. (1974). Potassium release in relation to crop production. In International Potash Institute (Ed.), *Potassium research and agricultural production* (pp. 131–136). Berne: Int. Potash Inst.
Grimme, H. (1979). Factors controlling potassium availability to plants. In A. Saurat, & M. M. El-Fouly (Eds.), *Procedding of International Potassium Workshop, Cairo, Egypt* (pp. 33–47).
Gull, M., Hafeez, F. Y., Saleem, M., & Malik, K. A. (2004). Phosphorus uptake and growth promotion of chickpea by co-inoculation of mineral phosphate solubilizing bacteria and a mixed rhizobial culture. *Australian Journal of Experimental Agriculture, 44*, 623–628.
Gunasekaran, S., & Pandiarajan, P. (1995). Dual inoculation of rhizobium and phosphobacteria with two forms of phosphorus in pigeonpea. In *Microbiology abstracts XXXVI. Annual conference of the Association of Microbiologists of India* (p. 111). Hissar, Haryana State, India. November 8–10.
Gupta, R. K., & Abrol, I. P. (2000). Salinity buildup and changes in the rice-wheat systems of the Indo-Gangetic plains. *Experimental Agriculture, 37*, 99–113.
Gupta, V. V. S. R., & Germida, J. J. (1988). Distribution of microbial biomass and its activity in different soil aggregate size classes as affected by cultivation. *Soil Biology and Biochemistry, 20*, 777–786.
Gupta, R. R., Singal, R., Shanker, A., Kuhad, R. C., & Saxena, R. K. (1994). A modified plate assay for screening phosphate solubilizing microorganisms Gen. *Applied Microbiology, 40*, 255–260.

Gupta, R. K., Naresh, R. K., Hobbs, P. R., Jiaguo, J., & Ladha, J. K. (2003). Sustainability of post-Green Revolution agriculture: The rice-wheat systems of the Indo-Gangetic plains and China. In J. K. Ladha, J. Hill, R. K. Gupta, J. Duxbury, & R. J. Buresh (Eds.), *Improving the productivity and sustainability of the Rice-Wheat systems: Issues and impact* (ASA Special Publication 65) (pp. 1–26). Madison: American Society of Agronomy.

Gyaneshwar, P., Naresh, K. G., & Parekh, L. J. (1998). Effect of buffering on the phosphate solubilizing ability of microorganisms. *World Journal of Microbiology and Biotechnology, 14*, 669–673.

Gyaneshwar, P., Parekh, L. J., Archana, G., Podle, P. S., Collins, M. D., Hutson, R. A., & Naresh, K. G. (1999). Involvement of a phosphate starvation inducible glucose dehydrogenase in soil phosphate solubilization by Enterobacter asburiae. *FEMS Microbiology Letters, 171*, 223–229.

Hajjar, R., Jarvis, D. I., & Gemmill-Herren, B. (2008). The utility of crop genetic diversity in maintaining ecosystem services. *Agriculture, Ecosystems & Environment, 123*, 261–270.

Halder, A. K., & Chakrabarty, P. K. (1993). Solubilization of inorganic phosphate by Rhizobium. *Folia Microbiologica, 38*, 325–330.

Halstead, E. H., Barber, S. A., Warncke, D. D., & Bole, J. (1968). Supply of Ca, Sr, Mn and Zn to plant roots. *Soil Science Society of America Proceedings, 32*, 69–72.

Hardin, G. (1968). Tragedy of the commons. *Science, 162*, 1243–1248.

Harris, H. C. (1991). Implications of climate variability. In H. C. Harris, P. J. M. Cooper, & M. Pela (Eds.), *Soil and crop management for improved water use efficiency in Rainfed areas* (Proceedings of International Workshop held in May 15–19, 1989) (pp. 179–198). Ankara/Turkey/Aleppo/Syria: International Center for Agricultural Research in Dry Areas s (ICARDA).

Harris, R. F., & Bezdicek, D. F. (1994). Descriptive aspects of soil quality/health. In J. W. Doran, D. C. Coleman, D. F. Bezdicek, & B. A. Stewart (Eds.), *Defining soil quality for a sustainable environment* (SSSA special publication 35) (pp. 23–35). Madison: American Society of Agronomy, Crop Science Society of America, and Soil Science Society of America.

Harris, R. F., Karlen, D. L., & Mulla, D. J. (1996). A conceptual framework for assessment and management of soil quality and health. In J. W. Doran & A. J. Jones (Eds.), *Methods for assessing soil quality* (SSSA special publication 49) (pp. 61–82). Madison: Soil Science Society of America.

Hartshorn, A. S., Chadwick, O. A., Vitousek, P. M., & Kirch, P. V. (2006). Prehistoric agricultural depletion of soil nutrients in Hawaii. *Proceedings of the National Academy of Sciences of the United States of America, 103*, 11092–11097.

Harwood, R. R. (1990). A history of sustainable agriculture. In C. A. Edwards, R. Lal, P. Madden, R. H. Miller, & G. House (Eds.), *Sustainable Agricultural Systems* (pp. 3–19). Ankey: Soil and Water Conservation Society.

Hazell, P. (1998). *Why invest more in sustainable development of less-favored lands* (IFPRI Report 20) (Vol. 2). Washington, DC: International Food Policy Research Institute.

Hegde, D. M., & Dwivedi, S. S. (1992). Nutrient management in rice-wheat cropping system in India. *Fertility News, 37*, 27–42.

Henao, J., & Banante, C. (1999). *Nutrient depletion in the agricultural soils of Africa* (Vision 2020 Brief 62). Washington, DC: International Food Policy Research Institute.

Hendriks, L., Claassen, N., & Jungk, A. (1981). Phosphatverarmung des wurzelnahen Bodens und Phosphataufnahme von Mais und Raps. *Zeitschrift für Pflanzenernährung und Bodenkunde, 144*, 486–499.

Herdt, R. W., & Steiner, R. A. (1995). Agricultural sustainability: Concepts and conundrums. In V. Barnett, R. Payne, & R. Steiner (Eds.), *Agricultural sustainability: Economic, environmental and statistical considerations* (pp. 1–17). New York: Wiley.

Herridge, D. F., Peoples, M. B., & Boddey, R. M. (2008). Global inputs of biological nitrogen fixation in agricultural systems. *Plant and Soil, 311*, 1–18.

Hicks, A. (1997). *Power and food security. Paper presented at the Solar World Congress, August 24–30*. International Solar Energy Society (ISES), Tejon, Republic of Korea

Highman, C. F. W., Kijngam, A., Manly, B. F. J., & Moore, S. J. E. (1981). The bovid third phalanx and prehistoric ploughing. *Journal of Archaeological Science, 8*, 353–365.

Hillel, D. (1994). *Rivers of Eden* (pp. 221–225). New York: Oxford University Press.

Hinsinger, P., Jaillard, B., & Dufey, J. E. (1992). Rapid weathering of a trioctahedral mica by roots of ryegrass. *Soil Science Society of America Journal, 56*, 977–982.

Ho, W. C., & Ko, W. H. (1985). Effect of environmental edaphic factors. *Soil Biology and Biochemistry, 17*, 167–170.

Hoagland, R., & Martin, J. C. (1933). Absorption of potassium by plants in relation to replaceable, non-replaceable and soil solution potassium. *Soil Science, 36*, 1–34.

Hobbs, P. R., Ken, S., & Raj, G. (2008). The role of conservation agriculture in sustainable agriculture. *Philosophical Transactions of the Royal Society B, 363*, 543–555.

Holben, W. E., Noto, K., Sumino, T., & Suwa, Y. (1998). Molecular analysis of bacterial communities in a three compartment granular activated sludge system indicates community-level control by incompatible nitrification process. *Applied and Environmental Microbiology, 64*, 2528–2532.

Holford, I. C. R. (1988). Buffering of phosphate in the soil solution during growth of ryegrass compared with buffering by sorption. *Plant and Soil, 111*, 3–9.

Holling, C. S., Schindler, D. W., Walker, B. H., & Roughgarden, J. (1995). Bio diversity in the functioning of ecosystems: An ecological synthesis. In C. A. Perrings, K. G. Maler, C. Folke, C. S. Holling, & B. O. Jansson (Eds.), *Biodiversity loss: Ecological and economic issues* (pp. 44–83). Cambridge: Cambridge University Press.

Hoon, H., Park, R. D., Kim, Y. W., Rim, Y. S., Park, K. H., Kim, T. H., Such, J. S., & Kim, K. Y. (2003). 2-ketogluconic acid production and phosphate solubilization by Enterobacter intermedium. *Current Microbiology, 47*, 87–92.

Horrocks, M., Shane, P. A., Barber, I. G., D'Costa, D. M., & Nichol, S. L. (2004). Microbotanical remains reveal Polynesian agriculture and mixed cropping in early New Zealand. *Review of Palaeobotany and Palynology, 131*, 147–157.

Howden, S. M., Sousana, J. F., Tubiello, F. N., Chhetri, N., Dunlop, M., & Meinke, H. (2007). Adapting agriculture to climate change. *PNAS, 104*, 19691–19696.

Hugenholtz, P., Goebel, B. M., & Pace, N. R. (1998). Impact of culture independent studies on the emerging polygenetic view of bacterial diversity. *Journal of Bacteriology, 180*, 4765–4774.

Hurrel, R. F. (2001). Modifying the composition of plant foods for better human health. In J. Nosberger, H. H. Geiger, & P. C. Struik (Eds.), *Crop science: Progress and prospects* (pp. 53–64). Wallingford: CAB International.

IFPRI. (1997a). *Water resources in the twenty-first century: Challenges and implications for action*. Washington, DC: International Food Policy Research Institute.

IFPRI. (1997b). *The world food situation: Recent developments, emerging issues and long term prospects* (Vision 2020 Food Policy Report). Washington, DC: International Food Policy Research Institute.

Illmer, P. A., Barbato, A., & Schinner, F. (1995). Solubilization of hardly soluble $AlPO_4$ with P-solubilizing microorganisms. *Soil Biology and Biochemistry, 27*, 260–270.

IPCC. (1996). In J. T. Houghton, L. G. M. Filho, B. A. Gallander, N. Harris, A. Kattenberg, & K. Maskell (Eds.), *Climate Change 1995: The science of climate change. Inter-Governmental Panel on Climate Change*. Cambridge: Cambridge University Press.

IPCC. (2001). In R. T. Watson (Ed.), *Climate change 2001: Synthesis report*. Cambridge: Cambridge University Press.

Izac, A. M. N. (1994). Ecological economic assessment of soil management practices in tropical countries. In D. J. Greenkand & I. Szabolcs (Eds.), *Soil resilience and sustainable land use* (pp. 77–96). Wallingford: CAB International.

Izac, A. M. M., & Swift, M. J. (1994). On agricultural sustainability and its measurement in small scale farming in sub Saharan Africa. *Ecological Economics, 11*, 105–125.

Jackson, R. B., IV. (1993). Greenhouse gases and agriculture. In R. A. Greyer (Ed.), *A global warming forum: Scientific, economic and legal overview* (pp. 417–444). Boca Raton: CRC Press.

Jardine, P. M., & Sparks, D. L. (1984). Potassium-calcium exchange in a multi-reactive soil system II. Thermodynamics. *Soil Science Society of America Journal, 48*, 45–50.

Jastrow, J. D., Boutton, T. W., & Miller, R. M. (1996). Carbon dynamics of aggregate-associated organic matter, estimated by C-13 natural abundance. *Soil Science Society of America Journal, 60*, 81–807.

Jeffries, P. (1987). Use of mycorrhizae in agriculture. *CRC Critical Reviews in Biotechnology, 5*, 319–357.

Jenny, H. (1941). *Factors of soil formation*. New York: McGraw-Hill.

Johri, J. K., Surange, S., & Nautiyal, C. S. (1999). Occurrence of salt, pH and temperature tolerant phosphate solubilizing bacteria in alkaline soils. *Current Microbiology, 39*, 89–93.

Kang, H. (1994). The Spark programme and rural township enterprises of China. In M. S. Swaminathan (Ed.), *Ecotechnology and Rural Employment* (pp. 142–146). Chennai: India Macmillan India.

Karlen, D. L., & Sharpley, A. N. (1990). Management strategies for sustainable soil fertility. In J. L. Hartfield & D. L. Karlen (Eds.), *Sustainable agricultural systems* (pp. 47–108). Boca Raton: Lewis Publications/CRC Press.

Karlen, D. L., & Stott, D. E. (1994). A framework for evaluating physical and chemical indicators of soil quality. In J. W. Doran, D. C. Coleman, E. D. Bezdicek, & B. A. Stewart (Eds.), *Defining soil quality for a sustainable environment* (SSSA special publication 35) (pp. 53–72). Madison: Soil Science Society of America.

Karlen, D. L., Eash, N. S., & Unger, P. W. (1992). Soil and crop management effects on soil quality indicators. *American Journal of Alternative Agriculture, 7*, 48–55.

Karlen, D. L., Mausbach, M. J., Doran, R. G., Cline, R. F., & Harris and Schuman G.E. (1997). Soil quality: A concept, definition and framework for evaluation. *Soil Science Society of America Journal, 61*, 4–10.

Kay, B. D., Rasiah, V., & Perfect, E. (1994). In D. J. Greenland & I. Szabolcs (Eds.), *Soil resilience and sustainable land use* (pp. 351–367). Wallingford: CAB International.

Keating, M. (1993). *The Earth Summit's Agenda for Change* (A Plain Language Version of Agenda 21 and the Other Rio Agreements). Geneva: Centre for our Common Future.

Keating, B. A., & Carberry, P. S. (1993). Resource capture and use in intercropping-solar radiation. *Field Crops Research, 34*, 273–301.

Keerthisinghe, G., & Mengel, K. (1979). Phosphatpufferung verschiedener Böden und ihre Veränderung infolge Phosphatalterung. *Mitteilungen der Deutschen Gesellschaft, 29*, 217–230.

Kennedy, A. C., & Papendick, R. I. (1995). Microbial characteristics of soil quality. *Journal of Soil and Water Conservation, 50*, 243–248.

Ketola, T. (1997). Ecological Eldorado: Eliminating excess over ecology. In R. Welfod (Ed.), *Hijacking environmentalism: Corporate response to sustainable development* (pp. 99–136). London: Earthscan.

Khan, M. S., Aamil, M., & Zaidi, A. (1997). Associative effect of Bradyrhizobium sp (vigna) and phosphate solubilizing bacteria on moongbean (Vigna radiata L.) wilczek. *Biojournal, 10*, 101–106.

Khan, M. S., Aamil, M., & Zaidi, A. (1998). Moogbean response to inoculation with nitrogen fixing and phosphate solubilizaing bacteria. In A. M. Deshmukh (Ed.), *Biofertilizers and biopesticides* (pp. 40–48). Jaipur: Technoscience Publications.

Kichey, T., Hirel, B., Heumez, E., Dubois, F., & Le Gouis, J. (2007). In winter wheat (Triticum aestivum L.) post-anthesis nitrogen uptake and remobilization to the grain correlates with agronomic traits and nitrogen physiological markers. *Field Crops Research, 102*, 22–32.

Kijne, J. W., Tuong, T. R., Bennett, J., Bowman, B., & Oweis, T. (2002). Ensuring food security via improvement in crop water productivity. In *Background paper 1, challenge program for water and food*. Washington, DC: Consultative Group for International Agricultural Research.

Kim, K. Y., Jordan, D., & McDonald, G. A. (1997). Solubilization of hydroxyapatite by Enterobacter agglomerans and cloned Escherichia coli in culture medium. *Biology and Fertility of Soils, 24*, 347–352.

Kim, K. Y., Jordan, D., & McDonald, G. A. (1998). Enterobacter aglomerans phosphate solubilizing bacteria and microbial activity in soil: Effect of carbon sources. *Soil Biology and Biochemistry, 30*, 995–1003.

Kimble, J. M. (1998). Minimum data sets needed for soil and climatic characterization. In R. Lal (Ed.), *Soil quality and agricultural sustainability* (pp. 44–50). Chelsea: Ann Arbor Press.

King, F. H. (1911). *Farmers of forty centuries*. Emmaus: P.A. Rodale Press.

Knorzer, K. N. (2000). 3000 years of agriculture in a valley of the high Himalayas. *Vegetation History and Archaeobotany, 9*, 219–222.

Koohafkan, A. P. (2000). Land resources potential and sustainable land management: An overview. In *Advances in land resources management for the 21st Century. Proceeding of international conference on land resource management for food, employment and environmental security (ICLRM), November 9–13, 2000* (Soil conservation society of India, New Delhi) (pp. 1–22). New Delhi: Angkor Publishers.

Kottgen, P., & Diehl, R. (1929). Lieber die anwendung der dialyse und electroultrafiltration zur bestimmung des nahvstoff bedurfnisses des bodens. *Zeitschrift für Pflanzenernährung und Bodenkdunde, 12*, 65–105.

Kovar, J. L., & Barber, S. A. (1988). Phosphorus supply characteristics of 33 soils as influenced by seven rates of phosphorus addition. *Soil Science Society of America Journal, 52*, 160–165.

Kucey, R. M. N. (1983). Phosphate solubilizing bacteria and fungi in various cultivated and virgin Alberta soils. *Canadian Journal of Soil Science, 63*, 671–678.

Kucey, R. M. N. (1987). Increased P uptake by wheat and field beans inoculated with a phosphorus solubilizing Penicillium bilaji strain and with vesicular arbuscular mycorrhizal fungi. *Applied and Environmental Microbiology, 53*, 2699–2703.

Kucey, R. M. N. (1988). Effect of Penicillium bilaji on the solubility and uptake of P and micronutrients from soil by wheat. *Canadian Journal of Soil Science, 68*, 261–270.

Kucey, R. M. N., Janzen, H. H., & Legget, M. E. (1989). Microbial mediated increases in plant available phosphorus. *Advances in Agronomy, 42*, 199–228.

Kuchenbuch, R., & Jungk, A. (1984). Wirkung der Kaliumdüngung auf die Kalium verfügbarkeit in der Rhizosphäre von Raps. *Zeitschrift für Pflanzenernährung und Bodenkdunde, 147*, 435–448.

Kuhlmann, H., & Wehrmann, J. (1984). Kali-Düngeempfehlung auf der Grundlage von 81K-Düngeversuchen zu Gertreide und Zuckerrüben. *Zeitschrift für Pflanzenernährung und Bodenkunde, 147*, 349–360.

Kumar, V., Behl, R. K., & Narula, N. (2001). Establishment of phosphate solubilizing strains of Azotbacter chrococcum in the rhizosphere and their effect on wheat cultivars under greenhouse conditions. *Microbiological Research, 156*, 87–93.

Kuo, S. (1990). Cadmium buffering capacity and accumulation in Swiss chard in some sludge-amended soils. *Soil Science Society of America Journal, 54*, 86–91.

Kyllingsbaek, A., & Hansen, J. F. (2007). Development in nutrient balances in Danish agriculture 1980–2004. *Nutrient Cycling in Agroecosystems, 79*, 267–280.

Ladha, J. K., Pathak, H., Padre, A. T., Dave, D., & Gupta, R. K. (2003). Productivity trends in intensive rice-wheat cropping systems in Asia. In J. K. Ladha, J. Hill, R. K. Gupta, J. Duxbury, & R. J. Buresh (Eds.), *Improving the productivity and sustainability of the Rice-Wheat systems: Issues and impact* (ASA Special Publication 65) (pp. 45–76). Madison: American Society of Agronomy.

Ladha, J. K., Pathak, H., Krupnik, T. J., Six, J., & van Kessel, C. (2005). Efficiency of fertilizer nitrogen in cereal production: Retrospects and prospects. *Advances in Agronomy, 87*, 85–156.

Lal, R. (1994). Sustainable land use systems and soil resilience. In D. J. Greenland & I. Szabolcs (Eds.), *Soil resilience and sustainable land use* (pp. 41–67). Wallingford: CAB International.

Lal, R., & Stewart, B. A. (1992). *Soil restoration vol 17 advances in soil science*. New York: Springer.

Lal, R., Kimble, J. M., Levine, E., & Whitman, C. (1995). World soils and green house effect: An overview. In R. Lal, J. M. Kimble, E. Levine, & B. A. Stewart (Eds.), *Soils and global change* (pp. 1–7). Boca Raton: Lewis Publishers.

Lal, R., Reicosky, D. C., & Hanson, J. (2007). History of plowing over 10,000 years. *Soil and Tillage Research, 93*, 1–12.

Lange, C. H. (1992a). Canal irrigation in prehistoric Mexico – The sequence of technological change. Doolittle, WE New Mexico. *Historic Review, 67*, 194–195.

Lange, L. (1992b). Microbes and microbial products in plant protection. *Progress in Botany, 53*, 252–270.

Langeveld, J. W. A., Verhagen, A., Neeteson, J. J., van Keulen, H., Conijin, J. G., Schils, R. L. M., & Oenema, J. (2007). Evaluating farm performance using agri-environmental indicators: Recent experiences for nitrogen management in The Netherlands. *Journal of Environmental Management, 82*, 363–376.

Larsen, S. (1967). Soil phosphorus. *Advances in Agronomy, 19*, 151–210.

Larson, W. E., & Pierce, F. J. (1991). Conservation enhancement of soil quality. In *Evaluation for sustainable land management in the developing world* (Vol. 2.. IBSRAM Proc. 12 (2)). Bangkok: International Board for Soil Research and Management.

Latham, M., & Syers, J. K. (1994). Collaborative research networks to promote sustainable land use. In D. J. Greenland & I. Szabolcs (Eds.), *Soil resilience and sustainable land use* (pp. 513–520). Wallingford: CAB International.

Lavy, T. L., & Barber, S. A. (1964). Movement of molybdenum in the soil and its effect on availability to the plant. *Soil Science Society of America Proceedings, 28*, 93–97.

Lemaire, G., Oosterom, E., Sheehy, J., Jeuffroy, M. H., Massignam, A., & Rossato, L. (2007). Is crop N demand more closely related to dry matter accumulation or leaf area expansion during vegetative growth? *Field Crops Research, 100*, 91–106.

Lepofsky, D. (1995). A radio carbon chronology of pre-historic agriculture in the Society Islands. *French Polynesia, Radio-carbon, 37*, 917–930.

Lewis, D. G., & Quirk, J. P. (1967). Phosphate diffusion in soil and uptake by plants III. 31P-movement and uptake by plants as indicated by 32P-autoradiography. *Plant and Soil, 26*, 445–453.

Liebman, M., & Dyck, E. (1993). Crop rotation and intercropping strategies for weed management. *Ecological Applications, 3*, 92–122.

Lloveras, J., Chocarro, C., Freixes, O., Arque, E., Moreno, A., & Santiver, F. (2008). Yield, yield components, and forage nutritive value of alfalfa as affected by seeding rate under irrigated conditions. *Agronomy Journal, 100*, 191–197.

Lopez-Bellido, R. J., Castillo, J. E., & Lopez-Bellido, L. (2008). Comparative response of bread and durum wheat cultivars to nitrogen fertilizer in a rainfed Mediterranean environment: Soil nitrate and N uptake and efficiency. *Nutrient Cycling in Agroecosystems, 80*, 121–130.

Lu, S., & Miller, M. H. (1994). Prediction of phosphorus uptake by field-grown maize with the Barber-Cushman model. *Soil Science Society of America Journal, 58*, 852–857.

Lumbanraja, J., & Evangelou, V. P. (1990). Binary and ternary exchange behavior of potassium and ammonium at low exchangeable fractional loads. *Soil Science Society of America Journal, 54*, 698–705.

Lumbanraja, J., & Evangelou, V. P. (1992). Potassium quantity–intensity relationships in the presence and absence of NH4 for three Kentucky soils. *Soil Science, 154*, 366–376.

Lynch, J. M. (1983). *Soil Biotechnology: Microbiological factors in crop productivity*. Oxford: Blackwell.

Maas, E. V. (1969). Calcium uptake by excised maize roots and interactions with alkali cations. *Plant Physiology, 44*, 985–989.

Maliha, R., Samina, K., Najma, A., Sadia, A., & Farooq, L. (2004). Organic acids production and phosphate solubilization by phosphate solubilizing microorganisms under in vitro conditions. *Pakistan Journal of Biological Sciences, 7*, 187–196.

Malthus, T. R. (1798). *An essay on the principle of population, as it effects the future improvement of society*. London: J.J. Johnson.

Mann, C. C. (2002). The real dirt on rainforest fertility. *Science, 297*, 920–923.

Mannering, J. L., & Fenster, C. R. (1983). What is conservation tillage? *Journal of Soil and Water Conservation, 38*, 141–143.

Marschner, H. (1994). Micronutr. *News Information, 14*, 3–5.

Marschner, H., & Römheld, V. (1994). Strategies of plants for acquisition of iron. *Plant and Soil, 165*, 261–274.

Martin, H. W., & Sparks, D. L. (1983). Kinetics of nonexchangeable potassium release from two coastal plain soils. *Soil Science Society of America Journal, 47*, 883–887.

Martinez-Murcia, A. J., Acinas, S. G., & Rodriguez-Valera, F. (1995). Evaluation of prokaryotic diversity by restrictase digestion of 16S rDNA directly amplified from hipersaline environments. *FEMS Microbiology Ecology, 17*, 247–256.

Martre, P., Semenov, M. A., & Jamieson, P. D. (2007). Simulation analysis of physiological traits to improve yield, nitrogen use efficiency and grain protein concentration in wheat. In J. H. J. Spiertz, P. C. Struik, & H. H. Van Laar (Eds.), *Scale and complexity in plant systems research: Gene-plant-crop relations* (Gene-plant-crop relations, Wageningen UR Frontis Series) (pp. 181–201). Berlin: Springer.

Masse, W. B. (1981). Prehistoric irrigation systems in the Salt River Valley. *Arizona Science USA, 214*, 408–415.

Massey, R. E., Myers, D. B., Kitchen, N. R., & Sudduth, K. A. (2008). Profitability maps as an input for site-specific management decision making. *Agronomy Journal, 100*, 52–59.

Matson, P. A., Parton, W. J., Power, A. G., & Swift, M. J. (1997). Agriculture intensification and ecosystem properties. *Science, 277*, 504–509.

Mattingly, C. E. G., Russell, R. D., & Jephcott, B. M. (1963). Experiments on cumulative dressings of fertilizers on calcareous soils in Southwest England. II. Phosphorus uptake by ryegrass in the green house. *Journal of the Science of Food and Agriculture, 14*, 629–637.

Mausbach, J. M., & Seybold, C. A. (1998). Assessment of soil quality. In R. Lal (Ed.), *Soil quality and agricultural sustainability* (pp. 33–43). Chelsea: Ann Arbor Press.

Max-Neef, M. (1995). Economic growth and quality of life: Threshold hypothesis. *Ecological Economics, 15*, 115–118.

McCoy, M. D., & Hartshorn, A. S. (2007). Wind erosion and intensive pre-historic agriculture: A case study from the Kalaupapa field systems. Molokai Island, Hawaii. *Geoarchaeology, 22*, 511–532.

McGinn, A. P. (1999). *Safeguarding the health of oceans* (Paper 145). Washington, DC: World Watch Institute.

McLean, E. O., & Watson, M. E. (1985). Soil measurement of plant-available potassium. In R. D. Munson (Ed.), *Potassium in agriculture* (pp. 277–308). Madison: ASA, CSSA, and SSA.

McNaughton, N. (1998). Earning carbon credits: A future market for farmers. *Holistic Management Quarterly* 59 (April).

Mehana, T. A., & Wahid, O. A. A. (2002). Associative effect of phosphate dissolving fungi, Rhizobium and phosphate fertilizer on some soil properties, yield components and the phosphorus and nitrogen concentration and uptake by Vicia fabia L., under field conditions. *Pakistan Journal of Biological Sciences, 5*, 1226–1231.

Mehlich, A. (1984). Mehlich 3 soil test extractant. A modification of the Mehlich 2 extractant. *Communications in Soil Science and Plant Analysis, 15*, 1409–1416.

Mengel, K. (1985). Dynamics and availability of major nutrients in soils. *Advances in Soil Sciences, 2*, 65–131.

Mengel, K. (1994). Iron availability in plant tissues—Iron chlorosis on calcavcous soils. *Plant and Soil, 165*, 275–283.

Mengel, K., & Busch, R. (1982). The importance of the potassium buffer power on the critical potassium level in soils. *Soil Science, 133*, 27–32.

Mengel, K., & Kirby, E. A. (1980). Potassium in crop production. *Advances in Agronomy, 33*, 59–110.

Mengel, K., & Steffens, D. (1985). Potassium uptake of ryegrass (Lolium perenne) and red clover (Trifolium pratense) as related to root parameters. *Biology and Fertility of Soils, 1*, 53–58.

Mengel, K., & Uhlenbecker, K. (1993). Determination of available interlayer potassium and its uptake by ryegrass. *Soil Science Society of America Journal, 57*, 761–766.

Meyer, D., & Jungk, A. (1993). A new approach to quantify the utilization of nonexchangeable soil potassium by plants. *Plant and Soil, 149*, 235–243.

Mitsios, I. K., & Rowell, D. L. (1987). Plant uptake from exchangeable and non-exchangeable potassium. I. Measuring and modelling for onion roots in a chalky boulder clay soil. *Journal of Soil Science, 38*, 53–63.

Molden, D. J. (1997). *Accounting for water use and productivity* (SWIM Paper 1). Colombo: International Irrigation Management Institute.

Morgan, M. F. (1935). *A universal soil testing system* (Vol. 372). New Haven: Connecticut Agricultural Experiment Station.

Morgan, R. P. C. (1986). *Soil Erosion and Conservation*. London: Longmans.

Motsara, M. R., Bhattacharya, P. B., & Srivastava, B. (1995). Biofertilizers, their description and characteristics. In *Biofertilizer technology, marketing and usage. A sourcebook – Cum Glossary* (pp. 9–18). New Delhi: Fertilizer Development and Consultation Organization.

MSSRF. (2000). In K. V. Raman (Ed.), *Biovillage oriented development plan for the Union Territory of Pondicherry*. Chennai: M.S.Swaminathan Research Foundation.

Mukherjee, P. K., & Rai, R. K. (2000). Effect of vesicular arbuscular mycorrhizae and phosphate solubilizing bacteria on growth, yield and phosphorus uptake by wheat (Triticum aestivum) and chickpea (Cicer arietinum). *Indian Journal of Agronomy, 45*, 602–607.

Myers, N. (1997). *Perverse Subsidies: Their Nature, Scales and Impacts*. Chicago: McArthur Foundation.

Nahas, E. (1996). Factors determining rock phosphate solubilization by microorganism isolated from soil. *World Journal of Microbiology and Biotechnology, 12*, 18–23.

Nair, K. P. P. (1984a). Towards a better approach to soil testing based on the buffer power concept. In *Proceeding of 6th International colloquium for the optimization of plant nutrition* (Vol. 4, pp. 1221–1228), 2–8 September, Pierre-Martin Prevel, Montpellier, France.

Nair, K. P. P. (1984b). Zinc buffer power as an important criterion for a dependable assessment of plant uptake. *Plant and Soil, 81*, 209–215.

Nair, K. P. P. (1989). Comments on "Phosphorus supply characteristics of 33 soils as influenced by seven rates of phosphorus addition". *Soil Science Society of America Journal, 53*, 984.

Nair, K. P. P. (1990). Quantification of P and Zn replenishment capacity of soils as a dependable alternative to routine soil tests in assessing their bio-availability to growing plants. In *Transactions of the 14th international congress of soil science* (Comm. IV, pp. 493–494), Kyoto, Japan, August 12–18.

Nair, K. P. P. (1992a). Measuring P buffer power to improve routine soil testing for phosphate. *European Journal of Agnon, 1*(2), 79–84.

Nair, K. P. P. (1992b). Soil testing procedure to economise on fertilizer use in land management. In *The spirit of enterprise — The 1993 rolex awards* (pp. 323–325).

Nair, K. P. P. (1996). The buffering power of plant nutrients and effects on availability. *Advances in Agronomy, 57*, 237–287.

Nair, K. P. P. (2004). Sustaining crop production in the developing world through "the nutrient buffer power concept" – A case study with Black pepper (Piper nigrum) growing at low pH in the laterite soils of India. In *Proceeding of 16th international symposium on plant-soil interactions at low pH* (p. 392), Sendai, Japan, August 1–5.

Nair, K. P. P. (2013). The buffer power concept and its relevance in African and Asian soils. *Advances in Agronomy, 121*, 447–529.

Nair, K. P. P. (2014, August 18). Green revolution or silent spring? *The Hindu Business Line*.
Nair, K. P. P. (2015a, February 13). The debt story less told. *The Hindu Business Line*.
Nair, K. P. P. (2015b, April 7). When there is no soil left in this world. *The Hindu Business Line*.
Nair, K.P.P. (2015c). The nutrient buffer power concept – A revolutionary soil management technique for sustainable agriculture. In *Keynote address at the National Seminar on "issues, challenges and strategies in sustaining soil health"*. Kerala Forest Research Institute, Peechi, Kerala State, India, December 10–11, 2015.
Nair, K. P. P., & Babu, G. R. (1975). Zinc–phosphorus–iron interaction studies in maize. *Plant and Soil, 42*, 517–536.
Nair, K. P. P., & Mengel, K. (1984). Importance of phosphate buffer power for phosphate uptake by rye. *Soil Science Society of America Journal, 48*, 92–95.
Nair, K. P. P., & Prabhat, G. (1977). Differential response of tropical maize genotypes to zinc and manganese nutrition. *Plant and Soil, 47*, 149–159.
Nair, K. P. P., & Sharma, P. B. (1979). Mineralization and field effectiveness of ordinary and coated urea, urea-aldehyde condensation product and urea treated with nitrification inhibitor. *The Journal of Agricultural Science, Cambridge, 93*(3), 623–627.
Nair, K. P. P., Ram, N., & Sharma, P. K. (1984). Quantitative relationship between zinc transport and plant uptake. *Plant and Soil, 81*, 217–220.
Nair, K. P. P., Sadanandan, A. K., Hamza, S., & Abraham, J. (1997). The importance of Potassium buffer power in the growth and yield of Cardamom. *Journal of Plant Nutrition, 20*, 987–997.
Narang, R. S., & Bhandari, A. L. (1990). Integrated nitrogen management in rice–wheat system. Paper presented at the rice-wheat workshop. Directorate for Cropping System in Punjab, India. In *Proceedings of symposium on maximum yield research*, November 16–18, 1990, New Delhi, India.
Narang, R. S., Tiwanria, U. S., and Dev, G. (1994). Maximum yield research studies in rice-wheat system and soil productivity—The Indian experience. Research for maximum yield in harmony with nature. Satellite symposium sponsored by PPI (pp. 46–55). XV World Congress of Soil Science, Acapulco, July 10–16.
Natarajan, T., & Subrammanian, P. (1995). Response of phospho-bacteria along with Rhizobium at two levels of phosphorus on groundnut. In *Microbiology abstracts XXXVI annual conference of the association of microbiologists of India* (p. 111), Hissar, Haryana State, India November 8–10
Neeteson, J. J., Schroder, J. J., & Ten Berge, H. F. M. (2002). A multiscale system approach to nutrient management research in the Netherlands, NJAS Wageningen. *Journal of Life Sciences, 50*, 141–151.
Nemeth, K. (1971). Die characteriesierung des K-haushaltes von boden mittels K-desorptionskurven. *Geoderma, 5*, 99–101.
Nemeth, K. (1972). The determination of desorption and solubility rates of nutrients in the soil by means of electroultrafiltration (EUF). *Proceedings of the ... Colloquium of the International Potash Institute, 9*, 171–180.
Nemeth, K. (1979). The availability of nutrients in the soil as determined by electroultrafiltration (EUF). *Advances in Agronomy, 31*, 155–188.
Nemeth, K., & Recke, H. (1982). Reproducibility of Ca, K, Na, and P contents in the different EUF fractions. *Plant and Soil, 64*, 25–33.
Nicholls, R. J., & Leatherman, S. P. (1995). Global sea levels rise. In K. M. Strzepek & J. B. Smith (Eds.), *As climate changes: International impacts and implications* (pp. 92–123). Cambridge: Cambridge University Press.
Nordt, L. F., Hayashida, H. T., & Crawford, C. (2004). Late prehistoric soil fertility, irrigation management, and agricultural production in northwest coastal Peru. *Geoarchaelogy, 19*, 21–46.
Norman, D., Janka, S., Freyenberger, B., Schurle, B., & Kok, H. (1997). *Defining and implementing sustainable agriculture* (Kansas Sustainable Agriculture Series Paper. I). Manhattan: Kansas State University.

NRC. (1999). *Our common journey: A transition towards sustainability*. Washington, DC: National Academy Press.

Nye, P. H. (1972). Diffusion of ions and uncharged solutes in soils and clays. *Advances in Agronomy, 31*, 225–272.

Oades, J. M. (1984). Soil organic matter and structural stability: Mechanisms and implications for management. *Plant and Soil, 76*, 319–337.

O'Connor, G. A., Lindsay, W. L., & Olsen, S. R. (1971). Diffusion of iron and iron chelates in soil. *Soil Science Society of America Proceedings, 35*, 407–410.

Oerlemans, J., Von Oberfeld, W. O., & Wolf, D. (2007). Impact of long-term nutrient supply on plant species diversity in grassland: An experimental approach on conventionally used pastures. *Journal of Applied Botany And Food Quality, 81*, 151–157.

Okajima, H., Kubota, H., & Sakuma, T. (1983). Hysteresis in the phosphorus sorption and desorption process of soils. *Soil Science & Plant Nutrition, 29*, 271–283.

Oldeman, L. R. (1994). The global extent of soil degradation. In D. J. Greenland & I. Szabolcs (Eds.), *Soil resilience and sustainable land use* (pp. 99–118). Wallingford: CAB International.

Olsen, S. R., Cole, C. V., Watanabe, F. S., & Dean, L. A. (1954). Estimation of available phosphorus in soils by extraction with sodium bicarbonate. USDA Cir. No. 939.

Olson, D. M., & Wackers, F. G. (2007). Management of field margins to maximize multiple ecological services. *Journal of Applied Ecology, 44*, 13–21.

Omar, S. A. (1998). The role of rock phosphate solubilizing fungi and vesicular arbuscular mycorrhiza (VAM) in growth of wheat plants fertilized with rock phosphate. *World Journal of Microbiology and Biotechnology, 14*, 211–219.

Orr, A., Islam, A. S. M. N., & Barnes, G. (1991). *Treadle Pump: Manual irrigation for small farmers in Bangladesh*. Dhaka: Dinajpur Rural Services.

Ozanne, P. G. (1980). Phosphate nutrition of plants – General treatise. In F. E. Khasawneh, E. C. Sample, & E. J. Kamprath (Eds.), *The role of phosphorus in agriculture* (pp. 559–589). Madison: American Society of Agronomy, Crop Science Society of America.

Page, A. L., Logan, T. J., & Ryan, J. A. (1987). *Land application of sludge—Food chain implications*. Chelsea: Lewis Publication.

Pagel, H., & van Huay, H. (1976). Wichtige Parameter der Phosphat-Sorptionskurven einiger Böden der Tropen und Subtropen und ihre Zeitliche Veränderung durch P—Düngung. *Archiv für Acker- und Pflanzenbau und Bodenkunde, 20*, 765–778.

PAI. (2002). *People in the balance: Population and natural resources at the turn of the Millennium*. New York: Population Action International.

Palacios-Fest, M. R., Mabry, J. B., Nialis, J. F., Holmlund, J. P., Miksa, E., & Davis, O. K. (2001). Early irrigation systems in southeastern Arizona: The ostracode perspective. *Journal of South American Earth Sciences, 14*, 541–545.

Park, C. C. (1983). Paleohydrologic reconstruction from stratigraphic evidence – A case study from prehistoric irrigation canals in Peru. *Journal of the Geological Society, 140*, 321.

Parks, E. J., Olson, G. J., Brinckman, F. E., & Baldi, F. (1990). Characterization by high performance liquid chromatography (HPLC) of the solubilization of phosphorus in iron ore by a fungus. *Journal of Industrial Microbiology & Biotechnology, 5*, 183–189.

Parr, J. F., Papendick, R. I., Hornick, S. B., & Meyer, R. E. (1992). Soil quality attributes and relationship to alternative sustainable agriculture. *American Journal of Alternative Agriculture, 7*, 5–11.

Patter, C. (1998). *Against the grain: Agricultural environment reform in the United States and European Union*. Wallingford: CAB International.

Peng, S., & Bouman, B. A. M. (2007). Prospects for genetic improvement to increase lowland rice yields with less water and nitrogen. In J. H. J. Spiertz, P. C. Struik, & H. H. Van Laar (Eds.), *Scale and complexity in plant systems research: Gene-Plant-Crop relations* (pp. 251–266). Berlin: Springer.

Piccini, A., & Azcon, R. (1987). Effect of phosphate solubilizing bacteria and vesicular arbuuscular mycorrhizal fungi on the utilization of Bayovar rock phosphate by alfalfa plants using a sand vermiculite medium. *Plant and Soil, 101*, 45–50.
Pierce, F. J., & Larson, W. E. (1993). Developing criteria to evaluate sustainable land management. In J. M. Kimble (Ed.), *Proceedings of the 8th international soil management workshop: Utilization of soil survey information for sustainable agriculture land use* (pp. 7–14). Lincoln: USDA-SCS National Soil Survey.
Pierce, F. P., & Nowak, P. (1999). Aspects of precision farming. *Advances in Agronomy, 67*, 1–85.
Pikovskaya, R. I. (1948). Mobilization of phosphorus in soil in connection with vital activity of some microbial species. *Microbiology, 17*, 362–370.
Pimentel, D., Hurd, E., Bellotti, A. C., Foster, M. J., Oke, I. N., Sholes, O. D., & And Whitman, R. J. (1973). Food production and the energy crisis. *Science, 182*, 442–449.
Pimentel, D. J., Allen, J., & Beers, A. (1993). Soil erosion and agricultural productivity. In D. J. Pimentel (Ed.), *World soil erosion and conservation*. Cambridge, MA: Cambridge University Press.
Pimentel, D. R., Doughty, R., Cowthers, C., Lamberson, S., Bora, N., & Lele, U. (2003). Energy impacts in crop production in developing and developed countries. In R. Lal, D. Hansen, N. Uphoff, & S. Slack (Eds.), *Food security and environmental quality in the developing world* (pp. 129–151). Boca Raton: Lewis Publishers.
Pingali, P., Hossein, M., & Gerpacio, R. V. (1995). *Asian rice bowls: The returning crisis*. Wallingford: CAB International.
Pinstrup, A. P. (2002). Towards a sustainable global food system: What will it take? *Keynote presentation for the Annual John Pesek colloquium in sustainable agriculture, March 26–27*, Iowa State University
Pinstrup, A. P., & Pandya-Lorsch, R. (1994). *Alleviating poverty and effectively managing natural resources* (FAO Discussion Paper I). Washington, DC: International Food Policy Research Institute.
Poi, S. C., Ghosh, G., & Kabi, M. C. (1989). Response of chickpea (Cicer arietinum L.) to combined inoculation with Rhizobium, phosphobacteria and mycorrhizal organisms. *Zentral fur Microbiology, 114*, 249–253.
Ponmurugan, P., & Gopi, C. (2006). In vitro production of growth regulators and phosphatase activity by phosphate solubilizing bacteria. *African Journal of Biotechnology, 5*, 348–350.
Postel, S. (1992). *The last oasis: Facing water scarcity*. London: Earthscan.
Postel, S. (1998). Water for food production: Will there be enough in 2025? *Bio Science, 48*, 629–637.
Postel, S. (1999). *Pillars of sand: Can the irrigation miracle last?* New York: W.W.Norton.
Postel, S. (2001). Drip irrigation for small farms: A new initiative to alleviate hunger and poverty. *Water International, 26*(1), 3–13.
Postel, S., & Carpenter, S. (1997). Freshwater ecosystem services. In G. C. Daily (Ed.), *Nature's services: Societal dependence on natural ecosystems* (pp. 195–214). Washington, DC: Island Press.
Prabhakaran, J., Ravi, K. B., & Srinivasan, K. (1996). Response of Vamban-I Blackgram to N2 (2 subscript please) fixer and P mobilizers in acid soil. In *Microbiology abstracts XXXVII. Annual conference of the microbiologists of India* (p. 120), IIT, Chennai, Tamil Nadu, India, December 4–6.
Prasad, A. S. (1982). History of zinc in human nutrition. In A. S. Prasad, I. E. Dreosti, & B. S. Hetzel (Eds.), *Clinical applications of recent advances in zinc metabolism* (pp. 1–17). New York: Alan R Liss Inc.
Pratt, P. F. (1965). Potassium. In C. A. Black et al. (Eds.), *Methods of soil analysis* (Part 2, pp. 1023–1031). Madison: Agron. Monogr. 9. ASA.
Premathilake, R. (2006). Relationship of environmental changes in central Sri Lanka to possible pre-historic land-use and climate changes. *Palaeogeography, 240*, 468–496.

Pretty, J. N. (1999). Value of natural and social capital for sustainable agriculture. In G. S. Dhaliwal, R. Arora, & A. K. Dhavan (Eds.), *Emerging trends in sustainable agriculture*. New Delhi: Commonwealth Publishers.

Pretty, J. (2008). Agricultural sustainability: Concepts, principles and evidence. *Philosophical Transactions of the Royal Society B, 363*, 447–465.

Prihar, S. S., Gajri, P. R., Benbi, D. K., & Arora, V. K. (Eds.). (2000). *Intensive cropping: Efficient use of water, nutrients and tillage*. New York: Food Products Press.

Raman, S. (1987). Mineralogy in relation to pesticide use and soil pollution. *Clay Research, 6*(2), 50–58.

Raman, S. (1989). Adsorption of metoxuron and tebuthiuron on model clay-organo complexes. *Toxicological and Environmental Chemistry, 24*, 207–213.

Raman, S. (1992). Mineral-microbe interface: A veritable biochemical factory. In *Invited lecture, annual convention of the clay mineral society of India, November 14–16, 1992, Hyderabad, India*

Raman, S. (2001). Soil quality audit for resource conservation in Indian agriculture. *Social Change, 31*, 75–86.

Raman, K. V. (2003). Multiple livelihood opportunities towards sustainable human development. In *Reshaping our earth view: Creative thoughts and alternative*. New York: Tata McGraw-Hill Pub. Co.

Raman, S. (2006). *Agricultural sustainability: Principles, processes and prospects* (pp. 13904–11580). Birmingham: The Hawoth Press Inc.

Raman, S., & Rao, P. C. (1987). Kinetics of extraction of soil-applied metoxuron and its biological implications. *Water, Air, and Soil Pollution, 38*, 217–223.

Raman, S., & Rao, P. C. (1989). Adsorption of metoxuron and tebuthiuron on model clay-organo complexes. *Toxicological and Environmental Chemistry, 24*, 20–213.

Raven, K. P., & Hossner, L. R. (1994). Sorption and desorption—Quantity–intensity parameters related to plant-available soil phosphorus. *Soil Science Society of America Journal, 58*, 405–410.

Reganold, J. P. (1989). Comparison of soil properties as influenced by organic and conventional farming systems. *American Journal of Alternative Agriculture, 3*, 144–154.

Reyes, I., Bernier, L., Simard, R. R., & Antoun, H. (1999). Effect of nitrogen source on the solubilization of different inorganic phosphates by an isolate of Penicillium rugulosum and two UV induced mutants. *FEMS Microbiology Ecology, 28*, 281–290.

Richards, R. A., Lopez-Castaneda, C., Gomez-Macpherson, H., & Condon, A. G. (1993). Improving the efficiency of water use by plant breeding and molecular biology Irrigation. *Science, 14*, 93–104.

Robison, L. R., Johnston, N. P., & Hill, J. M. (1981). Small-scale agriculture, an untapped giant. *Benson Institute, 4*(2), 4–7.

Rockstrom, J., Barron, J., & Fox, P. (2002). Water productivity in rainfed agriculture: Challenges and opportunities for smallholder farmers in drought-prone tropical agroecosystems. In J. K. Kijne, R. B. Barker, & D. Molden (Eds.), *Water productivity in agriculture: Limits and opportunities for improvement* (pp. 145–162). Wallingford: CAB International.

Rodale, R. (1983). Breaking new ground: The search for a sustainable agriculture. *Futurist, 1*, 15–20.

Romig, D. E., Garlynd, M. J., & Harris, R. F. (1996). Farmer-based assessment of soil quality: A Soil health storecard. In J. W. Doran & A. J. Jones (Eds.), *Methods for assessing soil quality* (SSSA publication 49) (pp. 30–60). Madison: Soil Science Society of America.

Roos, W., & Luckner, M. (1984). Relationships between proton extrusion and fluxes of ammonium ions and organic acid in Penicillium cyclopium. *Journal of General Microbiology, 130*, 1007–1014.

Rosengrant, M. W., & Ringler, C. (1999). *Impact on food security and rural development of reallocating water from agriculture*. Washington, DC: International Food Policy Research Institute.

Rosenzweig, C., & Hillel, D. (1998). *Climate change and the global harvest: Potential impacts of the greenhouse effect on agriculture*. Oxford: Oxford University Press.

Roy, H. K., Kumar, A., & Sinha, K. (1990). Response of wheat to K in red loam (Alfisols) soils of Ranchi. *Journal of Potassium Research, 6*, 23–28.

Russell, G., Jarvis, P. G., & Monteith, J. L. (1989). Absorption of radiation by canopies and stand growth. In G. Russell (Ed.), *Plant Canopy: Their growth, form and function* (pp. 21–39). Cambridge: Cambridge University Press.

Saber, M. S. M., & Kabesh, M. O. (1990). Utilization of biofertilizers in field crop production II. A comparison study on the effect of biofertilization or sulphur application on yield and nutrient uptake by lentil plants. *Egyptian Journal of Soil Science, 30*, 415–422.

Saito, K., Atlin, G. N., Linquist, B., Phanthaboon, K., Shiraiwa, T., & Horie, T. (2007). Performance of traditional and improved upland rice cultivars under nonfertilized and fertilized conditions in northern Laos. *Crop Science, 47*, 2473–2481.

Samonte, S. O. P. B., Wilson, L. T., Medley, J. C., Pinson, S. R. M., Clung, A. M., & Lales, J. S. (2006). Nitrogen utilization efficiency: Relationships with grain yield, grain protein and yield-related traits in rice. *Agronomy Journal, 98*, 168–176.

Sanchez, P. A. (1994). Tropical soil fertility research: Towards the second paradigm. In *Proceeding XV ISSS symposium, Acapulco, Mexico, July 10–16, 1994*," Vol. 1, pp. 89–104).

Sarojini, V., Verma, S., & Mathur, M. S. (1989). Biocoenotic association between nitrogen fixing and phosphate solubilizing microorganisms. *Current Science, 59*, 1099–1100.

Satizabal, E. J. H., & Saif, U. S. R. (1987). Interaction between vesicular arbuscular mycorrhiza and leguminous Rhizobium in an oxisol of the eastern plains of Colombia. *Acta Agronomica Journal, 7–21*.

Schachtschabel, P. (1937). Aufnahme von nicht austauschbarem Kalium durch die Pflanze. *Zeitschrift für Pflanzenernährung und Bodenkunde, 48*, 107–133.

Schachtschabel, P. (1961). Fixierung und Nachlieferung von Kalium-und Ammonium-Ionen. Beurteilung und Bestimmung des Kaliumversorgungs-grades von Böden. *Landwirtschaftliche Forschung, 14*, 29–47.

Scherer, S. J., & Yadav, S. (1996). *Land degradation in the developing world: Implications for food, agriculture and environment to 2020* (Food and Agriculture and the Environment Discussion Paper 14). Washington, DC: International Food Policy Research Institute.

Scherr, S. J. (1999). *Soil Degradation: A threat to developing-country food security by 2020?* (Vision 2020 Brief 58). Washington, DC: International Food Policy Research Institute.

Schofield, R. K. (1955). Can a precise meaning be given to "available" soil phosphorus? *Soil Fertility, 18*, 373–375.

Scholes, R. J. (1990). The influence of soil fertility on the ecology of south African dry savannas. *Journal of Biogeography, 17*, 415–419.

Scholes, M. C., Swift, M. J., Heal, O. W., Sanchez, P. A., Ingram, J. S. I., & Dalal, R. (1994). Soil fertility research in response to the demand for sustainability. In P. L. Woomer & M. J. Swift (Eds.), *The biological management of tropical soil fertility* (pp. 1–14). Chichester: John Wiley.

Schroder, J. J., Aarts, H. F. M., ten Berge, H. F. M., van Keulen, H., & Neeteson, J. J. (2003). An evaluation of whole-farm nitrogen balances and related indices for efficient nitrogen use. *European Journal of Agronomy, 20*, 33–44.

Schüller, H. (1969). Die CAL-Methode, eine neue Methode zur Bestimmung des pflanzen verfügbaren Phosphates in Böden. *Zeitschrift für Pflanzenernährung und Bodenkunde, 123*, 48–63.

Seckler, D., Molden, D., & Barker, R. (1998). *Water scarcity in the twenty-first century* (Water Brief 1). Colombo: International Water Management Institute.

Selim, H. M. (1992). Modeling the transport and retention of inorganics in soils. *Advances in Agronomy, 47*, 331–384.

Sen, A. (1981). *Poverty and famines: An essay on entitlement and deprivation*. Oxford: Clarendon Press.

Sharpley, A. N. I., Pierzynki, G., & Sims, J. T. (1993). Innovative soil phosphorus availability indices: Assessing inorganic phosphorus. In *Agronomy Abstracts*. Madison: American Society of Agronomy.
Sheridan, H., Finn, J. A., Culleton, N., & O'Donovan, G. (2008). Plant and invertebrate diversity in grassland field margins. *Agriculture, Ecosystems & Environment, 123*, 225–232.
Shuman, L. M. (1975). The effect of soil properties on zinc adsorption by soils. *Soil Science Society of America Proceedings, 39*, 454–458.
Shu-Yan, C., & Sposito, G. (1981). The thermodynamics of ternary cation exchange systems and the subregular model. *Soil Science Society of America Journal, 45*, 1084–1089.
Sibma, L., & Spiertz, J. H. J. (1986). Dry matter production and nitrogen utilization in cropping systems with grass, lucerne and maize I: Comparison of crop characteristics, growth and production. *Netherlands Journal of Agricultural Science, 34*, 25–35.
Sillanpaa, M. (1982). *Micronuttrients and the nutrient status of soils. A global study* (FAO Soils Bulletin, No. 48). Rome: FAO.
Sillanpaa, M., & Vlek, P. L. G. (1985). Micronutrients and the agroecology of tropical and Mediterranean regions. *Fertilizer Research, 7*, 151–167.
Simrad, R. R., & Tran, T. S. (1993). Evaluating plant available phosphorus with the electroultrafiltration technique. *Soil Science Society of America Journal, 57*, 404–409.
Simrad, R. R., Tran, T. S., & Zizka, J. (1991). Evaluation of the electro-ultrafiltration technique (EUF) as a measure of the K supplying power of Quebec soils. *Plant and Soil, 132*, 91–101.
Sims, G. K. (1990). Biological degradation of soil. In R. Lal & B. A. Stewart (Eds.), *Soil degradation vol II advances in soil science* (pp. 289–320). New York: Springer.
Sinclair, T. R. (1994). Limits to crop yield. In *Physiology and determination of crop yields*. Madison: American Society of Agronomy, Crop Science Society of America, and Soil Science Society of America.
Sinclair, T. R., Tanner, C. B., & Bennett, J. (1984). Water-use efficiency in crop production. *BioScience, 34*, 36–40.
Singal, R., Gupta, R., & Saxena, R. K. (1994). Rock phosphate solubilization under alkaline conditions by Aspergillus japonicus and A. foetidus. *Folia, 39*, 33–36.
Singh, H. P. (1990). Response of dual inoculation with Bradyrhizobium and VAM mycorrhiza or phosphate solubilizer on soybean in mollisol. In B. L. Jalali & H. Chand (Eds.), *Trends in mycorrhiza* (Research Proceedings of the National Conference on Mycorrhiza) (pp. 14–16). Hissar: Haryana Agricultural University.
Singh, K. P., & Kushwaha, C. P. (2000). Ecological aspect of soil fertility management in Indian agro ecosystem. In P. K. Jha, S. B. Karmacharya, S. R. Baral, & P. Lacoul (Eds.), *Environment and agriculture: At the cross road of the new millennium* (Proc. International conference on environment and agriculture, November 1–3, 1998, Kathmandu) (pp. 205–222). Kathmandu: Ecological Society (ECOS).
Singh, G. B., & Nambiar, K. K. M. (1986). Crop productivity and fertility under intensive use of chemical fertilizers in long-term field experiments. *Indian Journal of Agronomy, 31*, 115–127.
Sivanappan, R. K. (1994). Prospects of micro irrigation in India. *Irrigation and Drainage Systems, 8*, 49–58.
Smaling, E. M. A., Stoonvogel, J. J., & Windmeijer, N. P. (1993). Calculating soil nutrient balances in Africa at different scales. II. District Scale. *Fertilizer Research, 35*, 237–250.
Smil, V., Nachman, P., & Long, T. V. (1983). *Energy analysis in agriculture: An application to the US Corn production*. Boulder: Westview Press.
Soil Science Society of America (Committee Report). (1965). Glossary of soil science terms. *Soil Science Society of America Proceedings, 29*, 330–351.
Sparks, D. L. (1987). Potassium dynamics in soils. *Advances in Soil Sciences, 6*, 2–63.
Sparks, D. L. (1989). *Kinetics of soil chemical processes*. San Diego: Academic.
Sparks, D. L., & Huang, P. M. (1985). Physical chemistry of soil potassium. In R. D. Munson (Ed.), *Potassium in agriculture* (American Society of Agronomy) (pp. 201–276). WI: Madison.

Spencer, D. S. C., & Swift, M. J. (1992). Sustainable agriculture: Definition and measurement. In K. Mulongoy, M. Gueye, & D. S. C. Spencer (Eds.), *Biological nitrogen fixation and sustainability of tropical agriculture* (pp. 15–24). Ibadan: International Institute of Tropical Agriculture.

Spiertz, J. H. J., & Oenema, O. (2005). Resource use efficiency and management of nutrients in agricultural systems. In *ERSEC Ecological Book Series – 1. On sustainable agricultural systems* (pp. 171–183). Berlin: Tsinghua University Press/Springer.

Spiertz, J. H. J., & Sibma, L. (1986). Dry matter production and nitrogen utilization in cropping systems with grass, lucerne and maize 2. Nitrogen yield and utilization with various cropping systems and their after-effects. *Netherlands Journal of Agricultural Science, 34*, 37–47.

Sposito, G. (1981a). Cation exchange in soils. A historical and theoretical perspective. In R. H. Dowday (Ed.), *Chemistry in the soil environment*. Madison: Soil Science Society of America.

Sposito, G. (1981b). *The Thermodynamics of the soil solution*. Oxford: Clarendon Press.

Sposito, G., Holtzelaw, K. M., Johnston, C. T., & Le Vesque, C. S. (1981). Thermodynamics of sodium-copper exchange on Wyoming bentonite at 198°K. *Soil Science Society of America Journal, 45*, 1079–1084.

Sposito, G., Holtzelaw, K. M., Charlet, L., Jonany, C., & Page, A. L. (1983). Sodium-calcium and sodium magnesium exchange on Wyoming bentonite in perchlorate and chloride background ionic media. *Soil Science Society of America Journal, 47*, 51–56.

SSSA. (1995, June 7). Soil quality: A conceptual definition. *Agronomy News*.

Stammer, L. B. (1992, January). Study finds serious harm to 10 percent of world's best soil. *Los Angeles Times* (quotations from World Resources Institute of U.N. Environmental Program).

Steffens, D. (1994). Phosphorus release kinetics and extractable phosphorus after long-term fertilization. *Soil Science Society of America Journal, 58*, 1702–1708.

Steffens, D., & Mengel, K. (1979). Das Aneignungsvermögen von Lolium perenne im Vergleich zu Trifolium pratense für Zwischenschilht-Kalium der Touminerale. *Landw. Forsch. SH., 36*, 120–127.

Steffens, D., & Mengel, K. (1981). Vergleichende Untersuchungen zwischen Lolium perenne und Trifolium pratense über das Aneignungsvermögen von Kalium. *Mitt. Dtsch. Bodenk. Ges., 32*, 375–386.

Stevenson, C. M., Jackson, T. L., Meith, A., Bork, H. R., & Ladeffaged, T. N. (2006). Pre-historic and early historic agriculture at Maunga Orito, Easter Island (Rapa Nui). *Chile, Antiquity, 80*, 919–936.

Stockdale, E. A., Gaunt, J. L., & Vos, J. (1997). Soil-plant nitrogen dynamics: What concepts are required? *European Journal of Agronomy, 7*, 145–159.

Stocking, M. A. (1984). *Erosion and soil productivity: A review*. Rome: Soil Conservation Program. Land and Water Development Division, Food and Agriculture Organization.

Storkey, J., & Westbury, D. B. (2007). Managing arable weeds for biodiversity. *Pest Management Science, 63*, 517–523.

Struik, P. C., & Bonciarelli, F. (1997). Resource use at the cropping system level. *European Journal of Agronomy, 7*, 133–143.

Subba Rao, N. S. (1982). In N. S. S. Rao (Ed.), *Advances in agricultural microbiology* (pp. 229–305). New Delhi: Oxford and IBH Publ. Co.

Sudding, K. N., Collins, S. L., Gough, L., Clark, C., Cleland, E. E., Gross, K. L., Milchunas, D. G., & Pennings, S. (2005). Functional-and abundance-based mechanisms explain diversity loss due to N fertilization. *PNAS, 102*, 4387–4392.

Sulaiman, W., & Kay, B. D. (1972). Measurement of the diffusion coefficient of boron in soil using a single cell technique. *Soil Science Society of America Proceedings, 36*, 746–752.

Sullivan, L. A. (1990). Soil organic matter air encapsulation and water – Stable aggregation. *Journal of Soil Science, 41*, 529–534.

Swarup, A., & Singh, K. N. (1987). Movement of potassium in a sodic soil profile as influenced by long-term use of inorganic fertilizers under rice-wheat rotation. *Journal of the Indian Society of Soil Science, 35*, 253–256.

Swift, M. J. (1999). Integraing soils, systems and society. *Nature & Resources, 35*, 12–20.

Swift, M. J., & Anderson, J. M. (1994). Biodiversity and ecosystem functions in agricultural systems. In E. D. Schulze & H. Mooney (Eds.), *Biodiversity and ecosystem function* (pp. 15–42). New York: Springer.

Swift, M. J., Izac, A., & van Noordwijk, M. (2004). Biodiversity and ecosystem services in agricultural landscapes. *Agriculture, Ecosystems & Environment, 104*, 113–134.

Syers, J. K., & Latham, M. (1994). Sustainable land management for mitigating stresses. In S. M. Virmani, J. C. Katyal, H. Eswaran, & I. P. Ibrol (Eds.), *Stressed ecosystems and sustainable agriculture* (pp. 332–338). New Delhi: Oxford and IBH Publishing Company.

Sys, C. (1976). *Land evaluation: Part I*. Ghent: ITC State University Ghent.

Szabolcs, I. (1994). The concept of soil resilience. In D. J. Greenland & I. Szabolcs (Eds.), *Soil resilience and sustainable land use* (pp. 33–39). Wallingford: CAB International.

Ten Berge, H. F. M., & Kropff, M. J. (1995). Founding a systems research network for rice. In Bouma (Ed.), *Eco-regional approaches for sustainable land use and food production* (pp. 263–282). Dordrecht: Kluwer.

Tinker, P. B. (1980). In F. E. Khasawneh, E. C. Sample, & E. J. Kamprath (Eds.), *The role of phosphorus in agriculture*. Madison: Soil Science Society of America.

Tisdall, J. M., & Qades, J. M. (1982). Organic matter and water-stable aggregates in soils. *Journal of Soil Science, 33*, 141–163.

Tuong, T. P. (1999). Productive water use in rice production: Opportunities and limitations. *Journal of Crop Production, 2*(2), 241–264.

Tushar, S. (1993). *Ground water markets and irrigation development*. Bombay: Oxford University Press.

Tyagi, N. K. (2002). Managing saline and alkaline water for higher productivity. In J. W. Kijne, R. Barker, & D. Molden (Eds.), *Water productivity in agriculture: Limits and opportunities for improvement* (pp. 69–87). Wallingfrord: CAB International.

Umali, D. L. (1993). *Irrigation induced salinity*. Washington, DC: World Bank.

UN. (1999). *World population prospects: The 1998 revision*. New York: United Nations Publications Division.

UNCED. (1992). Chapter 24: Global action for women towards sustainable and equitable development. In *Agenda* (Vol. 21, pp. 349–354). Rio de Janeiro: United Nations Conference on Environment and Development.

UNDP. (1995). *The Aral in Crisis: United Nations development project, Tashkent, Russia*. New York: United Nations Publications Division.

UNDP/IEI/SEI. (1997). *Energy after Rio: Prospects and challenges*. New York: United Nations Development Program in collaboration with the International Energy Institute and Stockholm Environmental Institute.

UNEP. (2007). Reactive nitrogen in the environment. In *Too much or too little of a good thing* (Vol. 51). Paris: The Woods Hole Research Center, USA/UNEP DTIE Sustainable Consumption and Production Branch. ISBN 978 92 807 2783 8.

UNFCCC. (1997). *Kyoto protocol for and United Nations framework convention on climate change*. Kyoto, Japan.

UNMD. (2000). United National millennium declaration. UN General Assembly Resolution 55/2, document A/RES/55/2, September 8, New York: United Nations

Van Delden, A. (2001). Yield and growth components of potato and wheat under organic nitrogen management. *Agronomy Journal, 93*, 1370–1385.

van der Paauw, F. (1971). An effective water extraction method for the determination of plant-available soil phosphorus. *Plant and Soil, 34*, 467–481.

Van Elsas, J. D., Van Overbeek, L. S., & Fouchier, R. (1991). A specific marker pat for studying the fate of introduced bacteria and their DNA in soil using a combination of detection techniques. *Plant and Soil, 138*, 49–60.

Van Ginkel, M., Ortiz-Monasterio, I., Trethowan, R., & Hernandez, E. (2001). Methodology for selecting segregating populations for improved N-use efficiency in bread wheat. *Euphytica, 119*, 223–230.

Van Keulen, H., Aarts, H. F. M., Habekotte, B., van der Meer, H. G., & Spiertz, J. H. J. (2000). Soil-plant-animal relations in nutrient cycling: The case of dairy farming system "De Marke". *European Journal of Agronomy, 13*, 245–261.

Van Veen, J. A., Leonard, S., Van Overbeek, L. S., & Van Ellsas, J. D. (1997). Fate and activity of microorganisms introduced into soil. *Microbiology and Molecular Biology Reviews, 61*, 121–135.

Vandermeer, J. M. (1992). *Thoughts on agriculture and the environment in a post-modern world. Symposium on enhancing the future of the land grant system.* Irving: National Academy of Sciences Board on Agriculture.

Vanselow, A. P. (1932). Equilibria of the base-exchange reactions of bentonites, permutities, soil colloids, and zeolites. *Soil Science, 33*, 95–113.

Vassilev, N., Fenice, M., & Federici, F. (1996). Rock phosphate solubilization with gluconic acid produced by immobilized Penicillium variable P16. *Biotechnology Techniques, 20*, 585–588.

Vazquez, P., Holguin, G., Puente, M., ELopez Cortes, A., & Bashan, Y. (2000). Phosphate solubilizing microorganisms associated with the rhizosphere of mangroves in a semi arid coastal lagoon. *Biology and Fertility of Soils, 30*, 460–468.

Venkateswarlu, B., Rao, A. V., Raina, P., & Ahmad, N. (1984). Evaluation of phosphorus solubilization by microorganisms isolated from arid soil. *Journal of the Indian Society of Soil Science, 32*, 273–277.

Virmani, S. M., Katyal, J. C., Eswaran, H., & Abrol, I. P. (Eds.). (1994). *Stressed ecosystems and sustainable agriculture* (pp. 229–235). New Delhi: Oxford and IBH Publishing Company.

Visser, S., & Parkinson, D. (1992). Soil biological criteria as indicators of soil quality: Soil microorganisms. *American Journal of Alternative Agriculture, 7*, 33–37.

Vlek, P. L. G., & Vielhauer, K. (1994). Nutrient management strategies in stressed environments. In S. M. Virmani, J. C. Katyal, H. Eswaran, & I. P. Abrol (Eds.), *Stressed ecosystems and sustainable agriculture* (pp. 203–228). New Delhi: Oxford and IBH Publishing Company.

von Reichenbach, H. (1972). Factors of mica transformation. In *Potassium in Soil* (pp. 33–42). Berne, Switzerland: Int Potash Institute.

WCED. (1987). *Our common future.* Oxford: Oxford University Press.

WEC/FAO. (1999). *The challenge of rural energy poverty in developing countries.* London/Rome: World Energy Council and Food and Agriculture Organization.

Welch, R. M. (1993). Zinc concentrations and forms in plants for humans and animals. In A. D. Robson (Ed.), *Zinc in soils and plants* (pp. 183–195). Dordrecht: Kluwer Academic Publishers.

Whitelaw, M. A. (2000). Growth promotion of plants inoculated with phosphate solubilizing fungi. *Advances in Agronomy, 69*, 99–151.

Whitelaw, M. A., Harden, T. J., & Helyar, K. R. (1999). Phosphate solubilization in solution culture by the soil fungus Penicillium radicum. *Soil Biology and Biochemistry, 32*, 655–665.

Whitmore, A. P., & van Noordwijk, M. (1994). Bridging the gaps between environmentally acceptable and agronomically desirable nutrient supply. In D. Glen (Ed.), *Proceedings of the Long Ashton Symposium: Agriculture in the 21st century*. in press.

Wilkins, R. J. (2008). Eco-efficient approaches to land management: A case for increased integration of crop and animal production systems. *Philosophical Transactions of the Royal Society B, 363*, 517–525.

Wilkinson, H. F., Loneragan, J. F., & Quirk, J. P. (1968). The movement of zinc to plant roots. *Soil Science Society of America Proceedings, 32*, 831–833.

Wilkinson, P., Smith, K. R., Joffe, M., & Haines, A. (2007). Energy and Health 1: A global perspective on energy: Health effects and injustices. *The Lancet, 370*, 965–978.

Wood, L. K., & De Turk, E. E. (1941). The adsorption of potassium in soils in replaceable forms. *Soil Science Society of America Proceedings, 5*, 152–161.

Wood, S., Sebastian, K., & Scherr, S. J. (2000). *Pilot analysis of global ecosystems: Agroecosystems.* Washington, DC: International Food Policy Research Institute/World Resources Institute.

Woodruff, C. M. (1955a). Ionic equilibria between clay and dilute salt solutions. *Soil Science Society of America Proceedings, 19*, 36–40.

Woodruff, C. M. (1955b). The energies of replacement of calcium by potassium in soils. *Soil Science Society of America Proceedings, 19*, 167–171.

Woomer, P. L., & Swift, M. J. (Eds.). (1994). *The Biological management of tropical soil fertility.* Chicester: John Wiley.

World Bank. (2000). *World development indicators*. Washington DC: World Bank.

WRI. (1994). *World Resources 1994–95*. Washington, DC: World Resources Institute.

WRI. (1998). *World Resources 1998–99. Food production: Have yields stopped rising?* Washington, DC: World Resources Institute.

WRI. (2000). *World resources 2000–2001. People and ecosystems, the Fraying web of life.* Washington, DC: World Resources Institute.

WSSD. (2002). *World summit for sustainable development. United Nations conference on environment and development,* August 26 – September 4, Johannesburg, South Africa

Xu, M. L., & Liu, Z. Y. (1983). The nutrient status of soil-root interface. II. Potassium accumulation and depletion in rhizosphere soils. *Acta Pedologica Sinica, 20*, 295–302.

Yang, J. E., & Skogley, E. O. (1989). Influence of copper or cadmium on soil potassium availability properties. *Soil Science Society of America Journal, 53*, 1019–1023.

Yang, J. E., & Skogley, E. O. (1990). Copper and cadmium effects on potassium adsorption and buffering. *Soil Science Society of America Journal, 54*, 739–744.

Yang, X., Römheld, V., & Marschner, H. (1994). Effect of bicarbonate on root growth and accumulation of organic acids in Zn-inefficient and Zn-efficient rice cultivars (Oryza sativa L.). *Plant and Soil: An International Journal on Plant-Soil Relationships, 164*, 1–7.

Yates, R. A., & Kiss, A. (1992). Using and sustaining Africa's soils. In *Agricultural and rural development series* (Vol. 6). Washington: World Bank.

Yin, X., & Van Laar, H. H. (2005). *Crop systems dynamics: An ecophysiological simulation model for genotype-by-environment interactions* (p. 153). Wageningen: Wageningen Academic Publishers. ISBN 907 69 985 82.

Young, E. M. (1997). *World Hunger*. London: Routledge.

Zaidi, A., & Khan, M. S. (2005). Interactive effect of rhizospheric microorganisms on growth, yield and nutrient uptake of wheat. *Journal of Plant Nutrition, 28*, 2079–2092.

Zaidi, A., Khan, M. S., & Aamil, M. (2004). Bio-associative effect of rhizospheric microorganisms on growth, yield and nutrient uptake of greengram. *Journal of Plant Nutrition, 27*, 599–610.

Zhao, J., Luo, Q., Deng, H., & Yan, Y. (2008). Opportunities and challenges of sustainable agricultural development in China. *Philosophical Transactions of the Royal Society B, 363*, 893–904.

Printed by Printforce, the Netherlands